Communications
in Computer and Information Science 141

Sumeet Dua   Sartaj Sahni   D.P. Goyal (Eds.)

Information Intelligence, Systems, Technology and Management

5th International Conference, ICISTM 2011
Gurgaon, India, March 10-12, 2011
Proceedings

 Springer

Volume Editors

Sumeet Dua
Louisiana Tech University
Ruston, LA, USA
E-mail: sdua@coes.latech.edu

Sartaj Sahni
University of Florida
Gainesville, FL, USA
E-mail: sahni@cise.ufl.edu

D.P. Goyal
Management Development Institute
Sukhrali, Gurgaon, India
E-mail: dpgoyal@mdi.ac.in

ISSN 1865-0929   e-ISSN 1865-0937
ISBN 978-3-642-19422-1   ISBN 978-3-642-19423-8 (eBook)
DOI 10.1007/978-3-642-19423-8
Springer Heidelberg Dordrecht London New York

Library of Congress Control Number: 2011921552

CR Subject Classification (1998): I.2, I.2.11, H.3-4, C.2, D, H.5

Typesetting: Camera-ready by author, data conversion by Scientific Publishing Services, Chennai, India

Printed on acid-free paper

Springer is part of Springer Science+Business Media (www.springer.com)

Message from the Program Chair

It is my pleasure to welcome you to the 5th International Conference on Information Systems, Management and Technology (ICISTM 2011).

The proceedings of the 5th International Conference on Information Systems, Management and Technology contain 39 papers that were selected for presentation at the conference. The peer-review process for these papers was particularly rigorous this year due to the large number of submissions (106 total submissions), which originated from seven countries, primarily India, the USA, France, Nigeria, and Thailand, in response to the call for papers announced in May 2010.

Along with the influx of submissions, the 2011 conference featured some distinct changes from previous ICISTM conferences, including a new conference website, http://www.icistm.org, to effectively disseminate information and updates. In addition, new tracks were added to the conference, and researchers were invited to submit papers in each of the resulting seven tracks. An exceptional international multidisciplinary team of 13 scientists and academicians from seven countries including the USA, Taiwan, Singapore, Canada, the UK, India, and Chile served as Chairs for the seven tracks [(1) Information Systems, (2) Information Technology, (3) Business Intelligence, (4) Information Management, (5) Health Information Management and Technology, (6) Management Science and Education, and (7) Applications]. A team of five Publicity Chairs, a Tutorials Chair, and the Publications Chair assisted the Track Chairs and the Program Chair in the review process and the organization of the program.

The Track Chairs and the Program Chair established an international Program Committee consisting of 113 members from 21 countries (primarily from the USA, India, Singapore, Taiwan, and China) to assist in the review process. Seventeen external reviewers also assisted this team. The Track Chairs and the Program Chair made paper assignments to the Program Committee members so that the focus of each paper matched its reviewer's expertise. A total of 302 reviews and reviewer ratings were obtained for the submitted papers. Each reviewer's rating was weighted by his or her confidence in the domain area. The Program Chair, in coordination with the Track Chairs, made the final selection of the accepted papers. The papers were accepted into two categories: regular papers (up to 12 pages) and short papers (up to five pages).

The proceedings are published in Springer's *Communications in Computer and Information Science* (CCIS) series, and the papers have been reviewed by Springer's editorial team for final publication in the proceedings. The publication of the proceedings by the CCIS series offers a uniquely formulated and widely available medium for the dissemination of the selected papers.

The conference was a result of a harmonized effort between several people. I would like to thank these people, as their contributions to the conference organization and proceedings have been invaluable. The General Chairs, Sartaj Sahni and D.P. Goyal, offered their perpetual guidance throughout the process. This conference is a result of their enthusiastic vision and unwavering commitment to the conference. The expertise and leadership offered by our exceptional team of Track Chairs was crucial for the peer-review of the conference papers and, thus, for the final product that is included in this volume. The Publicity Chairs and the Tutorials Chair did a remarkable job in circulating the conference information and generating interest in the project.

I would like to issue a special thanks to our Publications Chair, Pradeep Chowriappa, for the timely and effective preparation of the proceedings, which involved the rigorous task of carefully checking each accepted paper and ensuring compliance with the Springer LNCS/CCIS standards. Stefan Goeller and Leonie Kunz from Springer and their editorial team skillfully assisted us with the compilation of these proceedings, and the organizing team at MDI, Gurgaon, provided the venue and made local arrangements for the conference. I sincerely thank each of the people involved in these processes. Finally, I would like to thank the authors and the Program Committee for making this year's program particularly notable and uniquely multidisciplinary.

March 2011   Sumeet Dua

Conference Organization

General Co-chairs

Sartaj Sahni	University of Florida, USA
D.P. Goyal	Management Development Institute, India
Renaud Cornu Emieux	Grenoble Ecole de Management, France

Program Chair

Sumeet Dua	Louisiana Tech University, USA

Track Chairs

Information Management	Rajanish Dass
	Indian Institute of Management-Ahemdabad, India
	K.B.C. Saxena
	Management Development Institute, India
Information Systems	James Courtney
	Louisiana Tech University, USA
	Han C.W. Hsiao
	Asia University, Taiwan
Information Technology	Yan Huang
	University of North Texas, USA
	Yu Lin
	Data Storage Institute, A*Star, Singapore
Healthcare Information Management and Technology	William J. Rudman
	American Health Information Association Foundation, USA
	Craig Kuziemsky
	University of Ottawa, Canada
Business Intelligence	J. Michael Hardin
	University of Alabama, USA
	Hamparsum Bozdogan
	The University of Tennessee, USA
Applications	Alan Hartman
	IBM Research, India
	Visakan Kadirkamanathan
	University of Sheffield, UK
Management Science and Education	Marcela González Araya
	Universidad de Talca, Chile

Publicity Co-chairs

Mario Dantas	Federal University of Santa Catarina, Brazil
Adagunodo E.R.	Obafemi Awolowo University, Nigeria
Ayad Salhieh	Jordan University of Science and Technology, Jordan
Prerna Sethi	Louisiana Tech University, USA
Payal Mehra	Indian Institute of Management, Lucknow, India

Publications Chair

Pradeep Chowriappa	Louisiana Tech University, USA

Tutorials and Workshop Chair

Krishna Karuturi	Genome Institute of Singapore, Singapore

ICISTM 2011 Program Committee

Information Management

Travis Atkinson	Louisiana Tech University, USA
S. K. Batra	Institute of Management Technology at Ghaziabad, India
Pradeep Chowriappa	Louisiana Tech University, USA
Parthasarathi Dasgupta	Indian Institute of Management at Calcutta, India
Mohit Jain	Louisiana Tech University, USA
Vishal Jha	Jaypee Institute of Information Technology, India
Rahul Kala	ABV-Indian Institute of Information Technology and Management at Gwalior, India
Yung-Chung Ku	National Taiwan University, Taiwan
Guangqin Ma	AT&T Research Labs, USA
Kamna Malik	U21Global, India
Hima Bindu Maringanti	Jaypee Institute of Information Technology University, India
Payal Mehra	Indian Institute of Management at Lucknow, India
Teng Moh	San Jose State University, USA
Vasanth Nair	Louisiana Tech University, USA
Kiyoshi Nosu	Tokai University, Japan

Olufade Falade W. Onifade   University of Ibadan Nigeria and
                             SITE-LORIA, Nancy Université, France
Adenike Oyinlola Osofisan   University of Ibadan, Nigeria
Prashant Patnaik   Konark Institute of Technolgy, India
Dhiren Patel   Indian Institute of Technology
                             at Gandhinagar, India
Sheetal Saini   Louisiana Tech University, USA
Anupam Shukla   ABV-Indian Institute of Information
                             Technology and Management at Gwalior,
                             India
Anongnart Srivihok   Kasetsart University, Thailand

Information Systems

Rizwan Ahmed   Anjuman College of Engineering
                             and Technology, India
Oluwatosin Akinsol Ayeni   Louisiana Tech University, USA
Veena Bhat   IBS - ICFAI Business School, India
A. J. Burns   Louisiana Tech University, USA
Chun-Ming Chang   Asia University, Taiwan
Feng-Cheng Chang   Tamkang University, Taiwan
Pradeep Chowriappa   Louisiana Tech University, USA
Liou Chu   Tamkang University, Taiwan
Fred Coleman   Louisiana Tech University, USA
Wafa Elgarah   Al Akhawayn University, Morocco
Selwyn Ellis   Louisiana Tech University, USA
Christie Fuller   Louisiana Tech University, USA
Adarsh Garg   Institute of Management Technology
                             at Ghaziabad, India
Lin Hui   Tamkang University, Taiwan
Virginia Ilie   Florida State University, USA
Yogesh Karunakar   Shree LRT College of Engineering and
                             Technology, India
Peeter Kirs   University of Texas at El Paso, USA
John Kuhn   University of Louisville, USA
Erwin Leonardi   Universiti Tunku Abdul Rahman,
                             Malaysia
Lars Linden   Georgia Southern University, USA
Ali Mirza Mahmood   Acharya Nagarjuna University, India
James Parrish   University of Arkansas at Little Rock, USA
Clay Posey   University of Arkansas at Little Rock, USA
Ravi Prakash   Mumbai University, India
Sandra Richardson   University of Memphis, USA
Bikramjit Rishi   Institute of Management Technology
                             at Ghaziabad, India

Tom Roberts   Louisiana Tech University, USA
Sheetal Saini   Louisiana Tech University, USA
Victor Hock Kim Tan   Universiti Tunku Abdul Rahman, Malaysia
Shih-Jung Wu   Tamkang University, Taiwan

Information Technoglogy

Joonsang Baek   Institute for Infocomm Research, Singapore
Robert Brazile   University of North Texas, USA
Leonard Brown   University of Texas at Tyler, USA
Bill Buckles   University of North Texas, USA
Alan Chiu   Louisiana Tech University, USA
Manoranjan Dash   Nanyang Technological University, Singapore
Wei Ding   University of Massachusetts at Boston, USA
Nigel Gwee   Southern University and A&M College, USA
Sridhar Hariharaputran   Bielefeld University, Germany
Luo Huaien   Genome Institute of Singapore, Singapore
Minhe Ji   East China Normal University, China
Rajaraman Kanagasabai   Institute for Infocomm Research, Singapore
Xiaoli Li   Institute for Infocomm Research, Singapore
Wei Li   IBM, USA
Patrick McDowell   Southeastern Louisiana University, USA
Teng Moh   San Jose State University, USA
Jihyoun Park   Seoul National University, Korea
Kavitha Ranganathan   Indian Institute of Management, India
Ayad Salhieh   Jordan University of Science and
                             Technology, Jordan
Sheetal Saini   Louisiana Tech University, USA
Yuni Xia   Indiana University-Purdue University
                             Indianapolis, USA
Yinping Yang   Institute of High Performance Computing,
                             Agency for Science, Technology,
                             and Research, Singapore
Yanjiang Yang   Institute for Infocomm Research, Singapore
Jin Soung Yoo   Indiana University-Purdue University
                             at Fort Wayne, USA
Jian Zhang   Texas Woman's University, USA
Ying Zhao   Tsinghua University, China
Zehua Zhang   School of Information Science and Engineering,
                             Yunnan University, China
Yu Zheng   Microsoft Research Aisa, China

Healthcare Information Management and Technoglogy

Sheetal Saini	Louisiana Tech University, USA
Vasanth Nair	Louisiana Tech University, USA
Mohit Jain	Louisiana Tech University, USA

Business Intelligence

Baisakhi Chakraborty	National Institute of Technology at Durgapur, India
Pradeep Chowriappa	Louisiana Tech University, USA
Demian Antony D'Mello	St. Joseph Engineering College, India
Tatiana Escovedo	Pontifcia Universidade Catlica do Rio de Janeiro, Brazil
Rahul Hakhu	Baddi University, India
Nazrul Islam	Department of Agriculture & Food, Western Australia, Australia
Marcus Perry	University of Alabama at Tuscaloosa, USA
Srinivas Prasad	Trident Academy of Technology, India
Holger Schroedl	University of Augsburg, Germany
Sheetal Saini	Louisiana Tech University, USA
Uday Kulkarni	Arizona State University, USA
Parijat Upadhyay	International School of Business and Media, India
Vadim Timkovsky	University of Sydney, Australia

Applications

Sunil Arvindam	SAP Labs India Pvt. Ltd., India
Yannis Charalabidis	National Technical University of Athens, Greece
Akshaye Dhawan	Ursinus College, USA
Phuong-Quyen Huynh	San Jose State University, USA
Mika Katara	Tampere University of Technology, Finland
Holger Kett	Fraunhofer Institute, Germany
Rajiv Narvekar	Infosys Technologies Ltd., India
Shrikant Parikh	S.P. Jain Institute of Management & Research, India
Siddhartha Sengupta	Tata Consultancy Services Ltd., India
Anupam Shukla	ABV-Indian Institute of Information Technology and Management at Gwalior, India
Renuka Sindhgatta	IBM Research, India
Christian Zirpins	Karlsruhe Institute of Technology, Germany
Jay Ramanathan	The Ohio State University, USA

Management Science and Education

Nair Abreu   Federal University of Rio de Janeiro, Brazil
Carmen Belderrain   Technological Institute of Aeronautics, Brazil
Mariana Funes   National University of Cordoba, Argentina
Richard Weber   University of Chile, Chile

Table of Contents

3. Information Technology

4. Healthcare Information Management and Technology

5. Business Intelligence

6. Applications

7. Management Science and Education

8. Short Papers

Selection of Outsourcing Agency through AHP and Grey Relational Analysis: A Case Analysis

Debendra Kumar Mahalik

PG Department of Business Administration
Sambalpur University
Debendra_mahalik@hotmail.com

Abstract. IT enabled services (ITeS), often outsourced by different utility industry. But the problem before the manager is to from where to outsource, wrong selection often blame for delay, partial success and failure. This selection is vital and is a multi criteria decision making process (MCDM), where the manager has to take a decision taking several factor in to consideration. Often these managers have to take on decision on some qualitative and quantitative data. This paper focuses on selection of outsourcing agency through an integrated Analytic Hierarchy Process (AHP) along with Grey Relational Analysis (GRA) approach for an organization for their e-procurement services. This paper proposes an integrated AHP and GRA method for multi criteria decision problem.

Keywords: Outsourcing, MCDM, AHP, GRA, ITeS.

1 Introduction

Information Technology enabled Services (ITeS) are often being outsourced from a third party for many reasons. The term Outsourcing generally used, when a company takes part of its business and gives that part to another company. In recent times, the terms have been most commonly used for technology related initiatives such as handing over the IT help-desk to a third party [1]. Webster's define the meaning of "Outsourcing" as "A company or person that provides information; to find a supplier or service, to identify a source". The term "outsourcing" has been defined in the information system literature in various ways, some of them are: "... turning over to a vendor some or all of the IS functions..." [2], "...the contracting of various information systems' sub-functions by user firms to outside information systems vendors"[3],"...the organizational decision to turn over part or all of an organization's IS functions to external service provider(s) in order for an organization to be able to achieve its goals" [4]. Some of the reason for outsourcing are not having sufficient expertise in developing solutions both in terms of manpower and know how, it saves Cost, saving in long run, internal People do not take active participation, less risk, better control, not core competency, Less time taking, faster delivery, more efficient, more focus on core competency and vested interested. This outsourcing also have some critic, some of them are risk of data/information miss utilization, lack of competent agency, fear of not supporting,

S. Dua, S. Sahni, and D.P. Goyal (Eds.): ICISTM 2011, CCIS 141, pp. 1–12, 2011.

fear of failure, vendor driven policy and rules and regulation. But Outsourcing is a buzz word, which is finding its increase in use due to its advantages.

Selection of outsourcing agency is a hard and vital task for success, as it is a multi criteria decision making problem, involves selection based on several criteria's, which depends upon the selector(s). Improper selection often blamed and affects the performances, resulted in more failure, partial success, delay, loss etc. So the selection process needs to carefully examine and executed.

2 Literature Survey

In fact, when Eastman Kodak announced that it was outsourcing its information systems (IS) function in 1989 to IBM, DEC and Businessland it created quite a stir in the information technology industry. Never before had such a well-known organization, where IS was considered to be a strategic asset, turned it over to third party providers [5]. Since then both large and small companies have found it acceptable, indeed fashionable, to transfer their IS assets, leases and staff to outsourcing vendors [6]. Kodak appears to have legitimized outsourcing, leading to what some have called "the Kodak effect"[7]. Senior executives at well known companies in the U.S. and abroad have followed Kodak's example and signed long term contracts worth hundreds of millions of dollars with outsourcing "partners". A number of high-profile multi-billion dollar "mega-deals" have been signed which has raised awareness even more. A Dataquest report (2000) notes that since 1989 there have been over 100 of these mega-deals [8].

Despite the numerous success stories illustrating the advantages of bringing information technology into organizations, it is broadly accepted that the processes of designing, developing and implementing are cumbersome and not straightforward. Recent and older reports show that IS projects frequently fail [9]. The broad and elaborate research on IS failures has been conducted for more than four decennia [10, 11, 12, 13, 14, 15, 16, 17, 18, 19]. As per literature that there is an extra dimension to information system failures i.e., the Outsourced Information System Failure (OISF). An OISF is a failure that occurs during an IS project in an outsourced environment. We use the taxonomy of Lacity and Hirschheim [20] of outsourcing options and focus on Project Management. Some academics have already pointed out that outsourcing increases risks leading to IS failures [21, 22].For decisions of whether a task can be outsourced, different approaches and criteria have been developed or have been worked out in the literature, irrespective of the potential of IT [23, 24, 25, 26, 27]. New Institutional Economics and in particular the transaction cost theory have gained importance in efficiency-oriented and costs-oriented decision processes about outsourcing [28, 29, 30].

There are several criteria for selection of outsourcing agency/supplier, twenty three criteria for supplier selection based on the extensive survey [31]; the result shows that quality is the most important parameter followed by delivery and performance history. A number of quantitative techniques have been used to supplier selection problem such as weighing method, statistical methods, Analytic Hierarchy Process (AHP), Data Envelopment Analysis (DEA) etc [32].Authors have examine the supplier selection and evaluation criterion in small and large electronic firms[33]. The results confirm the importance of the quality criteria in the supplier selection and evaluation. The other criteria found to be relatively important are speed to market, design capability and technology. The result shows that the nature of industry and its competitive environment may have a greater influence on selection criteria in comparison to the size of the firm [33]. [34] has used fuzzy multi-objective model with capacity, demand and budget constraint for supplier selection problem. They used two models to solve the problem. First Zimmerman's approach of symmetric model is used followed by Tiwari, Dharmar and Rao's weighted additive model as symmetric model. Analytical Hierarchy Process (AHP) has been applied for effective supplier selection in a leading automobile component manufacturing company [35]. The study shows that application of AHP enhances the decision making process and reduces the time taken to select the supplier. The paper uses Additive Normalization Method and Eigen vector Method to find priority vector. Some uses a novel model based on aggregation technique for combining group members' preferences into one consensus ranking for multi-criteria group decision making for supplier rating [36]. Some author have used activity based costing (ABC) approach under the fuzzy variables by considering multi period of supplier-purchaser relationship for vendor selection [37], use total cost of ownership and analytical hierarchy process for supplier selection problem and a comparison is made among different approaches [38], uses a non parametric technique called Data Envelopment Analysis for identifying and selecting vendors[39], this paper also makes a comparison between DEA approach with the current practices for vendor selection and superiority of DEA, some has used probabilistic hierarchical classification model for rating suppliers in context of software development[40], this approach, a positive probability is assigned to a supplier belonging to each class of supplier then this probabilistic classification is employed as an input to any decision making procedure that is used for actual supplier selection, some have used voting analytic hierarchy process (VAHP) for supplier selection [41], this method uses "vote ranking" rather than "paired comparison" for quantifying and measuring consistence. The study uses the vote ranking to determine the weights in the selected rank in place of the paired comparison method, some have used fuzzy goal programming for supplier selection [42] and used fuzzy multi-objective mathematical programming for supplier selection with three goals: cost minimization, quality maximization and on-time delivery maximization with constraints as demand, capacity, and quota flexibility [43].

3 Case Introduction

E-Procurement is the process of purchasing of goods and services using the internet and other ICT technology. It covers full life cycle of purchasing (indent to receipt of goods) and it connects buyers and suppliers through electronic media via exchange of Tenders, catalogs, contracts, POs, invoices etc. In the past, traditional methods of procurement offered little transparency and lesser satisfaction of negotiation with suppliers. It automates the manual buying processes, from the creation of the requisition through to payment of suppliers and brings more transparency and efficiency in the system. These reasons forced XYZ organization to adopt for an e-procurement system. The success of e-procurement depends upon the selection of a proper agency for carrying out development and maintenance of the system due to various reasons. Normally, traditional industry depends upon other outsourcing agency for the operational and maintenance for which reasons are many. These types of organization conduct a performance evaluation on each agency before selecting agency. The selection process is based on several parameter like cost, reputation, experience, quality, terms and conditions. A literature survey shoes there are various method for selection problem, the selection process are based on availability of qualitative and quantitative data and is multi criteria decision making problem (MCDM).

4 Methodology

The MCDM Unlike other methods that assume the availability of measurements, measurements in MCDM are derived or interpreted subjectively as indicators of the strength of various preferences. Preferences differ from decision maker to decision maker, so the outcome depends on who is making the decision and what their goals and preferences are [44]. In our case analysis, we will be using combination of AHP [45] and GRA (Grey Relational Analysis theory) [46]. Grey relational analysis in the Grey theory that was already proved to be a simple and accurate method for multiple attributes decision problems [47, 48, 49, 50], very unique characteristic [51, 52].

4.1 Analytic Hierarchy Process (AHP)

AHP, decision elements at each component are compared pair-wise with respect to their importance towards their control criterion, and the components themselves are also compared pair-wise with respect to their contribution to the goal. Decision makers are asked to respond to a series of pair-wise comparisons where two elements or two components at a time will be compared in terms of how they contribute to their particular upper level criterion [53, 54]. In addition, if there are interdependencies among elements of a component, pair-wise comparisons also need to be created, and an eigenvector can be obtained for each element to show the influence of other elements on it. The relative importance values are determined with a scale of 1 to 9, where a score of 1 represents equal importance between the two elements and a score of 9 indicates the extreme importance of one element (row component in the matrix) compared to the other one (column component in the matrix) [53, 54]. A reciprocal value is assigned to the inverse comparison; that is, $a_{ij}=1/ a_{ji}$; where a_{ij} (a_{ij}) denotes the importance of the ith (jth) element compared to the jth (ith) element. In AHP, pair-wise comparison is made in the framework of a matrix, and a local priority vector can be derived as an estimate of relative importance associated with the elements (or components) being compared by solving the following formulae:

$$A \cdot w = \lambda_{max} \cdot w \qquad (1)$$

where A is the matrix of pair-wise comparison, w is the eigenvector, and λ_{max} is the largest Eigenvalue of A. If A is a consistency matrix, eigenvector X can be calculated by

$$(A - \lambda_{max}I)X = 0. \qquad (2)$$

Consistency index (*C.I.*) and consistency ratio (*C.R.*) to verify the consistency of the comparison matrix. *C.I.* and *R.I.* are defined as follows:

$$C.I. = \frac{\lambda_{max} - n}{n - 1} \, , \tag{3}$$

$$C.R. = \frac{C.I.}{R.I.}, \tag{4}$$

where *R.I.* represents the average consistency index over numerous random entries of same order reciprocal matrices. If $C.R \leq 0.1$, the estimate is accepted; otherwise, a new comparison matrix is solicited until $C.R \leq 0.1$[44].

4.2 Grey Theory

The black box is used to indicate a system lacking interior information [55]. Nowadays, the black is represented, as lack of information, but the white is full of information. Thus, the information that is either incomplete or undetermined is called Grey. A system having incomplete information is called Grey system. The Grey number in Grey system represents a number with less complete information. The Grey element represents an element with incomplete information. The Grey relation is the relation with incomplete information. Those three terms are the typical symbols and features for Grey system and Grey phenomenon [56]. There are several aspects for the theory of Grey system [56]:

1. Grey generation: This is data processing to supplement information. It is aimed to process those complicate and tedious data to gain a clear rule, which is the whitening of a sequence of numbers.
2. Grey modeling: This is done by step 1 to establish a set of Grey variation equations and Grey differential equations, which is the whitening of the model.
3. Grey prediction: By using the Grey model to conduct a qualitative prediction, this is called the whitening of development.
4. Grey decision: A decision is made under imperfect countermeasure and unclear situation, which is called the whitening of status.
5. Grey relational analysis: Quantify all influences of various factors and their relation, which is called the whitening of factor relation.
6. Grey control: Work on the data of system behavior and look for any rules of behavior development predict future's behavior, the prediction value can be fed back into the system in order to control the system.

The concept of grey relational space was proposed by Deng based on the combined concepts of system theory, space theory and control theory [46]. It can be used to capture the correlations between the references factor and other compared factors of a system [57]. One of the features of GRA is that both qualitative and quantitative relationships can be identified among complex factors with insufficient information (relative to conventional statistical methods). Under such a condition, the results generated by conventional statistical techniques may not be acceptable without sufficient data to achieve desired confidence levels. In contrast, grey system theory can be used to identify major correlations among factors of a system with a relatively small amount of data. The procedure for calculating the GRA as followed.

(a). Calculate the Grey Relation Grade

Let X_0 be the referential series with k entities (or criteria) of $X_1, X_2, \ldots, X_i, \ldots, X_N$ (or N measurement criteria). Then

$$X_0 = \{x_0(1),\ x_0(2),\ \ldots,\ x_0(j),\ \ldots,\ x_0(k)\},$$

$$X_1 = \{x_1(1),\ x_1(2),\ \ldots,\ x_1(j),\ \ldots,\ x_1(k)\},$$

$$X_i = \{x_i(1),\ x_i(2),\ \ldots,\ x_i(j),\ \ldots,\ x_i(k)\},$$

$$X_N = \{x_N(1),\ x_N(2),\ \ldots,\ x_N(j),\ \ldots,\ x_N(k)\}.$$

The grey relational coefficient between the compared series X_i and the referential series of X_0 at the j-th entity is defined as:

$$\gamma_{0i}(j) = \frac{\Delta \min + \Delta \max}{\Delta_{0j}(j) + \Delta \max}. \tag{5}$$

where $\Delta_{0j}(j)$ is the absolute value of difference between X_0 and X_i at the j-th entity, that is $\Delta_{0j}(j) = |x_0(j) - x_i(j)|$, and $\Delta \max = \max\limits_{i} \max\limits_{j} \Delta_{0j}(j)$, $\Delta \min = \min\limits_{i} \min\limits_{j} \Delta_{0j}(j)$.

The grey relational grade (GRG) for series of X_i is given as:

$$\Gamma_{0i} = \sum_{j=1}^{K} w_j \gamma_{0i}(j). \tag{6}$$

where w_j is the weight of j-th entity. If it is not necessary to apply the weight, take $\omega_j = \dfrac{1}{K}$ for averaging.

(b). Data Normalization (or Data Dimensionless)

Before calculating the grey relation coefficients, the data series can be treated based on the following three kinds of situation and the linearity of data normalization to avoid distorting the normalized data [58]. They are:

(a). Upper-bound effectiveness measuring (i.e., larger-the-better)

$$x_i^*(j) = \frac{x_i(j) - \min\limits_{j} x_i(j)}{\max\limits_{j} x_i(j) - \min\limits_{j} x_i(j)}. \tag{7}$$

(b). Lower-bound effectiveness measuring (i.e., smaller-the-better)

$$x_i^*(j) = \frac{\max\limits_{j} x_i(j) - x_i(j)}{\max\limits_{j} x_i(j) - \min\limits_{j} x_i(j)} \tag{8}$$

(c). Moderate effectiveness measuring (i.e., nominal-the-best)

If $\min\limits_{j} x_i(j) \leq x_{ob}(j) \leq \max\limits_{j} x_i(j)$, then

$$x_i^*(j) = \frac{\left|x_i(j) - x_{ob}(j)\right|}{\max\limits_{j} x_i(j) - \min\limits_{j} x_i(j)}, \qquad (9)$$

If $\max\limits_{j} x_i(j) \leq x_{ob}(j)$, then

$$x_i^*(j) = \frac{x_i(j) - \min\limits_{j} x_i(j)}{x_{ob}(j) - \min\limits_{j} x_i(j)}, \qquad (10)$$

If $x_{ob}(j) \leq \min\limits_{j} x_i(j)$, then

$$x_i^*(j) = \frac{\max\limits_{j} x_i(j) - x_i(j)}{\max\limits_{j} x_i(j) - x_{ob}(j)}. \qquad (11)$$

where $x_{ob}(j)$ is the objective value of entity j.

5   Case Analysis

The model for Selection of Implementing Agency for implementation of the E-procurement through AHP and GRA, starts with Identify the criteria/parameters for establishing the evaluation structure. In order to find best agency for implementation agency four parameters are decided by the authority mainly cost (C1), reputation (C2), experience (c3), terms and condition (C4). The experts are asked to rate the above parameters with respect to each other and mutually agreed weight are represented in Table 1. Initially in order to find the priority weight AHP is used with Expert choice 4.0 as software and result is represented in result column in Table 1.

Table 1. Relative Weights of Different Parameters

Parameters	C1 (cost)	C2(Reputation)	C3(experience)	C4(terms and condition)	Results
C1	1	5	6	7	.636
C2	1/5	1	4	3	.211
C3	1/6	¼	1	2	.089
C4	1/7	1/3	½	1	.064

After getting the relative weights of different criteria, each of the four agencies A1-A4 are rated against the different four criteria's mention above by the same expert group. Since due to some confidentiality expert have given relative rank of different firm with respect to different criteria's are depicted in the Table 2. Finally all the results of the above table are summarized in Table 6. A group of assumptions are made for the following: (1) cost: the quantifying value for unit price (smaller-the-better) is 0; (2) Reputation (larger-the-better) is 5; (3) Experience: (larger-the-better) is 5; (4) terms and condition (larger-the-better) is 5.

Table 2. Measurement Values for Each Evaluation Attribute

Measure Vendors	C1 (cost)	C2(Reputation)	C3(experience)	C4(terms and condition)
A1	2	3	5	4
A2	4	2	4	5
A3	5	5	3	3
A4	3	4	2	2

Table 2 was rationalized using equations mention (b) above and resulted Table 3. According to our expected goal for each evaluation factor, an ideal standard series $(X_0 = 1)$ is established in the last line in Table 4.

Table 3. Data Rationalizing

Item Comparative series	C1 (cost)	C2(Reputation)	C3(experience)	C4(terms and condition)
X_1	1	.333	1	.667
X_2	.333	0.00	.667	1
X_3	0.00	1	.333	.333
X_4	.667	.667	0.00	0.00
Standard series (X_0)	1	1	1	1

Determination of Grey relational coefficient for each evaluation factor by using the Equation (1).

Table 4. The Grey Relational Coefficient

Item Comparative series	C1 (cost)	C2(Reputation)	C3(experience)	C4(terms and condition)
X_1	1	.333	1	.667
X_2	.333	0.00	.667	1
X_3	0.00	1	.333	.333
X_4	.667	.667	0.00	0.00

Determination of the relational grade for each agency using the equation-2, weight are taken from the Table 1.

$\Gamma_{01} = 0.745$; $\Gamma_{02} = 0.773$; $\Gamma_{03} = 0.856$; $\Gamma04 = 0.576$

This value of Grey relation is the overall performance that the enterprise requires.

Because of $\Gamma_{03} > \Gamma_{02} > \Gamma_{01} > \Gamma_{04}$, the ranking order for all candidate agency is: (1) A3; (2) A2; (3) A3; (4) A4 It is noted that the ranking order will change while we change the weighting value for each evaluation factor. In other words, the owner of an enterprise may select a suitable agency based on his own requirements.

6 Conclusions

E-procurement is a highly technical concept, requires expert hand for development, execution and maintenance. This resulted in outsourcing of such services, so selection of proper agency for implementation of e-procurement concept is vital for success. It depends upon several parameters and decision taker. It is important to develop an accurate evaluation method. This selection is a multi criteria decision making having both qualitative and quantitative data. Proposed methodology i.e. combine both AHP and GRA cater to both qualitative and quantitative nature of data in an integrated manner. This selection helps the decision maker for selecting an agency under multi criteria.

Acknowledgements

We acknowledge that we have used the following papers as source papers to write this paper. We further acknowledge that for literature review we have drawn heavily from these papers. We thank the researchers for making their research articles freely available in the Internet.

Training Program Selection for Life Insurers- By Using Analytic Network Process and Grey Relational Analysis, Da Han Chung, Yen Lin Hung, Yu Hsuang Lee, Jun Min Wang,Department of Risk management and Insurance, Shih Chien University, Taipei, Taiwan, *fbm.ypu.edu.tw/ezfiles/10/1010/img/698/08.doc on 12.10.2010*

Applying Grey Relational Analysis to the Vendor Evaluation Model, Chih-Hung Tsai, Ching-Liang Chang, and Lieh Chen, Department of Industrial Engineering and Management, Ta-Hwa Institute of Technology,1 Ta-Hwa Road, Chung-Lin, Hsin-Chu, Taiwan, ROC,E-mail: ietch@thit.edu.tw , International Journal of The Computer, The Internet and Management, Vol. 11, No.3, 2003, pp. 45 - 53

A fuzzy multi-criteria decision making approach for supplier selection in supply chain management Sreekumar and S. S. Mahapatra, African Journal of Business Management Vol.3 (4), pp. 168-177, April, 2009 Available at http://www.academicjournals.org/AJBM ISSN 1993-8233 © 2008 Academic Journal.

References

1. http://www.mariosalexandrou.com/definition/outsourcing.asp
2. Apte, U.M., Sobol, M.G., Hanaoka, S., Shimada, T., Saarinen, T., Salmela, T., Vepsalainen, A.P.J.: IS outsourcing practices in the USA. Japan and Finland: A Comparative Study. Journal of Information Technology 12, 289–304 (1997)

3. Chaudhury, A., Nam, K., Rao, H.R.: Management of information systems outsourcing: a bidding perspective. Journal of Management Information Systems 12(2), 131–159 (1995)

4. Cheon, M.J., Grover, V., Teng, J.T.C.: Theoretical perspectives on the outsourcing of information systems. Journal of Information Technology 10, 209–210 (1995)

5. Applegate, L., Montealegre, R.: Eastman Kodak Company: managing information systems through strategic alliances. Harvard Business School Case 9-192-030. Harvard Business School, Boston (1991)

6. Arnett, K.P., Jones, M.C.: Firms that choose outsourcing: a profile. Information and Management 26(4), 179–188 (1994)

7. Caldwell, B.: Farming out client-server. Information Week 510(5) (1994), http://www.informationweek.com/510/05mtcs.html

8. Young, S.: Outsourcing: lessons from the literature. Labour and Industry 10(3), 97–118 (2000)

9. Devos, J., Van Landeghem, H., Deschoolmeester, D.: Outsourced information systems failures in SMEs: a multiple case study. The Electronic Journal Information Systems Evaluation 11(2), 73–82 (2002), http://www.ejise.com

10. Ackoff, R.L.: Management misinformation systems. Management Science 14(4), 147–156 (1967)

11. Lucas, H.C.: Why information systems fail. Columbia University Press, New York (1975)

12. Lyytinen, K., Hirschheim, R.: Information systems failures-a survey and classification of the empirical literature. In: Zorkoczy, P.I. (ed.) Oxford Surveys in Information Technology, vol. 4, pp. 257–309. Oxford University Press, Oxford (1987)

13. Sauer, C.: Why information systems fail: a case study approach. Alfred Wailer, Henley-on-Thames (1993)

14. Keil, M.: Pulling the plug: software project management and the problem of project escalation. MIS Quarterly 19(4), 421–431 (1995)

15. Beynon, D.: Information systems 'failure': the case of the London ambulance service's computer aided dispatch project. European Journal of Information Systems 4(3), 171–184 (2002)

16. Schmidt, R., Lyytinen, K., Keil, M., Cule, P.: Identifying software project risks: an international Delphi study. Journal of Management Information Systems 17(4), 5–36 (2001)

17. Ewusi-Mensah, K.: Software development failures. MIT Press, Cambridge (2003)

18. Iacovou, C.L., Dexter, A.S.: Surviving IT project cancellations. Communications of the ACM 48(4), 83–86 (2005)

19. Avison, D., Gregor, S., Wilson, M.: Managerial IT unconsciousness. Communications of the ACM 49(7), 89–93 (2006)

20. Dibbern, J.: The sourcing of application software services: empirical evidence of cultural. In: Industry and Functional Differences. Physika-Verlag, Reihe, Information Age Economy, Berlin (2004)

21. Dibbern, J., Goles, T., Hirschheim, R.: Information systems outsourcing: a survey and analysis of the literature. Database for Advances in Information Systems 35(4), 6–102 (2004)

22. Natovich, J.: Vendor related risks in IT development: a chronology of an outsourced project failure. Technology Analysis & Strategic Management 15(4), 409–419 (2003)

23. Aubert, B.A., Patry, M., Rivard, S.: A tale of two outsourcing contracts - an agency-theoretical perspective. Wirtschaftsinformatik 45(2), 181–190 (2003)

24. Savas, E.S.: Privatization: the key to better government. Chatham House, Chatham (1987)

25. Young, S.: Outsourcing: uncovering the complexity of the decision. International Public Management Journal 10(3), 307–325 (2007)
26. Farneti, F., Young, D.: A contingency approach to managing outsourcing risk in municipalities. Public Management Review 10(1), 89–99 (2008)
27. National Research Council Committee on Outsourcing Design and Construction-related Management Services for Federal Facilities and National Research Council Board on Infrastructure and the Constructed Environment and the National Research Council Commission on Engineering and Technical Systems (2000)
28. Preker, A.S., Harding, A., Travis, P.: Make or buy decisions in the production of health care goods and services: new insights from institutional economics and organizational theory. Bulletin of the World Health Organization 78, 779–790 (2000)
29. Lane, J.E.: New public management. Routledge, London (2000)
30. Savas, E.S.: Privatization: the key to better government. Chatham House, Chatham (1987)
31. Dickson, G.: An analysis of vendor selection systems and decisions. Journal of Purchasing 2(1), 5–17 (1996)
32. Sreekumar, M.S.S.: A fuzzy multi-criteria decision making approach for supplier selection in supply chain management. African Journal of Business Management 3(4), 168–177 (2009)
33. Pearson, J.N., Ellram, L.M.: Supplier selection and evaluation in small versus large electronics firms. Journal of Small Business Management 33(4), 53–65 (1995)
34. Kagnicioglu, C.H.: A Fuzzy multiobjective programming approach for supplier selection in a supply chain. The Business Review 6(1), 107–115 (2006)
35. Sasikumar, G., Selladurai, V., Manimaran, P.: Application of analytical hierarchy process in supplier selection: an automobile industry case study. South Asian Journal of Management 13(4), 89–100 (2006)
36. Muralidharan, C., Anantharaman, N., Deshmukh, S.G.: Multi-criteria group decision making model for supplier rating. Journal of Sup. Chain Management 38(4), 22–33 (2002)
37. Ibrahim, D., Ugur, S.: Supplier selection using activity based costing and fuzzy present-worth techniques. Logistics Information Management 16(6), 420–426 (2003)
38. Bhutta, K.S., Faizul, H.: Supplier selection problem: a comparison of the total cost of ownership and analytical hierarchy process approaches. Supply Chain Management - An International Journal 7(3), 126–135 (2002)
39. Al-Faraj, T.N.: Vendor selection by means of data envelopment analysis. Bus. Rev. Cambridge 6(1), 70–77 (2006)
40. Singpurwalla, N.D.: A probabilistic hierarchical classification model for rating suppliers. Journal of Quality Technology 31(4), 444–454 (1999)
41. Fuh-Hwa, L., Hai, F., Lin, H.: The voting analytic hierarchy process method for selecting supplier. Int. J. Prod. Eco. 97, 308–317 (2005)
42. Kumar, K., Vart, P., Shankar, R.: A fuzzy goal programming approach for vendor selection problem in supply chain management (2004)
43. Kumar, M., Vrat, P., Shankar, R.: A fuzzy programming approach for vendor selection in a supply chain. Int. J. Prod. Eco. 101(2), 273–285 (2006)
44. Saaty, T.L.: Theory and applications of the analytic network process, p. ix. RWS Publications, Pittsburgh (2005) ISBN1-888603-06-2
45. Saaty, T.L.: The Analytic Hierarchy Process. McGraw-Hill, New York (1980)
46. Deng, J.L.: Control problems of grey system. Systems and Control letters 1, 288–294 (1982)
47. Hsu, T.H.: The study on the application of grey theory to the evaluation of service quality. In: The Proceeding of Conference of Grey System and Application, pp. 1–15 (1997)

48. Chang, Y.C.: The application of grey theory to purchasing-decision. Knowledge of Military Supplies 94 (1996)
49. Wen, K.L.: The quantitative research on the grey relational grade. Journal of Grey System 2(2), 117–135 (1999)
50. Luo, M., Kuhnell, B.T.: Fault detection using grey relational grade analysis. The Journal of Grey System 5(4), 319–328 (1993)
51. Chiang, K.S.: The introduction of grey theory. Gauli Publishing Co., Taipei (1997)
52. Wu, H.H.: The method and application of grey decisions. Chien-Kuo Institute of Technology, Chang-Hwa (1998)
53. Meade, L.M., Sarkis, J.: Analyzing organizational project alternatives for agile manufacturing processes-an analytical network approach. International Journal of Production Research 37, 241–261 (1999)
54. Saaty, T.L.: Rank from comparisons and from ratings in the analytic hierarchy/network processes. European Journal of Operational Research 168, 557–570 (2006)
55. Ashby, W.R.: Effect of control on stability. Nature (London) 155(3933), 242–243 (1945)
56. Wu, H.H.: The introduction of grey analysis. Gauli Publishing Co., Taipei (1996)
57. Deng, J.L.: Properties of relational space for grey system. In: Deng, J.L. (ed.) Essential Topics on Grey System—Theory and Applications, pp. 1–13. China Ocean, Beijing (1988)
58. Hsia, K.H., Wu, H.: A study on the data preprocessing in Gery relation analysis. J. Chin. Gery Syst. Assoc., 147–153 (1998)

A Study on XML Data Indexing for Web Based Application

Puspha Rani Suri[1] and Neetu Sardana[2]

[1] Computer Science and Application Department
Kurukshetra University, Kurukshetra
[2] Apeejay School of Management
Sector-8, Institutional Area, Dwarka, New Delhi
pushpa.suri@yahoo.com, sardana.neetu@gmail.com

Abstract. XML is a standard representation format for data operated in web environment, a variety of XML data management systems are becoming available for maintaining web information residing in web pages of their respective websites. There are various indexing techniques under each database system for transferring the data between the XML documents and the database system. This paper tries to sum some of the techniques up, classify them and to evaluate their advantages and disadvantages. Our purpose is to contribute to a better understanding of these indexing techniques and explore the suitability of these indexing techniques for web applications.

Keywords: Native database, XML enabled database, web applications, indexing.

1 Introduction

The simplicity of XML in combination with the Web has opened up new possibilities for moving data and for building new application architectures centered on common Internet protocols. Some of the changes brought about by XML include [1]

• Reduced dependence on proprietary data formats for applications.
• A new way for data exchange using XML instead of the formats defined by traditional EDI systems.
• The emergence of Web services as technology for discovering and connecting to Internet-based services.

XML based web applications are popular due to the ubiquity of web browsers, and the convenience of using a web browser as a client, sometimes called a thin client. The ability to update and maintain web applications without distributing and installing software on potentially thousands of client computers is a key reason for their popularity. Web revolution has exposed more and more developers to database issues because of the desire for ever more dynamic web sites. And the crown prince of web technologies, XML, has had the effect of increasing awareness of data design in general. Direct XML support in the database means there is no need for any other tool to translate data from an external provider into something that can be used immediately.

An important task of XML document storage in a database is indexing. Indexing the XML data improves the retrieval speed when querying. There are several methods for indexing in literature. In this paper we are presenting some of the indexing techniques and are compared and discussed their suitability for web application.

S. Dua, S. Sahni, and D.P. Goyal (Eds.): ICISTM 2011, CCIS 141, pp. 13–22, 2011.
© Springer-Verlag Berlin Heidelberg 2011

In the rest of the paper, Section 2 draws a comparison between two categories of databases- XML Enabled and Native XML. Section 3 evaluates various indexing techniques. Section 4 then describes suitability of indexing techniques for web applications. We then conclude in Section 5 summarizing the contents of the paper.

2 XML and Databases

2.1 XML Document

We can look at XML documents in two ways. In one view they serve as containers for data transferred over the web. In this sense they are data centric, since their meaning depends only on the structured data represented inside them. Typically data centric documents are regular in structure and homogeneous in content. They occur in business to business applications such as buyer-supplier trading automation, inventory database access and sharing, integration of commercial transactions and work flow. In another view, XML documents are application relevant objects – that is new data objects to be stored and managed by a Database management system. In this sense they are document-centric, since their meaning depends on the document as a whole. Their structure is more irregular. Examples include books, email messages, weather forecast etc.

Most of the web contents have prose oriented XML documents that changes over time in real world business systems. As information is communicated daily and is being created every day and hence becomes the part of web. Changing the structure of the XML document (it's possible for attributes to become elements and vice versa) - which is easy, natural, and useful to do in XML - forces a redesign of the database. Access management of XML documents and querying the XML database need special consideration. There is no one size fits all storage model to give the best performance and flexibility for each of these cases.

2.2 XML Enabled System and Native XML Database System (NXD) [2]

Presently there exist two approaches for storing and accessing XML documents in databases. The simplest approach to storing XML data in a relational database system is to use a long, character, string data type, such as CLOB. The major disadvantage of this approach is that though this preserves the structure of the XML document, but it fails to take advantage of the structural information available in the markup. Also, separate indexes need to be created to locate the required document based on one or more keys. Another method for storing XML documents in relational databases, called shredding, maps the information in XML document to specify columns of one or more tables. Shredding leads to the loss of details like element order, processing instructions, comments, white space, and other elements. Nature of the document becomes more irregular and less predictable; the mapping becomes inefficient and generic.

Another approach is to storage XML data in Native XML database (NXD). NXD is a collection of various documents arranged in hierarchical form. NXD stores the XML document in its intact form that is in original tree structure. NXD understands the structure of the XML document and is able to perform queries and retrieve the data taking advantage of knowledge of XML structure. Query languages like Xpath or Xquery are used to extract data residing in these documents. Both of these languages use path expression to navigate through the logical, hierarchical structure of an XML document.

The emerging trend towards XML applications creates the need for persistent storage of XML documents in databases. To efficiently query documents, XML databases require indices on the content and structure of frequently queried document fragments.

3 Indexing in Databases

Indexing plays an important role in any database to retrieve requested data without scanning all data. Index is a data structure that provides a fast access to XML Documents. To conform to the specific requirements of the XML data model, various XML index structures have been proposed recently. Basically indexing system uses numbering scheme to identify various XML nodes and the relationship like parent-child, ancestor-descendent or sibling between XML nodes. Numbering scheme assigns a unique identifier to each XML node. Identifiers are used as indexes as a reference to the actual node.

3.1 Indexing Techniques Classification

Recently an adopted classification for indexing methods is done on the basis of their structure and query processing strategy. The three classes of indexing are namely summary, structural join and sequence-based indexes [3].

3.1.1 Summary Based Indexing
Summary indexes also known as path based index identifies nodes of XML document using path. Path also helps in identifying parent-child relationship. Each element is associated with its position w.r.t. a path that reaches in the document from the root to that element having some label name. To know ancestor- descendent relationship, we need to merge independent executed path queries.

Summary indexes can be partial or total. As name suggests partial summary indexes, it contains only a subset representing the most common paths or path templates corresponding to frequent queries. Another summary index total consists of complete path of an element from root. With this, only single look up will help in retrieving the element.

Pros n cons:
More complex data structures are required, however, to cope up with path expressions.

Require more storage space to store the path as an index from root to leaf, if using partial summary index then it will save some storage space up to certain extent.

3.1.2 Structural Join Indexing
Structural join indexes index nodes consisting attributes and elements with a particular name or those whose content satisfies a given condition. In general, developers use these indexes in the context of join-based processing, in which the query processor first determines elements that match query tree nodes. It then joins the obtained sets together with a structural join algorithm, using specific location schemes to improve processing.

Each element is defined by index consisting of <doc-id, pre, post, level>. Structural join will consist of pair of acendent-decendent or parent-child as (a, b) where a ,b are part of same doc-id, a.level < b.level or a.level=b.level-1 and a.pre < b.pre < a.post.

Pros n cons:

Reduce querying overhead.

Problem occurs while insertion, as we are following pre and post order but can be reduces by introducing gaps.

Save storage space in comparison to summary based indexes.

3.1.3 Sequential Based Indexing

The sequence-based indexes, encodes XML data as well as queries as sequences. A query is processed by sequence matching. It avoids expensive join operations. It uses structural encoded sequences.

Pros n cons:

Content based querying requires additional processing.

Most indexing methods that disassemble a structured query into multiple sub-queries, and then join the results of these sub-queries to provide the final answers, our method uses tree structures as the basic unit of query to avoid expensive join operations.

3.2 Indexing Techniques

3.2.1 DataGuide

DataGuide [4] uses summary based indexing scheme that means that it is possible to answer queries directly by the index without consulting the original XML data. The DataGuide index is a structural summary of all elements that can be reached in the documents. This summary contains all unique labeled paths. That means that if two different paths in the document have the same labels, they are put together as one path in the structural summary.

DataGuide consist of graph-based structure and a hash-table. The graph contains the structural summary and the hash table the content of the XML data. While creating the index the XML document is traversed in a depth-first way. All target sets are collected. A target set is the collection of leaf elements that are at the end of a path. When a target set is not in the index the node of the set will be added to the path in the structural summary and the target set will be added to the hash table. A DataGuide describes every unique label path of a source exactly once.

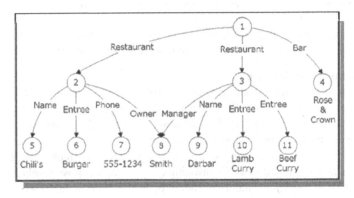

Fig. 1. Restaurant XML Document

A legal label path for extracting name of restaurant we use "Restaurant/Name" then for this we will have target set (Ts) as shown in Fig. 2.

If e=Restaurant/Entree then Ts(e) = {6,10,11}.

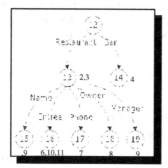

Fig. 2. Target Set in DataGuide

3.2.2 Index Fabric

Index Fabric [5] is a type of a summary based / path based indexing for XML documents. Index fabric encode path in the data as strings and insert these strings into an index that is highly optimized for string searching. Index Fabric defines the concept of raw paths and refined paths. Raw paths are the paths that exist in raw data and, thereby, help in answering adhoc queries over data with irregular structure. Refined paths help in optimizing frequently occurring and important patterns. This method is based on prefix encoding of strings and make use of specialized structure known as Patricia trees for storing encodings, look ups and path traversals. A Patricia trie is a simple form of compressed trie which merges single child nodes with their parents.For querying key lookup operator search for the path key corresponding to the path expression.

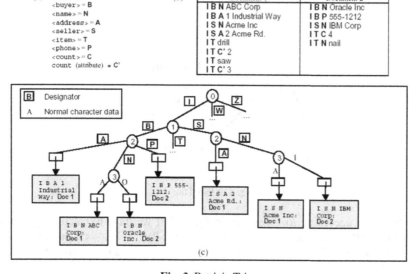

Fig. 3. Patricia Trie

3.2.3 XISS

XISS (XML Indexing and Storage System) [6] is structural join based indexing. It is based on extended preorder scheme. In this scheme each node is assigns a pair of numbers<order, number> to each node. This scheme determines fast ancestor-descendant relationship between nodes in the hierarchy of XML data. Ordering and numbering is done in such a way that for a tree node Y, X will be parent if

i) order(x) < order(y)
ii) order(y)+size(y) <= order(x)+size(x)   (3)

X will be sibling of Y if

i) X is predecessor of Y in preorder traversal and
ii) order(x)+size(x) <order(y)   (4)

Ordering is done as traversal per preorder of node. Size is total number of descendant of current node. X will be ancestor of Y if and only if

$$Order(x) < order(y) <<= order(x) +size(x)$$ (5)

The ancestor-descendant relationship for a pair of nodes can be determined by examining their order and size values

3.2.4 Signature File System

It interprets each document as k ary tree where k is the equal number of children of each node of a tree. It uses unique element identifier (uid) for identification of element in xml document. UID'S helps in for identifying structural relationship between the nodes in xml document. Formula used for calculating parent and child wrt to certain node are parent-id= [(i-2)/k +1] child(i,j)=k(i-1)+j+1 it occupies very less storage space as number of index entries are reduced [7].

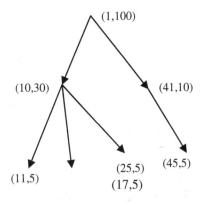

Fig. 4. XISS Tree

For Example according to given tree in figure 5 if we have to calculate parent of node having UID as 2. Parent (node (2)) =floor [(2-2)/2 + 1]   =1 and we have to calculate the second child of node 2, it will come out as Child (2nd child of node 2) = 2(2-1) +2 +1 = 5.

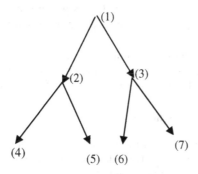

Fig. 5. Tree Structure Having UID

3.2.5 PRIX

The prüfer sequences indexing method (PRIX) [8] is sequence based indexing scheme. The encoding is based on deleting one element at a time. A tree has 0.. n-1 nodes, which are labeled in postorder started by 1. It works as follows: First delete the node with the smallest label, and put the parent of that node in the sequence. After that, delete the next node with the smallest label and also put the parent of that node in the sequence. This must be done until the whole tree has been processed. The sequence that is been made by putting each parent in it is called the prüfer sequence. The length of a prüfer sequence of a tree with n nodes is n-2. The original tree can be reconstructed from this sequence. Prüfer sequences can be represented as NPS (numbered prüfer sequence) and LPS (labeled prüfer sequence). In Fig. 6 each node has a pair of the form <label, postorder>. To reduce the size of the example the original labels has been replaced by the first letter of it and only the structure of the tree is used. The NPS of the tree in Figure 6 is <5, 4, 4, 5, 10, 9, 8, 9, 10 >. The way the NPS of a tree can be found is discussed earlier in this subsection. To get the LPS the only thing to do is to replace the numbers by the labels of the nodes. The LPS is then <B , A, A, B, BD, B, A, B, BD>.

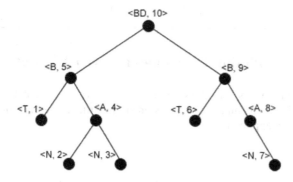

Fig. 6. Tree Consisting <label, postorder > for PRIX

When querying, the query is transformed to its prüfer sequence. Then there will be searched in the prüfer sequence of the tree for a subsequence that is the prüfer sequence of the query.

Complex queries are supported without breaking them into individual paths. The length of a prüfer sequence is linear to the number of elements that are in the original tree. The index size is also linear to the total number of tree nodes.

3.3 Evaluation of Indexing Schemes

I have drawn comparison among various indexing scheme on four parameters like Indexing type, how do we indentify nodes of document that is Identifier, how to calculate structural relationship like parent child or ancestor-descendent among nodes and querying.

Table 1. Evaluation of Various Indexing Schemes

Indexing Technique	Indexing Type	Identifier	Structural Relationship Search	Querying
DataGuide	Summary/path Based	One per element	Graph Based Traversal	Single path traversal
IndexFabric	Summary/path Based	One per documents	String search, do not maintain order of element	
XISS	Structural join	<order, size> One per element	Using numbering scheme	Complex queries are handled by joining simple path expressions
Signature file scheme(by Lee)	Structural join based	Unique element identifier	Calculation using formula	Regular path expression
Prix	Sequence based	<label, postorder> one per node	Using prüfer sequence	Suitable for simple and complex queries

4 Suitability of Indexing Techniques for Web Applications

There are different ways of classifying indexing used in databases. We are differentiating indexing techniques on basis on storage time, reference time and query time for various web applications.

4.1 Data Guide

Dataguide deals with the applications which have got very large repositories and their emphasis is maintenance of these storage systems. Basically warehouses come under this category, dataguide can handle such application efficiently.

4.2 Index Fabric

Application has information taken by multiple suppliers where buyers can query about some product. In these application data have irregularity or changing organization. Some web application that can have catalog of different vender data like deals2buy.com comes under this indexing scheme.

4.3 Signature File Scheme

It is widely recognized that throughput is the most important performance parameters for any web application. Throughput is defined as the number of requests served per unit of time—typically requests per second. Less is the response time more effective is the site to work with. Signature file scheme is suitable for web applications where response time is most important factor. Any ecommerce application like Amazon, eBay that requires less response time while processing of documents/objects (access/search of web pages, music/video files, directory/phone book type of data etc).

4.4 XISS

Web traffic is one of the dominant components of internet traffic. It is closely tied to contents of web pages. Web application that could handle heavy traffic like CBSE result site, which is accessed my multiple people, when result is out.

4.5 Prix

Prix indexing technique can be used to handle complex queries over large repositories. Web repositories, such as the Stanford WebBase repository, manage large heterogeneous collections of Web pages and associated indexes. For effective analysis and mining, these repositories must provide a query interface that supports "complex expressive Web queries". Different customers may exhibit different navigational patterns and hence invoke different functions in different ways and with different frequencies.

Prix queries contain branches and correspond to a small tree, called a twig. We can also make use of this indexing technique for sites like "makemytrip.com" in which databases are large and that can handle complex queries.

5 Conclusions

We studied five indexing schemes that are used in XML databases for storage. Every indexing technique has its own query type that it can handle the best. We had explored there suitability of indexing for web applications. We found that the choice of indexing scheme depends upon the purpose of a web application we are dealing with.

Web application where emphasis is query should be answered quickly, there we found signature file system is the best scheme as it uses formula to calculate the structural relationship between nodes. If the application has heavy requests most of the time that means application has to handle multiple queries at a time then XISS is preferred. Web application used for complex queries basically used for analysis, mining then PRIX is efficient. Most of web applications are getting its data from multiple sources with respect to different vendors on a one platform then in such application index fabric scheme is found to suitable. Dataguide is preferred for maintaining large warehouse.

References

1. Moller, A., Shwartzbach, M.I.: An introduction to XML and web technologies. Addison-Wesley, Reading (2006)
2. Bourret, R.: XML and databases (1999-2005),
   http://www.rpbourret.com/xml/XMLAndDatabases.htm
3. Catania, B., Maddalena, A., Vakali, A.: XML document indexes: a classification, pp. 64–71. IEEE Computer Society, Los Alamitos (2005)
4. Weigel, F.: Content-aware dataguides for indexing semi-structured data. University of Munich, Germany (2003)
5. Cooper, B., Sample, N., Franklin, M.J., Hjaltason, G.R., Shadmon, M.: A fast index for semistructured data. In: The Proceedings of the Twenty-Seventh International Conference on Very Large Data Bases, pp. 341–350 (2001)
6. Li, Q., Moon, B.: Indexing and querying XML data for regular path expressions. In: The Proceedings of the Twenty-Seventh International Conference on Very Large Data Bases (VLDB), Rome, Italy, pp. 361–370 (2001)
7. Lee, Y.K., Yoo, S.-J., Yoon, K.: Index structures for structured documents. ACM, New York (1996)
8. Raw, P.R., Moon, B.: PRIX: indexing and querying XML using Prüfer sequences. In: The Proceedings of the International Conference on Data Eng. (ICDE), pp. 288–300. IEEE CS Press, Los Alamitos (2004)

Knowledge Sharing Intentions among IT Professionals in India

Binoy Joseph[1] and Merin Jacob[2]

[1] Rajagiri College of Social Sciences, Kochi, 683 104 Kerala, India
[2] IAL Shipping Company Limited, Kochi 682 014 Kerala, India
binoyjoseph@rajagiri.edu, mjacob@ial.com

Abstract. In today's knowledge-based economy, an organization's ability to strategically leverage knowledge has become a crucial factor for global competitiveness. Knowledge is considered an important source of establishing and maintaining competitive advantage. Specifically, knowledge sharing and resultant knowledge creation are crucial for organizations to gain and sustain competitiveness. The aim of this study is to develop an understanding of the factors supporting or inhibiting individuals' knowledge-sharing intentions. Theory of Reasoned Action is employed and augmented with extrinsic motivators, social-psychological forces and organizational climate factors that are believed to influence individuals' knowledge sharing intentions. The study proved that organizational climates and social-psychological factors have a positive effect on the intension for knowledge sharing. Additionally it was found that the attitude towards knowledge sharing and subjective norm affects individual's intension to share knowledge. Contrary to the common belief, it was found that anticipated extrinsic rewards exert a negative effect on individuals' knowledge-sharing attitudes.

Keywords: Knowledge Sharing Intentions, Information Technology Employees, India.

1 Introduction

Knowledge is commonly acknowledged as a critical economic resource in the present global economy and it is progressively becoming evident that organizations should possess the right kind of knowledge in the desired form and context to be successful. Knowledge can be defined as a combination of experience; values, contextual information and expert insight that help evaluate and incorporate new experience and information [9].Knowledge sharing is critical to a firm's success [7] as it leads to faster knowledge deployment to portions of the organization that can greatly benefit from it [34].

For countries like India, which have joined this process of liberalization and globalization recently, this brings big issues. Now the platform of competition for Indian firms is global, because even a purely domestic Indian firm can face competition from multinational corporations or from imports. To survive and grow, Indian firns have to develop knowledge assets as well as to take better care of their already existing knowledge assets.

S. Dua, S. Sahni, and D.P. Goyal (Eds.): ICISTM 2011, CCIS 141, pp. 23–31, 2011.

The objective of this study is to deepen the understanding of the factors that increase or lessen employees' tendencies to engage in knowledge sharing behaviors. Since knowledge sharing behaviors are likely to be influenced not only by personal motivations but also by contextual forces [41]. A theoretical frame is applied in which extrinsic motivators, psychological forces and organizational climate are integrated with the Theory of Reasoned Action (TRA) [1].

2 Knowledge Sharing

Knowledge sharing is the process intended at exploiting existing and accessible knowledge, in order to transfer and apply this knowledge to solve specific tasks better, faster and cheaper than they would otherwise have been solved. Usually sharing knowledge can be defined as the dissemination of information and knowledge through the whole department and/or organization. [19] describes the process of "knowledge sharing" as enabling sharers to guide sharees through sharers' thinking and/or using their insights to assist sharees to examine their own situations. Those who share knowledge can refine their shared knowledge by the interactive dialogue process; those who shared knowledge can obtain knowledge from sharers. Knowledge sharing occurs when an individual is willing to assist as well as to learn from others in the development of new competencies.

2.1 Bridging Intention and Knowledge Sharing Behavior

The means by which knowledge is shared within organizations is a core issue in KM and is considered a firm's most valuable resource because it embodies intangible assets and creative processes that are difficult to imitate [12, 18]. Knowledge sharing has been explained on the basis of exchange theory [5] and the notion that a successful exchange process creates an obligation to reciprocate anticipated future monetary and non-monetary benefits [11]. Knowledge is one of the fundamental factors in an enterprise's momentum and goes part and parcel with the effectiveness and competitiveness of its operations and behavior [42]. The concept of tacit knowledge was defined by [28, 29]. He posited that all knowledge is either tacit or explicit knowledge and stressed the importance of a "personal" way of communicating knowledge. While explicit knowledge is codified and transferred mostly by technology, tacit knowledge is more embedded in social relations and transferred primarily through direct contact and observation of behavior.

2.2 Organizational Innovation

Continuous knowledge management (KM) can promote organizational innovation and play a key role in the organization's success [24]. In addition, problems such as maintaining, locating and applying knowledge have intensified the importance of organizational KM [3]. Innovation is defined as the generation, acceptance and implementation of new ideas, processes, products or services [39]. [4] defined innovation as "the successful implementation of creative ideas within the organization". Knowledge dissemination and responsiveness to knowledge, in other words knowledge sharing, have been put forward as the two most important components impacting upon innovation due to their ambiguous and unique nature within the firm [35, 12]. Therefore, knowledge creation is hard to achieve and other individuals randomly benefit from this donated knowledge [26, 40].

2.3 Knowledge Management

Knowledge sharing is the corner-stone of many organizations' knowledge-management (KM) strategy. Knowledge sharing practices and initiatives often form a key component of knowledge management programs, in terms of organizational and individual learning (e.g. [3,8,21,23,33]. Knowledge sharing (KS), the process by which an individual imparts his or her knowledge (e.g., expertise, insight, or understanding in a tacit or explicit format) to a recipient[12], is a key activity of effective knowledge management (the other activities include: creating, using, storing and identifying, [14]). KS among employees represents attempts and contributions towards creating an organizational knowledge database - and is attracting growing interest on the part of both practitioners and researchers alike [6]. The means by which knowledge is shared within organizations is a core issue in KM and is considered a firm's most valuable resource because it embodies intangible assets and creative processes that are difficult to imitate [12,18]. [25] indicated that knowledge management could be defined as a method for simplifying and improving the process of sharing, distributing, creating, and understanding company knowledge. Within Ajzen's theory, attitude toward behavior "refers to the degree to which a person has a favorable or unfavorable evaluation appraisal of the behavior in question" [2]. For regarding attitude toward a given behavior, behavioral beliefs describe subjective probability that the behavior in question would result in a particular outcome, and evaluations describe the implicit valuation or payoff that an individual associates with the outcome.

2.4 Organizational Learning and Knowledge Sharing

[32] claims "organizational learning and knowledge sharing are intimately connected (p. 189)". The knowing process is composed of sharing, thinking and learning components that have a reciprocity relationship. Knowledge sharing enables managers to keep the individual learning flowing throughout the company and to integrate it for practical applications. A common classification of organizational knowledge [22] comprises explicit knowledge, which can be documented and shared, and implicit or tacit knowledge, which resides in the minds, cultures, and experiences within the organization [30]. Turnover intentions is seen by [31] as a mental decision intervening between an individual's attitude regarding a job and the stay or leave decision and that can be regarded as an immediate antecedent to stay, or leave [31].

2.5 Organizational Climate and Culture

A climate which is mobilized for sharing and learning enables employees to acquire knowledge and skills, and to replenish creativity, imagination, exploration, discovery, and intentional risk taking [20].A constructive organizational climate will encourage individuals to have a positive attitude to their own learning, and to be willing to develop, overcoming their resistance to learning. According to [13], organizational culture involves six major categories: information systems, people, process, leadership, and reward system and organization structure. Organizational culture can be defined as the shared, basic assumptions that an organization learnt while coping with the environment and solving problems of external adaptation and internal integration that are taught to new members as the correct way to solve those problems [27].

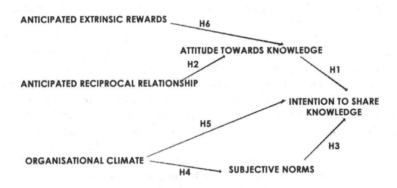

Fig. 1. Research Model

3   Research Methodology

The study was conducted among 125 IT knowledge workers in India. The Research design is Explanatory and the sampling design used for the study was Multi Stage Random Sampling. Companies with at least ten employees were selected and data was collected by distributing a questionnaire used in a study by Gee-Woo Bock and Robert W. Zmud. In this study, Ordinal scale with 5 point Likert scale has been used,. The questionnaire contained nine constructs. They are anticipated extrinsic rewards (with two items), anticipated reciprocal relationships (with five items), sense of self – worth (with five items), affiliation (with four items), innovativeness (with three items), fairness (with three items), subjective norm (with six items), attitude towards knowledge sharing (with five items) and intention to share knowledge with five items). The reliability value was 0.778. The data was analyzed using Regression.

3.1   Objective of the Study

3.1.1   Main Objective
To study the effect of extrinsic motivators, socio-psychological forces and organizational climate on the intention for knowledge sharing among knowledge workers.

3.1.2   Secondary Objectives
1. To study the relationship between attitude toward knowledge sharing and intention to share knowledge.
2. To study the relationship between anticipated reciprocal relationships and the attitude toward knowledge sharing.

3. To study the relationship between subjective norm to share knowledge and intention to share knowledge. To study the relationship between subjective norm to share knowledge and attitude toward knowledge sharing.
4. To study the relationship between organizational climate and subjective norm to share knowledge.
5. To study the relationship between organizational climate and the intention to share knowledge.

3.2 Hypotheses

S.No	Hypotheses	R^2 value	Decision
1	The more favorable the attitude toward knowledge sharing is, the greater the intention to share knowledge will be.	0.314	Supported
2	The greater the anticipated reciprocal relationships are, the more favorable the attitude toward knowledge sharing will be.	0.234	Supported
3	The greater the subjective norm to share knowledge will be, the greater the intention to share knowledge will be.	0.375	Supported
4	The greater the extent to which the organizational climate is perceived to be characterized by fairness, innovativeness and affiliation, the greater the subjective norm to share knowledge will be.	0.337	Supported
5	The greater the extent to which the organizational climate is perceived to be characterized by fairness, innovativeness and affiliation, the greater the intention to share knowledge will be.	0.255	Supported
6	The greater the extent to which the organizational climate, extrinsic motivators and social psychological forces, the greater the intention to share knowledge will be.	0.449	Supported

4 Discussion of Findings

The following insights were drawn from the above findings:

- Greater the attitude towards knowledge sharing greater the intension for sharing knowledge. If a person intends to do a behavior then it is likely that the person will do it. According to [1], Behavioral intention is a function of both attitudes toward a behavior and subjective norms toward that behavior, which has been found to predict actual behavior. Thus the relationship between attitude towards knowledge sharing and intension to share knowledge is established.
- It was also proved that the greater the anticipated reciprocal relationships more favorable the attitude toward knowledge sharing. In Soonhee Kim Hyangsoo Lee's research titled "Organizational Factors Affecting Knowledge Sharing Capabilities

In Electronic Government", Socio-psychological forces are pervasive within both organizational and societal cultures as they are seen as a key to "good attitudes" and knowledge worker motivational issues. [17]

- The relationship between attitude towards knowledge sharing and subjective norm regarding knowledge sharing was disproved. As per Theory of Reasoned Action, a person's intentions are guided by two things: the person's attitude towards the behavior and the subjective norm. So far no relation has been established between subjective norm and attitude towards performing a behavior in the existing literature.

- An individual's sense of self-worth through knowledge sharing does not enhance the salience of the subjective norm regarding knowledge sharing. Subjective norms look at the influence of people in one's social environment on his/her behavioral intentions; the beliefs of people, weighted by the importance one attributes to each of their opinions, will influence one's behavioral intention. The subjective norm as proved in the study can generate an intension in performing a task, but not necessarily generate self – worthiness.

- It was found that there exists no relationship between sense of self- worth and attitude towards knowledge sharing. [6] has identified that self-efficacy as a determinant to knowledge sharing. However an individual with a sense of self – worth need not have self – efficacy. Thus the relationship between attitude towards knowledge sharing and sense of self worth cannot be proved.

- An organizational climate conducive to knowledge sharing(operationalised here as fairness, innovativeness, and affiliation) exerts a strong influence on the formation of subjective norms regarding knowledge sharing; Salient aspects of organizational climate which helps in understanding individuals' tendencies toward knowledge sharing are a climate in which individuals are highly trusting of others and of the organization [15].

- Organizational climate also directly affects individuals' Intentions to engage in knowledge sharing behaviors. An individual' intensions toward knowledge sharing can be intensified in a climate in which individuals are highly trusting of others and of the organization [15].

- When the behavior being studied is strongly reflective of collective action. Subjective norms are likely to affect behavioral intentions. As per Theory of Reasoned Action, a person's intentions are guided by two things: the person's attitude towards the behavior and the subjective norm.

4.1 Suggestions

- Approach known knowledge hoarders and those unwilling to participate in the knowledge sharing initiative in a diplomatic and non-confrontational manner. Employees should not be tackled with a "do it or else" attitude.

- Implement a peer-to-peer support center for your knowledge community so that individual knowledge bears won't feel as though they're alone in their efforts. Put a "face" to the knowledge community by creating an "About Us" section profiling key knowledge contributors and also implement a knowledge contributors "Hall of Fame" or "Contributor of the Month" to highlight workers who go above-and-beyond.

- Create a mentoring program so newcomers don't feel estranged from the established knowledge community. And also conduct "Social Network Analysis" wherein key knowledge sources, sinks, and relationships (links or ties) in the organization can be identified.
- Develop learning and knowledge sharing competencies as part of the annual performance plans and organize knowledge fairs/ knowledge exchanges with the external community on selected topics of interest.

4.2 Limitations

- The concept of organizational unlearning, that is the elimination of old, obsolete organizational knowledge in order to make room for the development of new adaptive capacities was not taken in to consideration in this study.
- The procedures for identifying candidate antecedent motivating factors might have overlooked barriers of knowledge sharing acknowledged by others, that is, natural barriers of time and space.

5 Conclusion

Effective knowledge sharing cannot be forced or mandated. Firms desiring to institutionalize knowledge- sharing behaviors must foster facilitative work contexts. More open meetings are to be conducted. Discourage carrot and stick approach towards unwilling employees. A peer-to-peer support center can be implemented for the knowledge community so that individual knowledge bears won't feel as though they're alone in their efforts. A "Social Network Analysis" can be conducted wherein key knowledge sources, sinks, and relationships (links or ties) in the organization can be identified. Contributions should be acknowledged so that active knowledge bearers have a sense of recognition for their hard work. For example reward innovative ideas by introducing "Contributor of the Month". Conduct knowledge fairs/ knowledge exchanges with the external community on selected topics of interest. Changes won't happen overnight; it takes time for cultural behaviors to evolve. The true quality of a knowledge sharing initiative lies not so much with the tools, but with those who nourish it; and a cooperative knowledge community can form the backbone of many successful IT systems. Bertrand Russell approves this notion when he quotes, "The only thing that will redeem mankind is cooperation."

References

1. Ajzen, I., Fishbein, M.: Understanding Attitudes and Predicting Social Behavior. Prentice Hall, Englewood Cliffs (1980)
2. Ajzen, I.: The Theory of Planned Behavior. Organizational Behavior and Human Decision Processes 50(2), 179–211 (1991)
3. Alavi, M., Leidner, D.E.: Review: Knowledge Management and Knowledge Management Systems: Conceptual Foundations and Research Issues. MIS Quarterly 25(1), 107–136 (2001)

4. Amabile, T.: Motivating Creativity in Organizations: on Doing What You Love and Loving What you Do. California Management Review 40(1), 39–57 (1997)
5. Blau, P.M.: The Dynamics of Bureaucracy, 2nd rev. edn. University of Chicago Press, Chicago (1963)
6. Cabrera, A., Collins, W., Salgado, J.: Determinants of Individual Engagement in Knowledge Sharing. International Journal of Human Resource Management 17(2), 245–264 (2006)
7. Davenport, T.H., Prusak, L.: Working Knowledge: How Organizations Manage What They Know. Harvard Business School Press, Boston (1998)
8. Earl, M.: Knowledge Management Strategies: Toward a Taxonomy. Journal of Management Information Systems 18, 215–233 (2001)
9. Gammelgaard, J., Ritter, T.: The Knowledge Retrieval Matrix: Codification and Personification as Separate Strategies, working paper No. 10, Department of Management, Politics and Philosophy. Copenhagen (2004)
10. Grant, R.M.: The Resource-Based Theory of Competitive Advantage; Implications for Strategy Formulation. California Management Review 33, 114–135 (1991)
11. Gouldner, A.W.: The Norm of Reciprocity: A Preliminary Statement. American Sociological Review 25(2), 161–178 (1960)
12. Grant, R.M.: Toward a Knowledge-Based Theory of the Firm. Strategic Management Journal 17(special issue), 109–122 (1996)
13. Gupta, A.K., Govindarajan, V.: Knowledge Flows within Multinational Corporations. Strategic Management Journal 21, 473–496 (2000)
14. Heisig, P.: Harmonisation of Knowledge Management – Comparing 160 KM Frameworks Around the Globe. Journal of Knowledge Management 13(4), 4–31 (2009)
15. Hinds, P., Pfeffer, J.: Why organizations Don't Know What They Know: Cognitive and Motivational Factors Affecting the Transfer of Expertise. In: Ackerman, M., Pipek, V., Wulf, V. (eds.) Beyond Knowledge Management: Sharing Expertise. MIT Press, Cambridge (2003)
16. Heisig, P.: Harmonisation of Knowledge Management – Comparing 160 KM Frameworks Around the Globe. Journal of Knowledge Management 13(4), 4–31 (2009)
17. Huber, G.P.: Transfer of Knowledge in Knowledge Management Systems: Unexplored Issues and Suggested Studies. European Journal of Information Systems (2001)
18. Liebeskind, J.P.: Knowledge, Strategy, and The Theory of the Firm. Strategic Management Journal 17(special issue), 93–108 (1996)
19. McDermott, R.: Why Information Technology Inspired but Cannot Deliver Knowledge Management. California Management Review 41(4), 103–117 (1999)
20. McGill, M.E., Slocum Jr., J.W.: Management Practices in Learning Organizations. Organizational Dynamics 21(1), 5–17 (1992)
21. Nahapiet, J., Ghoshal, S.: Social Capital, Intellectual Capital, and the Organizational Advantage. Academy of Management Review 23(2), 242–266 (1998)
22. Nonaka, I.: The Knowledge-Creating Company. Harvard Business Review 69(6), 96–104 (1991)
23. Nonaka, I.: A Dynamic Theory of Organizational Knowledge Creation. Organizational Science 5(1), 14–37 (1994)
24. Nonaka, I., Takeuchi, H.: The Knowledge-Creating Company-How Japanese Companies Create the Dynamic of Innovation. Oxford University Press, New York (1995)
25. Nonaka, I., Konno, N.: The Concept of "Ba": Building a Foundation for Knowledge Creation. California Management Review 40(3), 40–55 (1998)

26. Nonaka, I., Peltokorpi, V.: Objectivity and Subjectivity in Knowledge Management: A Review of 20 Top Articles. Knowledge and Process Management 13(2), 73–82 (2006)
27. Park, H., Ribière, V., Schulte, W.: Critical Attributes of Organizational Culture that Promote Knowledge Management Technology Implementation Success. Journal of Knowledge Management 8(3), 106–117 (2004)
28. Polanyi, M., Walshe, F.: The Tacit Dimension. Doubleday, Garden City (1966)
29. Polanyi, M.: Personal Knowledge: Towards a Post-Critical Philosophy. University of Chicago Press, Chicago (1958)
30. Rowley, J.: Knowledge Management-The New Librarianship? From Custodians of History to Gatekeepers to the Future. Library Management 24(8), 433–440 (2003)
31. Sager, J.K., Griffeth, R.W., Hom, P.W.: A Comparison of Structural Models Representing Turnover Cognitions. Journal of Vocational Behaviour 53(2), 254–273 (1998)
32. Spinello, R.A.: The Knowledge Chain. In: Cortada, J.W., Woods, J.A. (eds.) The Knowledge Management Yearbook 2000-2001, pp. 189–207. Butterworth-Heinemann, Woburn (2000)
33. Sveiby, K.: The New Organizational Wealth: Managing & Measuring Knowledge Based Assets. Berrett-Koehler Publishers (1997)
34. Syed-Ikhsan, S.O.S., Rowland, F.: Knowledge Management in a Public Organization: A Study on the Relationship Between Organizational Elements and the Performance of Knowledge Transfer. Journal of Knowledge Management 8(2), 95–111 (2004)
35. Teece, D.J.: Strategies for Managing Knowledge Assets: The role of Firm Structure and Industrial Context. Long Range Planning 33(1), 35–54 (2000)
36. Takeishi, A.: Bridging Inter- and Intra-firm Boundaries: Management of supplier Involvement in Automobile Product Development. Strategic Management Journal 22, 403–433 (2001)
37. Teece, D.J.: Profiting from Technological Innovation: Implications for Integration, Collaboration, Licensing and Public Policy. Research Policy 15, 285–305 (1986)
38. Thompson, P., Fox-Kean, M.: Patent Citations and the Geography of Knowledge Spillovers: A Reassessment. American Economic Review 95, 450–460 (2005)
39. Thompson, A.: Organization in Action. Prentics-Hall, Englewood Cliff (1965)
40. Von Krogh, G.: The Communal Resource and Information Systems. The Journal of Strategic Information Systems 11, 85–107 (2002)
41. Yoo, Y., Torrey, B.: National Culture and Knowledge Management in a Global Learning Organization. In: Choo, C.W., Bontis, N. (eds.) The Strategic Management of Intellectual Capital and Organizational Knowledge, pp. 421–434. Oxford University Press, Oxford (2002)
42. Wiig, K.: People-Focused Knowledge Management: How Effective Decision-Making Leads to Corporate Success. Elsevier, Burlington (2004)

Critical Success Factors for CRM Implementation: A Study on Indian Banks

Sarika Sharma[1] and D.P. Goyal[2]

[1] JSPM Eniac Institute of Computer Application, Pune, India
[2] Management Development Institute, Gurgoan, India
sarika4@gmail.com, dpgoyal@mdi.ac.in

Abstract. Many organizations have identified the need of becoming more customer-oriented with increased global competition. Services sectors including airlines, hotel, and banking have taken efforts to implement customer relationship management (CRM), as a step towards achievement of success in their business. Technological improvements are responsible for majority of innovations in banking over the past few decades. It started with the introduction of personal computers and has come a long way with computerized banking, internet banking, ATMs, and CRM as the latest buzz. CRM systems are implemented by banks to gain competitive edge over their competitors. Also, business strategies may be formulated based on the predictions given by the intelligent data mining tools. The paper intends to identify the critical factors which affect the successful implementations of CRM systems.

Keywords: CRM implementation, intelligent data mining tools, critical success factors, Indian banks.

1 Introduction

To survive in today's competitive environment, an organization needs to be customer-focused, as it is the customer who makes or breaks the organization success. Customer Relationship Management (CRM) is a highly fragmented environment, and has come to mean different things to different people. As per definition in [1], CRM is a management approach that enables organization to identify, attract, and increase retention of profitable customers by managing relationships with them.

CRM is a broad approach to doing business. It is holistic in that it encompasses all aspects and functions of a company, focusing on managing the relationship between customer and company just as much between company and customer. According to [2], the financial services sector has been perceived by many as leading the adoption of CRM, and defining many leading practices.

Over the past few years, many companies have implemented Customer Relationship Management Systems and have turned their attention to "mining" their customer databases. Now they can learn more about customers, use that information to make appropriate offers to customers, and understand which offers succeed. Customer information can be distributed across an enterprise for use in analysis and marketing.

S. Dua, S. Sahni, and D.P. Goyal (Eds.): ICISTM 2011, CCIS 141, pp. 32–40, 2011.

Data mining, data exploration and knowledge discovery are all terms that create an image of the demanding and sometimes tedious search to uncover insights that are neither obvious to competitors nor easy for competitors to duplicate. CRM depends on data analysis activities to uncover directions and opportunities and highlight warning indicators for CRM initiatives. The combination of good customer information, data mining, and technology enables companies to better understand their customer base and communicate with them more effectively. Once a firm is actively using customer information to make decisions about how, when and what to market to customers, they often increase the volume of targeted customer contacts.

1.1 Customer Relationship Management Systems

CRM systems include data warehouse as its backbone and when intelligent data mining tools are applied on the data warehouse maintained by the company (to store all the details about the customer behavior), hidden patterns are generated and predictions can be made.

CRM is more than just a set of technologies: it is a process. This fact will be of significant importance to information technology professionals who will be asked to support CRM with information and applications. Furthermore, it is intended to be a repeatable process to ensure ongoing, continually improving, and consistent results.

1.2 Major Components of CRM Systems [3]

A study from available CRM vendors indicates that the following four software modules are available:

1.2.1 Sales Automation Module
This module is designed to automate sales related tasks such as sales-customer interaction, contact scheduling, sales campaign and promotion activities, sales lead tracking, sales trends, etc.

1.2.2 Marketing Automation Module
This module is designed to automate marketing related tasks such as marketing analysis and planning, marketing campaign activities, products promotion and scheduling, etc.

1.2.3 Customer Service and Support Module
This module is designed to document and manage customer information and activities thereby building strong and long lasting customer relationships.

1.2.4 Reporting and Analysis Tools
These tools are a set of software and technology to enable cross-channel, complete view of sales, marketing, and customer service information that stored on company's databases for analytic reporting and analysis.

In [4], the author points out that CRM software can only be an enabler to implement a customer strategy but it does not ensure a successful customer relationship. However there are various other factors, which strongly affect the success of a CRM initiative taken by the firm.

2   Review of the Literature

In [5], it was noted that CRM offers so much more than the contact management systems of the past - but only for those organizations that can successfully embrace an open culture underpinned by access to the key business systems. Visibility across finance, sales, service and marketing provides an organization with the joined up business understanding required to achieve return on investment, whether from credit control or stock reduction, and drive operational strategy, while supporting the CRM objective of enabling improved customer value. CRM implementation is a not off-risk and needs a customer focus organization; it may also need recognizing the current business processes to support the implementation [6, 7].

According to [8], the financial services sector has been perceived by many as leading the adoption of CRM, and defining many leading practices. However, while financial services companies, have pioneered in some cases, in others they have fallen victim to common mistakes in their approach to CRM investments. These shortcoming he related to "how" they went about applying CRM services to their CRM program, and on "what" objectives they focused these activities. Data mining software can help find the "high-profit" gems buried in mountains of information. However, merely identifying the best prospects is not enough to improve customer value. The Data Mining results should fit into the execution of marketing campaigns that enhance the profitability of customer relationships. Unfortunately, Data Mining and Campaign Management technologies have followed separate paths – until now. Your organization stands to gain a competitive edge by understanding and utilizing this new union [9]. The author of [10] discusses how even with the finest marketing organizations, the success of marketing ultimately comes down to the data.

A successful CRM must start from the top-down getting involvement from upper management as well as all key areas of the business will reduce risks and ensure success [11, 12]. In [13], it is suggested that retaining the right customers or winning back defecting customers can be accomplished by following these steps: identify potential defectors, communicate with customers, listen to front-line people, treat valuable customers well, be fair, even when you do not have to, use exit barriers, carefully, win the right customers back. The author of [14] recommends the capability to integrate two or more delivery channels through shared technology has only recently been deployed in any significant way. Today, a handful of retail banks can boast of globally integrated delivery channels that are built on standard technology principles.

Challenges like these require a complete, customer-focused banking solution that improves customer service and responsiveness, provides an open cross-bank platform for collecting all customer information from a myriad of systems and delivering a unified view of each customer, boosts the ability to target marketing programs based on comprehensive analytical tools, enhances flexibility to adapt to processes and working methods to meet rising customer expectations as given by [15] in their white paper. In [16], it is emphasized that CRM is essential to compete in the today's market place and the more effectively the information about customers can be used to meet their need, more success is in the business. But the route to successful business requires that you understand your customers and their requirements essentially. In a CRM system, there's a range of allowable (and even expected) data quality that depends on the specific data element. Some things *do* have to be perfect, such as unique keys, internal security information, order quotes, order history, and anything that is subject to an audit. But, other things can be just an approximation or can be missing altogether which can lead to failures [17].

In [18], the author tried to find out the answer of the question, "How do you make CRM work?" The answers were interesting:

- The company needs to have a clear sense of journey. CRM is not a destination; it's merely some of the tools that facilitate the business becoming more customer-centric;
- There needs to be a clear customer strategy that is implemented across the company end executed consistently though all points of customer engagement;
- The organization should be designed around execution of the customer strategy, not around business functions;
- The information architecture should be just that, providing the right information to support every customer engagement and supporting rapid change.

The authors of [19] and [20] have mentioned the critical success factors of CRM implementation from strategic and tactical perspectives but have not cited the level of project implementation, whereas [21] argues that one study of 202 CRM projects found that only 30.7% of the organizations said that they have achieved improvements in the way they sell to and service customers. Most implementations fail because the organizations fail to adopt a clear strategy and fail to make appropriate changes to its business processes. Thus, without comprehensive understanding of customer focused objectives, the company will find it difficult to really leverage this cutting-edge technology [22].

In [23] it is revealed that the actual results being measured and achieved in organisations are consistently and significantly falling short of expectations. This finding, in fact, supports the general criticisms that CRM is only producing disappointing results in practice. It was recommended that organisations concentrate on implementation issues very carefully, ensuring that their adoption of CRM is well executed and therefore successful in terms of delivering real results. It is advocated in [24] that while companies continue to invest in CRM enhancements, the technology is still under-delivering in the call center. Customer service staffers are still grappling with disparate systems that are often not underpinned by a clear customer centric strategy.

From the above survey of literature it can be seen that there are various methods to implement the CRM systems and technology supports these in form of data mining and warehousing. It is also revealed that there is a lack of research with regard to some critical factors of CRM implementation. In general CRM project is complex and not easy to adopt. Many companies are failing with the CRM solutions, which indicates that there are other hidden factor determining the success or failure of these systems are needed to be identified. Hence, we take the study with following objective.

3 Research Objectives and Methodology

The main objective of this research was to identify the hidden factors that affect the success of CRM systems. Two leading private banks: an Indian bank (ICICI Bank) and a foreign bank (Standard Chartered Bank) were selected for the purpose of the study. These banks have already implemented CRM and are operating in India. The banks needed to manage and analyze the huge volumes of data captured by its OLTP systems. In order to carry out a number of critical business activities and provide excellent customer service, the selected organizations had to get the right information, to the right people, at the right time.

- Secondary data for research was collected from related books, publications, annual reports and records of the organization under study.
- Primary data for research was collected using questionnaire cum interview methodology. The questionnaire was developed on the basis of existing literature and was pre-tested on 30 managers from the actual number to be interviewed for checking its reliability and content validity. The suggestions collected during pre-testing of the questionnaires were incorporated in the questionnaire. Thus, pre-tested and modified questionnaire was administered to all the respondents. 140 experienced managers (ICICI bank: 70, Standard Chartered Bank: 70), who were working for the customer relationship management, participated in the study.
- Data was collected on 5 point scale. Pearson's Correlation and average scores were computed using SPSS package. For identifying the success and hindrance factors extensive literature review was done. Table 1 illustrates the various factors.

4   Empirical Data Analysis

A total of 50 factors were arranged in the questionnaire to identify them as success or hindrance factors. Respondents were asked to rate each factor on a five point scale. Eleven factors are extracted using factor analysis. The summary of total variances of the factors is given in Table 1.

Table 1. Total Variance Explained

Component	Extraction Sums of Squared Loadings			Rotation Sums of Squared Loadings		
	Total	% of Variance	Cumulative %	Total	% of Variance	Cumulative %
1	14.424	29.437	29.437	6.550	13.368	13.368
2	6.678	13.628	43.065	5.934	12.111	25.479
3	2.518	5.138	48.203	5.646	11.523	37.002
4	2.198	4.485	52.688	3.232	6.597	43.599
5	1.835	3.744	56.433	2.915	5.949	49.548
6	1.659	3.386	59.818	2.903	5.925	55.473
7	1.593	3.251	63.070	2.217	4.524	59.997
8	1.438	2.934	66.003	1.839	3.753	63.749
9	1.248	2.547	68.551	1.583	3.230	66.980
10	1.123	2.293	70.843	1.477	3.015	69.994
11	1.023	2.087	72.930	1.439	2.936	72.930

Extraction Method: Principal Component Analysis.

From the above table eleven factors are extracted as given below. Score for these factors is given in Table 2:

Factor 1:   CRM implementation issues,
Factor 2:   Implementation of CRM through employees,
Factor 3:   CRM Business Strategies of Bank,
Factor 4:   Interpretation and synchronization of service,
Factor 5:   Customer centric approach,
Factor 6:   Differences between IT and CRM,
Factor 7:   Right Method to interact with customer,
Factor 8:   Right tool of data mining,
Factor 9:   Clear methodology and steps,
Factor 10: Measurable goals and ROI of Bank, and
Factor 11: Training of the front people.

Table 2. Mean Score (N = 70-ICICI, N=70-SCB)

	Name of Bank	Mean Score	Std. Deviation	Std. Error Mean
CRM Implementation Issues	ICICI	2.0750	.87195	.10422
	SCB	2.3607	.66025	.07892
Implementation of CRM through Employees	ICICI	1.9429	.72915	.08715
	SCB	2.2810	.73160	.08744
CRM Business Strategies of Bank	ICICI	4.1079	.62967	.07526
	SCB	3.6857	.69220	.08273
Interpretation and Synchronization of Service	ICICI	4.119	.75934	.09076
	SCB	3.9048	.67422	.08058
Customer Centric Approach	ICICI	4.0679	.51061	.06103
	SCB	3.9214	.66423	.07939
Differences between IT and CRM	ICICI	2.5607	.78992	.09441
	SCB	2.5786	.78198	.09346
Right Method to Interact with Customer	ICICI	4.1000	.70733	.08454
	SCB	3.6667	.56180	.06715
Right Tool of Data Mining	ICICI	4.1714	.74155	.08863
	SCB	3.8571	.76681	.09165
Clear methodology and Steps	ICICI	4.1714	.74155	.08863
	SCB	3.8571	.76681	.09165
Measurable Goals and ROI of Bank	ICICI	4.0643	.69115	.08261
	SCB	3.7429	.71598	.08558
Training of the Front People	ICICI	3.8857	1.05697	.12633
	SCB	3.6286	.96566	.11542

5   Interpretations from the above Tables (1 and 2)

- From the empirical data analysis listed in Table 2, it is revealed that
- Improper CRM implementation issues are hindrances and can lead to the failure of the whole system. This should be addressed rightly in synchronization of the technology and the bank policies.
- CRM implementation through employees, require the right kind of offers given to the customer while interacting without forcing it otherwise it may lead to irritating the customer. Internet banking, e-mails, phone calls, face-to-face is the various methods used by the banks to give offers. It when asked by the managers most of them agreed that face-to-face method is considered as the best way to interact with the customers.
- CRM strategies of the bank play most important role in the success of the whole setup. When interpreted correctly and synchronization of the services is there it may enhance the customer relations as well as business.
- Customer centric approach should be adopted by the bank, where customers' benefits and interests should be put before the banks own interests. Training of the front people while implementing the CRM plays very important role in the success of the implementation and it should be done properly.
- Technology plays vital role in the whole CRM process, as every stage can be automated. There are various CRM vendors who provide the required software. The selection of the software should be done with utmost care keeping the requirements of the bank in mind as at wrong selection may be disastrous.
- While using the data mining tools for generation of the prediction model selection of the right tool for a particular problem should be there.
- CRM is implemented for the marketing department through the IT department of the bank. The differences between the two can be the hindrances and may affect the success of the CRM. It is recommended that while applying the data mining tools IT department should consult the marketing people.
- Though CRM is implemented in the banks most of the time they are not clear about the expectations from the whole initiative. If the banks have measurable goals and Return on Investment (ROI), the performance of the CRM can be measured and analyzed.

6   Conclusions

The most important factors for successful CRM implementation are the interpretation of the outcome of a data-mining tool, synchronization of IT with marketing, customer centric approach, and clear methodology to implement CRM and defined steps, and training of front office people. Hindrances are difference between   IT and CRM departments. Both the banks on which the study was carried out have same views on these factors. However there are some factors on which there is difference of the opinions. Further research can be done with more number of respondents to clearly identify these factors.

Including more number of banks as well as the public sector banks, the research can be carried forward which, can give better insight of the whole scenario. There are other services sectors like Hotel, Airlines, etc. who have also implemented the CRM. The study can be carried out for these sectors, also.

References

1. Bradshaw, D., Brash, C.: Managing customer relationships in e-business world: how to personalize the computer relationships for increased profitability. International Journal of Retail 29(12) (2001)
2. Popli, G., Rao, D.N.: Customer relationship management in Indian banks (2009), http://ssrn.com/abstract=1373592
3. Lin, M.C.: A study of main stream features of CRM system and evaluation criteria. In: The Proceedings of the 2003 American Society for Engineering Education Annual Conference & Exposition (2003)
4. Baumeister, H.: Customer relationship management for SMEs. In: The Proceedings 2002 (2002)
5. Cole, R.: Hidden benefits are key to CRM (2005), http://www.crmtoday.com (retrieved on March 2006)
6. Bolton, M.: Customer centric business processing. International Journal of Productivity and Performance Management 53(1) (2004)
7. Xu, Y., Yen, D.C., Lin, B., Chou, D.: Adopting customer relationship management technology. Industrial Management & Data Systems 102(8), 442–452 (2002)
8. Walters, S.: What are the main weaknesses in the financial sector concerning the application of CRM services? (2004), http://www.crmtoday.com (retrieved on March)
9. Thearling, K.: Increasing customer value by integrating data mining and campaign management software. Direct Marketing Magazine (1999)
10. Dravis, F.: Data quality: a survival guide for marketing. Business Objects (2008), http://www.techlinks.net/345-data-quality-a-survival-guide-for-marketing.html
11. Buggy, D.J.: A CRM strategy for distribution. Industrial Distribution 90(3), 96 (2001)
12. Chen, I.J., Popovich, K.: Understanding customer relationship management (CRM): people, process, and technology. Business Process Management Journal 9(5) (2003)
13. Thompson, B.: The loyalty connection: secrets to customer retention and increased profits (2005), http://www.crmguru.com
14. Eckenrode, J.: Improving customer relationship management in banking with integrated delivery channels (2005), http://www.microsoft.com/industry/financialservices/banking/businessvalue/tgcrmarticle.mspx
15. SAP white paper.: mySAP™ Crm for the banking sector (2003), http://www.sap.com/industries/banking/pdf/BWP_CRM_banking
16. Edelstein, H.: Building profitable customer relationships with data mining. Two Crows Corporation (2006), http://www.twocrows.com
17. Taber, D.: What price CRM data quality? (2009), http://www.cio.com/article/497574
18. Rance, D.: Customer centricity: a home run for customers (2005), http://defyingthelimits.com

19. Kotorov, R.: Customer relationship management: strategic lessons and future directions. Business Process Management Journal 9(5) (2003)
20. Radcliffe, J.: Eight building blocks of CRM: a framework for success. Gartner Research (2001), http://www.gartnergroup.com
21. Bull, C.: Strategic issues in customer relationship management (CRM) implementation. Business Process Management Journal 9(5) (2003)
22. Cuthberston, R., Laine, A.: The role of CRM within retail loyalty marketing. Journal of Targeting Measurement and Analysis of Marketing 12(3), 59–61 (2004)
23. Winston, J.: CRM in financial services: are companies realising the benefits of CRM in practice, and how is the strategy being implemented in organisations? Interactive Marketing 5(4), 329–344 (2004)
24. Hop, M., Bullock, J.: Why is customer service still failing to benefit from CRM investment (2005), http://www.customerthink.com

Issues before Indian Small and Medium Scale Enterprises Opting for ERP Implementation

Parijat Upadhyay[1], Rana Basu[2], and Pranab K. Dan[2]

[1] Globsyn Business School, EN-22, Sector-V, Salt Lake City, PIN-700091, India
[2] West Bengal University of Technology, Sector-I, Salt Lake City, PIN-700064, India
{parijat.upadhyay,rbasu004}@gmail.com, dan1pk@hotmail.com

Abstract. Enterprise Resource Planning (ERP), a comprehensive integrated packaged software system, have been used by many companies. Such big ticket investment decisions are driven by growth strategies, customer service expectations and pressures to reduce costs. The purpose of this paper is to present the findings of a study which is based on the results of a comprehensive compilation of literature and subsequent analysis of ERP implementation success factors in context to Indian Micro, Small and Medium scale Enterprises (MSMEs). The paper attempts to empirically assess which factors are most critical in the ERP implementation process from the perspective of the Indian MSMEs. This research is potentially aimed to being useful to MSMEs as a guideline so as to ensure a positive outcome of the implementation process. The result of this study highlights four crucial factors which influences the ERP implementation process in Indian MSME segment.

Keywords: ERP, MSME, Pareto analysis, implementation.

1 Introduction

Organizations of any magnitude have implemented or in the process of implementing Enterprise Resource Planning (ERP) systems in order to remain competitive in business. It is regarded as one of the novel solutions for the organizations of any magnitude to reap benefits like improving productivity, efficiency and overall business efficiency. More and more organizations across the globe have chosen to build their information technology (IT) infrastructure around this class of off-the-shelf applications and at the same time there has been and should be a greater appreciation for the challenges in implementing these complex technologies. Although ERP systems can bring competitive advantages to organizations of any magnitude but the high failure rate in implementing such systems is a major concern [1] particularly for micro, small and medium scale enterprises (MSMEs).

MSMEs in particular, face difficulty while focusing on business drivers impacting ERP strategies. The Indian small and medium scale enterprises are eyeing the enterprise system solutions for their business in this changing business scenario. Although this shift has been gradual but it is picking up as Indian small and medium business are now competing globally and they need to gain a competitive edge to win amidst global competition. However, the initial experience of MSMEs in this regard is not very encouraging. Numerous cases can be cited where in organizations have

S. Dua, S. Sahni, and D.P. Goyal (Eds.): ICISTM 2011, CCIS 141, pp. 41–50, 2011.

spent substantial amount on implementation have not been able to derive any benefit at all out the implementation. A number of publications have highlighted the failures and the frustrations that enterprises go through in implementing ERP systems. Some of the notable failures are Dell, Fox-Meyer and Hershley. This paper tries to explore the experience of the Indian MSME sector and highlights those factors or issues that can ensure positive outcome of ERP implementation process.

2 Literature Review

ERP system implementation issues have been subjected to substantial research under various theoretical perspectives dealing with aspects like: characteristics, adoption and implementation process [2, 3], project design or execution [4], organizational impacts [5], forecasting of the probability of success [6], extension towards e-commerce [7, 8, 9]. For developing countries like India ERP system is in its early stage. It has been found that reasons like limit of capital, non-availability of resources, poor management base, and absence of IT expertise are seriously affecting the implementation and adaptation of enterprise system in India and other similar Asian developing countries compared to the developed countries. SMEs either do not have sufficient resources or are not willing to commit significant portion of resources to complex ERP implementation process [10], SMEs are more fragile than large companies [11], and the adoption of ERP is no longer limited to large scale enterprises [12]. According to Huin unless the differences between SME and large companies are clearly conceived, the ERP project will not reach to positive outcome [13]. So, deeper understanding of ERP implementation in SMEs needs to ensure a strong impact. Those fact holds that SME niche definitely require specific research and analysis other than previous investigations primarily targeting larger enterprises. This need has been understood so some researchers have started addressing specifically ERP implementation at SMEs. In this section some similar developing countries like India are selected to understand the characteristics of ERP in those countries and also to investigate the issues that are affecting the implementation of ERP system.

Several academics and practitioners have tried to capture the main reasons for failure or success of ERP implementations [14, 15, 16, 17, 18, 19, 20]. Most of this analyses and lists focuses on the factors that contributed to failures than those contributed to success. The factors that appeared mostly on the lists include: Top level management support, User training and education, Project management, clearly defined goals and objectives, Project team competence, and Change management.

A study in context to SMEs and MSMEs points that since such organisations do not have sufficient resources they are not willing to commit a substantial portion of their resources. Such kind of attitude may be due to implementation time being longer and high fees associated with it [21]. It is also studied and concluded that cost of package is a crucial issue towards successful implementation of ERP. The compatibility of technology and company's need must be carefully addressed as ERP project involves a complex transition from legacy information systems and business processes to an integrated IT infrastructure [22].

In a research study by Nah et al. [23] ten success factors are identified for successful ERP implementation that includes ERP teamwork and composition, Top management support, Business plan and vision, Effective communication, Project management, Appropriate business and legacy system, Software development, Testing and troubleshooting, Effective – decision making, Effective training [23].

In a research study by Jafari et al. [24] after reviewing 28 articles identified ten success factors or issues in implementing ERP system in Malaysian companies which are: Support from top management, clear goals and objectives, communication, effective project management, business process engineering, data accuracy and integrity, suitability of software and hardware, support from vendor, education and training and user involvement [24].

In case of small and medium scale enterprises of India a study conducted by researchers for ensuring successful implementation of ERP six major factors have been identified that includes: clarity in goals and objectives behind implementation, adequate user training, competent project implementation team, acceptance of change brought about by implementation, proper vendor support, participation of external consultant [25]. Further to this the researchers in context to implementation of IT projects revealed seven factors that have been found to be crucial that are: Top management support, properly defined goals and objectives, user knowledge, project champion, project team competency, improve work efficiency, scalability & scope and ERP importance [26].

Similar study being conducted by creating interactive structural model and four critical factors are identified which are: fund support, departmental participation, training and service of ERP which is having impact on successful implementation of ERP [27]. Based on the analysis of conditions of SMEs of China six critical success factors have been recognized for ensuring successful ERP implementation: support from top management, competent project team, right implementation scope, change management, data accuracy, education and training [28].

A study conducted on the SMEs of Taiwan to investigate how the role of external consultant affects the ERP implementation project. The study explored that the involvement of external consultant along with management consultant in ERP implementation have significant impact on the ERP implementation practice. Also, the researchers have identified that the relationship between external consultants with SME managers highly affect the implementation project [29].

Proper package selection plays a crucial role in successful implementation of ERP. Normally the organization selects a package which is most user friendly, has adequate scope for scalability and covers an array of business processes where organization experiences problem. The selection of the specific ERP package is one that requires careful attention [30, 31, 32, 33, 34].

By conducting and assessing important success factors and failure factors in Malaysian SME companies and thus from the research the critical issues identified for successful implementation of ERP are proper team composition and effective training of users. In addition to this two issues are responsible for failures of the implementation are poor project planning and inappropriate training method, thus need to be effectively normalized so that they are phrased as positive factors that contribute to success [35].

3   Research Methodology

Based on the literature review certain issues and factors leading to successful ERP implementation and on the basis of frequency of citations made by several authors in context to small and medium scale enterprises, of similar developing countries like India, 29 issues have been identified. The different issues were identified after doing an extensive literature study (further Pareto analysis done to extract key issues contributing 80% of total percentage contribution) and from the questionnaire survey of the organizations that have gone through the implementation process. Out of those key issues few were found inapplicable to the context of implementation of ERP systems in micro, small and medium scale industries in particular. They were more generic in number. The researchers therefore created items to measure the constructs, and used 5-point multi-item, Likert-type scales for each item where "1" meant "strongly disagree" and "5" meant "strongly agree."

The data collection phase involved exhaustive search of many of the prime Management Information System (MIS) journals that the researchers could access. In addition to, the preceding journals, some conference papers, articles, were also accessed as well as ,the following databases were searched like Emerald, Science Direct, Proquest, EbscoHost, Springer, J Stor, etc.

Data were collected from the users of the leading ERP vendors: SAP, Oracle, Peoplesoft and Microsoft. The researchers contacted the organizations those are relevant for the study. The respondents were briefed about the scope and purpose of the questions that is contained in the questionnaire and the feedbacks were obtained. However in some cases, the questionnaire were distributed with a covering letter explaining the scope and purpose of the study and sent via e-mail. Some leading ERP vendors were also contacted, who provided the list of users, where they have implemented ERP and they also assisted in collecting data by forwarding the questionnaire on behalf of the researcher. Overall, 102 responses from 50+ organizations were obtained for analysis. Multiple responses were allowed from an organization. Because the survey questionnaire were also distributed by the vendor officials to the members of their respective user lists through e-mails, it was not possible to determine the number of recipients and the response rate. It was deliberate attempt on the part of the researcher not to collect any response in his presence so as to minimize the chance of any biasness.

4 Data Analysis

For identifying the key issues for small-medium scale enterprises in the developing countries like India Pareto analysis has been applied (see Table 1). For this purpose, each issues mentioned by authors at least once in the literature is listed down with the frequency of each issues. From the frequency we calculated the percentage contribution of each issue. Hence, the cumulative percentage contribution of issues is calculated. It has been found that only 11 issues out of 29 are contributing 80% of the total percentage contribution. Hence, those 11 issues are regarded as the key issues being extracted from the review section in context to Indian MSMEs. The key issues identified are: education and training, top management support, clearly defined goals and objectives, project team, proper project management, change management, proper package selection, external consultant role, data accuracy, effective communication among department, proper vendor support.

Table 1. Pareto Analysis (80-20 Rule)

Issues Mentioned in the Literature	Number of Instances Cited in the Literature	Percentage Contribution of Each Issue	Cumulative Percentage of Issues
Education & Training	11	12.22%	12.22%
Top Management Support	11	12.22%	24.44%
Clearly Defined Goals & Objectives	10	11.11%	35.55%
Project Team	10	11.11%	46.66%
Proper Project management	9	10.00%	56.66%
Change Management	9	10.00%	66.66%
Proper Package Selection	4	4.44%	71.10%
External Consultant	2	2.22%	73.32%
Data Accuracy	2	2.22%	75.54%
Effective Communication	2	2.22%	77.76%
Vendor Support	2	2.22%	79.98%
Improve Work Efficiency	1	1.11%	81.09%
IT Infrastructure	1	1.11%	82.20%
Project Champion	1	1.11%	83.31%
Scalability and Scope	1	1.11%	84.42%
User Knowledge	1	1.11%	85.53%
Package Cost	1	1.11%	86.64%
Fund Support	1	1.11%	87.75%
Departmental Participation	1	1.11%	88.86%
Right Implementation Scope	1	1.11%	89.97%
Appropriate Business & Legacy System	1	1.11%	91.08%
Right Implementation Scope	1	1.11%	92.19%
Relationship between External Consultant & SME Managers	1	1.11%	93.30%
S/W Development	1	1.11%	94.41%
BPR	1	1.11%	95.52%
Effective Decision Making	1	1.11%	96.63%
User Involvement	1	1.11%	97.74%
Testing & Troubleshooting	1	1.11%	98.85%
ERP Importance	1	1.11%	99.96%
Total	90	100%(approx)	100%(approx)

Table 2. Construct, Indicator, Factor Loadings, Chronbachs Alpha and Comment

Construct[3]	Indicator[2]	Factor Loadings[1]	Chronbach's Alpha	Comment
Technical Perspective	Project Team Composition	0.639	0.656	Reliable and Acceptable
	User Knowledge	0.692		
	Project Team Competence	0.747		
	Minimum Customization	0.728		
Organizational Climate	Top Management Support	0.843	0.717	Reliable and Acceptable
	Inter Departmental Communication	0.887		
	External Consultant	0.707		
Project Execution Competency	Improve Work Efficiency	0.780	0.606	Reliable and Acceptable
	Project Management	0.706		
	ERP Importance	0.631		
Package and Vendor Perspective	Proper Package Selection	0.778	0.634	Reliable and Acceptable
	Vendor Staff Knowledge	0.716		

[1] Extraction method used is: Principal Component Analysis; Rotation method used: Varimax with Kaiser Normalization
[2] Prevailing IT infrastructure does not satisfy required loading and could be excluded from analysis
[3] Construct is treated as a factor which has been identified through exploratory factor analysis

We have conducted factor analysis on the explanatory variables with the primary objective to determine minimum number of factor explaining the maximum variance in data. Initially we performed the factor analysis on SPSS with principal component method using varimax rotation technique and to validate our result we recheck the result with the help of SAS using PROC factor method with same varimax rotation technique. From the SPSS output according to Eigen value and SCREE plot we retain the following factors as some of the critical factors.

    a. Project Execution Competency
    b. Package and Vendor Selection
    c. Management Perspective

The confirmatory factor analysis procedure was started with the entire measurement model including all the latent factors and all the items in each factor (except an overall summary item for each which might be used for additional validation if necessary). The model fit statistics were improved step by step through the item deletion approach, deleting one item from the entire model at a time. The item deletion process was carried out in two phases. First, the items that did not load significantly on their respective factors were deleted one by one until all of them showed significant loadings. Next, the Lagrange multiplier test was carried out to identify unrelated pairs of items and factors between which an added path would reduce the Chi-square value significantly. These items were subsequently deleted one by one until acceptable model fit indices were achieved.

As can be seen in Table 2, to extract the factor from the rotation table we consider the value more than 0.5 so according to that we have found that user knowledge, project team composition, project team competence, minimum customization contributing towards the Success Factor 1. (We name this dimension as Technical Perspective).

Similarly for Success Factor 2 the contribution came from improve work efficiency, project management, ERP importance (we name this factor as Project Execution Competency).

For Success Factor 3 Top management, interdepartmental communication, external consultants have contribution more than 0.5 in rotation table. (We name this factor as Organizational Climate).

And for Success Factor 4 the contributing factors are proper package selection, vendor's staff knowledge (we name this factor Package and Vendor Perspective).

5 Discussion

India has more than 35 million MSMEs according to various industry estimates. They generate more than 60 percent of the country's Gross Domestic Product (GDP) and they are increasingly becoming a part of the global supply chain. Although MSMEs form the backbone of India's economy, the technological advancement or development of this segment has been slow. There is absence of a business continuity model support system in case of any unforeseen contingency and such limitations have affected the MSMEs repeatedly. Koh et.al. [36] in their study have observed that they need a model to deal with uncertainty and business contingency. While the biggest advantage that MSMEs have over their larger counter parts is business agility where they can respond quicker to market changes, they need to sharpen their business edge by having a secure and reliable information system that provides superior business intelligence, collaboration power, and transactional efficiency. Mid-sized companies need to choose smart and affordable IT solutions that are easy to buy, install, and maintain, but that still give them the power, security, and functionality of enterprise-strength systems—at a fraction of the cost. They need to deploy the most powerful technologies while keeping IT costs as low as possible.

MSMEs face pressures from various corners that include competition from global vendors, new regulatory compliance requirements that would come up every year requiring complying with and providing information to government agencies, needing to strengthen products and services, at the same time, physical expansion need to be taken care. With exposure to global economies, customers' preferences are changing drastically and accordingly, organizations need to enhance their products / services. Internally, within the organizations, they need to take care of risks due to employee churn, data security against competition, copyright protection, duplication of data, silos of applications and data etc.

The findings hold significance for these MSME's which wishes to leverage the benefits of integration of business processes by implementing an ERP system in their organization. Literature shows many instances of ERP implementation which failed to deliver the business returns and in some cases the entire project implementation cost has been a sunk cost for the organization getting no return on their investment.

6 Conclusions

The study provided partial support for the explanation of success in implementing ERP projects in small and medium scale enterprises as reported by various researchers. However, in context to a country like India, the initial approach of various vendors in the form of 'one size fits all' approach has resulted in failures in many small and medium scale enterprises.

Analyzing the results, puts forwarding some interesting insights. As per the respondents, the ERP vendors have mainly concentrated on ensuring the competency in project execution and emphasized more on the technical perspective, quantity and quality of available environmental information. What the vendors fail to take into consideration is the implication of a negative organizational climate in the form of building a consensus amongst the target users regarding the purpose and scope of the implementation and the eventual benefits to the employees and to the organizational as a whole. However in Indian context to MSMEs, the vendors have a very crucial role to play. They should have agility and required flexibility in implementation requirements of MSMEs with a localized approach.

References

1. Davenport, T.H.: Mission critical: realizing the promise of enterprise systems. Harvard Business School Press, Boston (2000)
2. Nandhakumar, J., Talvinen, J.M., Rossi, M.: ERP revelations: the dynamics of contextual forces of ERP implementation. In: Leino, T., Saarinen, T., Klein, S. (eds.) The Proceedings of the Twelfth European Conference on Information Systems, Turku, Finland (2004)
3. Butler, T., Pyke, A.: Examining the influence of ERP systems on firm-specific knowledge and core capabilities: a case study of SAP implementation and use. In: Ciborra, C.U., Mercurio, R., de Marco, M., Martinez, M., Carignani, A. (eds.) The Proceedings of the Eleventh European Conference on Information Systems, Naples, Italy (2003)
4. Laframboise, K.: Business performance and enterprise resource planning. In: Wrycza, S. (ed.) The Proceedings of the Tenth European Conference on Information Systems, Gdansk, Poland (2002)
5. Westrup, C., Knight, F.: Consultants and enterprise resource planning (ERP) systems. In: Hansen, H.R., Bichler, M., Mahrer, H. (eds.) The Proceedings of the Eighth European Conference on Information Systems, Wien, Austria (2000)

6. Magnusson, J., Nilsson, A., Carlsson, F.: Forecasting ERP implementation success - towards a grounded framework. In: Leino, T., Saarinen, T., Klein, S. (eds.) The Proceedings of the Twelfth European Conference on Information Systems, Turku, Finland (2004)

7. Schubert, P., Leimstoll, U.: Personalization of e-commerce applications in SME's: conclusions from an empirical study in Switzerland. Journal of Electronic Commerce in Organizations 2(3), 21–39 (2004)

8. Kemppainen, I.: Change management perspectives in an ERP implementation. In: Leino, T., Saarinen, T., Klein, S. (eds.) The Proceedings of the Twelfth European Conference on Information Systems, Turku, Finland (2004)

9. Schubert, P.: Personalizing e-commerce applications in SME's. In: The Proceedings of the Ninth Americas Conference on Information Systems, AMCIS (2003)

10. Buonanno, G., Faverio, P., Pigni, F., Ravarini, A., Sciuto, D., Tagliavini, M.: Factors affecting ERP system adoption: a comparative analysis between SMEs and large companies. Journal of Enterprise Information Management 18(4), 384–426 (2005)

11. Rao, S.S.: Enterprise resource planning: business needs and technologies. Industrial Management & Data Systems 100(2), 81–88 (2000)

12. Loh, T.C., Koh, S.C.L.: Critical elements for a successful enterprise resource planning implementation in small-and medium-sized enterprises. International Journal of Production Research 42(17), 3433–3455 (2004)

13. Huin, S.F.: Managing deployment of ERP systems in SMEs using multi-agents. International Journal of Project Management 22(6), 511–517 (2004)

14. Ewusi-Mensah, K.: Critical issues in abandoned information systems development projects. Communications of the ACM 40(9), 74–80 (1997)

15. Stapleton, G., Rezak, C.J.: Change management underpins a successful ERP implementation at Marathon Oil. Journal of Organization Excellence 23(4), 15–21 (2004)

16. Weightman, C.: The top 10 ERP mistakes. Business Management, 36–40 (2004)

17. Anexinet, R.B.: Top 10 ERP implementation pitfalls (2006),
   http://www.anexinet.com/pdfs/ERP_top10pitfalls3-2006.pdf

18. Kimberling, E.: Seven critical success factors to make your ERP or IT project successful (2006), http://it.toolbox.com/blogs/erp-roi/7-criticalsuccess-factors-to-make-your-erp-or-it-project-successful-12058

19. Ibrahim, A.M.S., Sharp, J.M., Syntetos, A.A.: A framework for the implementation of ERP to improve business performance: a case study. In: Irani, Z., Sahraoui, S., Ghoneim, A., Sharp, J., Ozkan, S., Ali, M., Alshawi, S. (eds.) The Proceedings of the European and Mediterranean Conference on Information Systems (EMCIS) (2008)

20. Lindley, J.T., Topping, S., Lindley, L.: The hidden financial costs of ERP software. Managerial Finance 34(2), 78–90 (2008)

21. Chan, R.: Knowledge management for implementing ERP in SME's. Paper presented at the 3rd Annual SAP Asia Pacific, Institute of Higher Learning Forum, and Singapore (1999)

22. Al-Mashari, M.: Enterprise resource planning (ERP) systems: a research agenda. Industrial Management & Data Systems 103(1), 165–170 (2002)

23. Nah, F.F., Lau, J.L.: Critical success factors for successful implementation of enterprsie systems. Business Process Management, 285–296 (2001)

24. Jafari, S.M., Osman, M.R., Yusuf, R.M., Tang, S.H.: ERP systems implementation in Malaysia: the importance of critical success factors. Engineering and Technology 3, 125–131 (2006)

25. Upadhyay, P., Dan, P.K.: An explorative study to identify the critical success factors for ERP implementation in Indian small and medium scale enterprises -978-0-7695-3513-5/08. IEEE, Los Alamitos (2008)

26. Upadhyay, P., Dan, P.K.: ERP in Indian SME's: a post implementation study of the underlying critical success factors. International Journal of Management Innovation System 1(2), E1 (2009)

27. Jing, R., Qiu, X.: A study on critical success factors in ERP systems implementation. IEEE, Los Alamitos (2007)

28. Xia, Y., Lok, P., Yang, S.: The ERP implementation of SME in China. In: The Sixth International Conference on Service Systems and Service Management, pp. 135–140 (2009)

29. Chen, R., Sun, C., Helms, M.M., Jih, W.: Role negotiation and interaction: an exploratory case study of the impact of management consultants on ERP system implementation in SMEs in Taiwan. Information Systems Management 25(2), 159–173 (2008)

30. Kraemmergaard, P., Rose, J.: Managerial competences for ERP journeys. Information Systems Frontiers 4, 199 (2002)

31. Yusuf, Y., Gunasekaran, A., Abthorpe, M.S.: Enterprise information systems project implementation: a case study of ERP in Rolls-Royce. International Journal of Production Economics 87, 251–266 (2004)

32. Al-Mashari, M.: Enterprise resource planning (ERP) systems: a research agenda. Industrial Management & Data Systems 103(1), 22–27 (2003)

33. Somers, T.M., Nelson, K.: The impact of critical success factors across the stages of enterprise resource planning implementations. In: The Proceeding of the Thirty-Fourth Hawaii International Conference on System Sciences, Hawaii (2001)

34. Somers, T.M., Nelson, K.G.: A taxonomy of players and activities across the project life cycle. Information & Management 41, 257–278 (2004)

35. Noudoostbeni, A.: To investigate the success and failure factors of ERP implementation within Malaysian small and medium enterprises, pp. 157–160. IEEE, Los Alamitos (2009)

36. Koh, S.C.L., Simpson, M., Padmore, J., Dimitriadis, N., Misopoulos, F.: An exploratory study of enterprise resource planning adoption in Greek companies. Industrial Management & Data Systems 106(7), 1033–1059 (2006)

Steganalysis of Perturbed Quantization using HBCL Statistics and FR Index

Veena H. Bhat[2], S. Krishna[1], P. Deepa Shenoy[1], K.R.Venugopal[1], and L.M. Patnaik[3]

[1] Department of Computer Science and Engineering
University Visvesvaraya College of Engineering, Bangalore, India
[2] IBS Bangalore (Research Scholar, Bangalore University), India
[3] Defence Institute of Advanced Technology, Pune, India
{veena.h.bhat,krishna.somandepalli,venugopalkr,
kisna.philly}@gmail.com, shenoypd@yahoo.com

Abstract. Targeted steganalysis aims at detecting hidden data embedded by a particular algorithm without any knowledge of the 'cover' image. In this paper we propose a novel approach for detecting Perturbed Quantization Steganography (PQ) by HFS (Huffman FR index Steganalysis) algorithm using a combination of Huffman Bit Code Length (HBCL) Statistics and File size to Resolution ratio (FR Index) which is not yet explored by steganalysts. JPEG images spanning a wide range of sizes, resolutions, textures and quality are used to test the performance of the model. In this work we evaluate the model against several classifiers like Artificial Neural Networks (ANN), k-Nearest Neighbors (k-NN), Random Forests (RF) and Support Vector Machines (SVM) for steganalysis. Experiments conducted prove that the proposed HFS algorithm can detect PQ of several embedding rates with a better accuracy compared to the existing attacks.

Keywords: Steganography, classifiers, Huffman coding, perturbed quantization.

1 Introduction

Digital data hiding techniques have considerably influenced covert communication in the recent years. Steganography is one such application which can be defined as the art or practice of concealing a message, image or file within another image or file in such a way that the sender would be able to communicate to the intended recipient covertly. Steganalysis is the science of detecting messages (payload) hidden using steganography. The invasive nature of steganography leaves detectable traces within the stego-image which allows adversary to use such characteristic features to reveal any secret communication taking place. According to the terminology as agreed in the First International Workshop on Information Hiding [1], the embedding of a text or 'payload' in a 'cover image' gives a 'stego-image'.

In general steganalysis can be categorized into two types - targeted steganalysis and blind steganalysis. While blind steganalysis is a generic technique to detect secret information embedded by unknown steganographic schemes, targeted steganalysis is designed specifically to detect one specific embedding algorithm.

S. Dua, S. Sahni, and D.P. Goyal (Eds.): ICISTM 2011, CCIS 141, pp. 51–60, 2011.
© Springer-Verlag Berlin Heidelberg 2011

The JPEG format being the most widely used image format receives acute interest from the stego community. One of the JPEG steganographic schemes that has got significant attention in the recent years can be generally termed as 'Minimum Distortion Embedding' and Perturbed Quantization is one such embedding scheme which is relatively secure compared to older techniques such as LSB (Least Significant Bit) encoding and MBS (Model Based Steganography).

Statistical steganalysis deals with extraction of certain features from the image and using a classifier to build the prediction model - a binary classifier can be used in case of targeted steganalysis. This paper is focused onto such feature-based steganalysis to detect PQ.

A brief description of the existing steganalytic techniques is given in Section 2. To make the paper self-contained, Section 3 summarizes Perturbed Quantization. Section 4 describes JPEG preliminaries and feature extraction; Section 5 illustrates the implementation model and details the classifiers selected. Section 6 discusses the experimental results and performance analyses. Conclusions and the future work is outlined in Section 7.

2   Related Work

A stego process is defined as a ε-secure process if the Kullback-Leibler divergence ∂ between the probability density functions of the cover document p_{cover} and those of this very same content embedding a message p_{stego} is less than ε:

$$\partial\left(p_{cover}, p_{stego}\right) \leq \varepsilon. \tag{1}$$

The process is called 'secure' if ε = 0, and in this case the steganography is perfect, creating no statistical differences by embedding of the message [2], steganalysis would then be impossible.

Perturbed quantization steganography (PQ) is one of the relatively recent approaches to passive-warden steganography where the sender hides data while processing the cover object with an information reducing operation that involves quantization, such as lossy compression (JPEG), downsampling, or A/D conversion [3]. According to the heuristics presented in [4], the PQ steganographic scheme is relatively secure and significantly less detectable than other JPEG steganographic methods, however several models have been proposed to detect PQ [5, 6, 7]. The features extracted in these works are described in section 6.

Data mining and soft computing techniques are effectively used for pattern recognition in statistical steganalysis. These techniques are employed to build binary or multi-class classifiers to classify the images as genuine or stego. Artificial Neural Networks can be efficiently used as binary classifiers in detecting steganography [8]. Support Vector Machines (SVM) is one of the most effective methods for pattern recognition and classification [9]. A SVM classifier can be used to predict steganography over the features extracted [5, 6]. A statistical approach for steganalysis method using SVM is given in [10]. Instance based learning techniques such as k-Nearest Neighbors (k-NN) are effectively used for prediction in several steganalyzers [11] in combination with feature selection methods [12]. Ensemble classifiers like Random Forests [13] are yet to be explored by steganalysts. In our previous work [14] we have used HBCL statistics and FR Index for blind steganalysis to detect amongst LSB encoding, JPEG Hide & Seek and Model Based Steganography.

3 Perturbed Quantization

In the steganographic scheme proposed by Fridrich et al, the sender is presented with a new grayscale image X. Perturbed Quantization (PQ) being a minimum distortion embedding scheme hides the data (payload) in lossy compression used in JPEG. Discrete Cosine Transforms (DCT) is applied during JPEG compression, the DCT coefficients are divided by quantization steps from the quantization table and then rounded to integer and finally encoded to a JPEG image Y using the JPEG standard [15]. The DCT coefficients before and after rounding are denoted as d_i and D_i respectively. The d_i coefficients whose fractional part is less than 0.5 are identified as 'changeable coefficients', these coefficients are such that $d_i - floor(d_i) \in [0.5-\delta, 0.5+\delta]$ where δ is called tolerance and $\delta \leq 0.1$. The $floor(x)$ gives the largest integer smaller than or equal to x [3].

Let $S = \{i_1, ..., i_k\}$ be the set of indices of all changeable coefficients. During compression these coefficients d_i are rounded up or down to d_j where $j \in S$ and thus can encode upto $k = |S|$ bits to obtain a compressed and embedded image Y'. This method is called Perturbed Quantization (PQ) because during compression we slightly perturb the quantizer (the process of rounding to integers) for a certain subset of changeable coefficients in order to embed message bits. To detect PQ, steganalysts will have to find statistical evidence that some of the values of D_i were quantized incorrectly and certain features can reflect this anomaly to predict if an image is not genuine.

4 Feature Extraction

In this work, we propose to use Huffman Bit Code Length (HBCL) statistics and the image File size to Resolution ratio (FR index) for PQ targeted steganalysis. This section describes some of the salient characteristics of these features.

4.1 JPEG Preliminaries

The JPEG (Joint Photographic Experts Group) image format uses a discrete cosine transform (DCT) to transform successive 8X8-pixel blocks of the image into 64 DCT coefficients each. The DCT coefficients $F(u,v)$, of an 8X8 block of image pixels $f(x,y)$, are given by:

$$F(u,v)=\frac{1}{4}C(u)C(v)\left[\sum_{x=0}^{7}\sum_{y=0}^{7}f(x,y) * \cos\frac{(2x+1)u\pi}{16} * \cos\frac{(2y+1)v\pi}{16}\right]. \qquad (2)$$

Here, $C(u), C(v) = 1/\sqrt{2}$ when $u,v = 0$ and $C(u), C(v) = 1$ otherwise. After this step, the coefficients are quantized by the following operation:

$$F^Q(u,v) = IntegerRound\left(\frac{F(u,v)}{Q(u,v)}\right) \qquad (3)$$

where $Q(u, v)$ is a 64-element Quantization table.

The resulting matrix after quantization is re-ordered to a 64-element vector in a zig-zag pattern, the matrix is essentially sorted from low frequency components to high frequency components. In the following steps, Huffman encoding of the pixels is used that result in the JPEG format's unique lossy compression. In general, sequential Huffman coding is employed in JPEG and grayscale images use two Huffman tables, one each for class AC and DC. The JPEG decompression works in the opposite order. The JPEG bit-stream is decompressed using the Huffman decoder which generates details about the Huffman code lengths in bits. In general the average code length is defined as:

$$\bar{l} = \sum\nolimits_{x=\bar{X}} (p\,X(x)\,l(x)) \tag{4}$$

where X is the set of possible values of x. It can be shown that the Huffman code $H(X)$ is

$$H(X) < l \leq H(X) + 1. \tag{5}$$

4.2 Huffman Bit Code Lengths: HBCL Statistics

JPEG, a lossy compression format employs sequential Huffman Encoding for data compression wherein symbols (DCT coefficients in this case) are encoded with variable length codes that are assigned based on statistical probabilities. A grayscale image employs 2 Huffman tables, one each for the AC and DC portions. The number of particular lengths of the Huffman codes is unique for a given image of certain quality and resolution; we call these numerical statistics of the DC portion of the Huffman table as Huffman Bit Code Length Statistics (H), these features along with the file size to resolution ratio (FR index denoted by F) are used in the proposed HFS algorithm. When a certain JPEG image is embedded with a payload using PQ, as described in section 3, the changeable DCT coefficients d_i before embedding are rounded to d_j after embedding. Further, since Huffman coding is applied on these DCT coefficients during JPEG compression, considerable variations can be observed on the nature of the Huffman codes before and after embedding.

Table 1. HBCL Statistics for an Image of Resolution, 640X480

Feature	Huffman Table of class DC		Feature	Huffman Table of class DC	
	Bit Codes	*Number*		*Bit Codes*	*Number*
L_0	Number of HBCL of 1 bit	0	L_5	Number of HBCL of 6 bits	57
L_1	**Number of HBCL of 2 bits**	492	L_6	Number of HBCL of 7 bits	7
L_2	**Number of HBCL of 3 bits**	3492	L_8 . . L_{16}	Number of HBCL of 8 bits . . Number of HBCL of 16 bits	0 . . 0
L_3	**Number of HBCL of 4 bits**	463			
L_4	**Number of HBCL of 5 bits**	289			

Let us denote the Huffman codes of d_i coefficients as $H(d_i)$ and those of d_j coefficients as $H(d_j)$. The number of particular code lengths of $H(d_i)$ are denoted as L_m and that of $H(d_j)$ as L_n where $m, n = \{1, 2, ..., 16\}$. These L_n statistics show a significant difference compared to the statistics of L_m and hence used as features for steganalysis.

One of the scoring features of Huffman coding algorithm is its 'unique prefix property' that is no code is a prefix to any other code, making the codes assigned to the symbols unique. This fact further supports our choice of the HBCL statistics as evaluation features to be efficient as the anomalies introduced by PQ on an image are reflected in these features. Experiments have proven that these features evaluate steganography with nearly the same performance over several classifiers demonstrating the uniqueness of HBCL statistics. In our work we extract HBCL statistics using Huffman decoding for a JPEG image in Matlab.

4.3   File Size to Resolution Ratio: FR Index

When an image is compressed by JPEG compression, based on the resolution of the image, quality, the compression ratio and few other factors the resulting JPEG file takes up a particular file size. This indicates that the file size and the resolution of an image and further its quality are interrelated. Thus the ratio of file size of the image in bytes to that of its resolution is found to be unique and in a certain range for a given resolution, this functional termed '*FR Index*' is used as one of the input features to build the prediction model.

Thus, five features - FR Index and HBCL statistics - L_1, L_2, L_3, L_4, are extracted from both cover images and stego-images embedded with different payloads. Table 2 shows the correlation between features illustrating the nature of the data.

Table 2. Nature of Features Extracted for Steganalysis

Correlation	FR Index	L_1	L_2	L_3	L_4
FR Index	1				
L_1	-0.3068	1			
L_2	-0.3714	-0.1888	1		
L_3	0.4211	-0.3105	0.1016	1	
L_4	0.6207	-0.2293	-0.1866	0.8342	1

5   Implementation

5.1   Image Database

One of the important aspects of performance evaluation is the nature of image database employed in implementation. To evaluate the performance of our model we used the image database provided by Memon et al. [15]. The image dataset consists of over 160,000 JPEG cover images spanning various textures, quality, resolutions and sizes and corresponding PQ stego-images of over 95,000 each for four different embedding payloads; 0.05, 0.1, 0.2 and 0.4 bpac (Bits Per non zero AC DCT coefficients). Embedding a payload of more than 0.4 bpac with PQ is an unachievable embedding rate.

We chose grayscale images because it is harder to detect hidden data in these compared to color images where steganalysis can utilize dependencies between color channels. In this work we test the performance of our model against these four embedding rates. Feature vectors are extracted from images and a binary classifier is built for steganalysis. The entire data extracted from the images is further used for training and testing the classifiers to evaluate the performance of the steganalyzer, as described in Section 6.

5.2 Model

The model is designed to extract the HBCL statistics from a grayscale JPEG image using a Huffman decoder, the number of two bit codes, three bit codes, four bit codes and five bit codes i.e., L_1, L_2, L_3 and L_4 respectively are extracted, these functionals along with the file size to resolution ratio (FR Index) are further used to build the classifier. A fast Huffman decoder is used to extract the two Huffman tables, one each of AC and DC portions, the code buffer holds all the codes decoded from which the required statistics are selectively separated and fed to the functional space of the classifier.

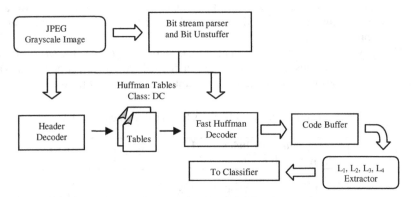

Fig. 1. Model Illustrating Extraction of HBCL Statistics for Classification

5.3 Classification

As described in the earlier sections, the selective feature vectors extracted from the model are used to train several classifiers to evaluate steganalysis. Four different techniques are used for classification; artificial neural networks (ANN), k-nearest neighbors (k-NN), Random Forests (RF) and support vector machines (SVM). The functional space for these classifiers consists of five variables, and is designed for binary classification to distinguish stego images from genuine ones.

To train and test each classifier the following steps are followed:

- The image database is divided into several combinations of training and testing sets for each of the four payloads tested. 60% of the data is used as the training set in each trial.

- To evaluate the accuracy of the model, the minimal total average probability of error (P) is computed which is given by:

$$P = \frac{P_{FP} + P_{FN}}{2} \tag{6}$$

where P_{FP} and P_{FN} are the probability of false positives and false negatives of the test images respectively.

- Given the predicted values for the test sets, Receiver Operating Characteristics curves (ROCs) are obtained to illustrate the performance of the classifier [16].

- The Area Under the ROC curve (AUR) is calculated to evaluate the accuracy of the designed classifier against previously unseen images in the test set.

Artificial Neural Network (ANN) is a computational model that simulates an interconnected group of artificial neurons to process the data [17]. In this work we use a feed forward back-propagation neural network with a single hidden layer of 10 neurons with radial basis activation function. Softmax function is used in the output layer to aid binary classification.

***k*-Nearest Neighbors** (*k*-NN) is an instance-based learning method. The training instances are taken as vectors to its feature space and stored along with its labels. The user defined constant k, is assigned based on a 10-fold cross validation technique where $k \in \{2, 3, ..., 12\}$. For an unlabelled vector or a test point to be classified, the label which is most frequent among the k training samples nearest to that test point is assigned to it. Hamming distance is the distance metric used.

Random Forests (RF) is an ensemble classifier that comprises of many decision trees and outputs the class that is the mode of the classes output by individual trees [18]. The number of trees is kept constant at 500 and the maximum number of nodes is varied from 3 to 9 for a sufficiently low Out-of-Bag (OOB) error with 2 variables tried at each split, these values are computed and assigned by a tune function proposed in R [19].

Support Vector Machines (SVM) are a set of related supervised learning methods used for classification and regression [20]. In this work we employ a C type binary class SVM with a Gaussian kernel; the two hyper-parameters of the C-SVMs; penalty parameter C and kernel width γ are estimated using a 10-fold cross-validation employing grid-search over the following multiplicative grid.

$$(C, \gamma) \in [(2^i, 2^j) | i \in \{1, ..., 10\}, j \in \{-10, ..., 2\}] \tag{7}$$

These techniques are selected to evaluate the proposed model as they can classify non-linear data with a huge number of instances effectively. Section 6 describes the performance of each of these classifiers across different payloads. Further the AUR and the error probability (P) are computed, evaluated and compared with existing works in Section 6.

6 Results and Performance Analysis

The parameters used to evaluate the performance of the classifiers are the error probability (P) and Area under ROC (AUR). An AUR of 1 represents a perfect test; an area of 0.5 represents a worthless test. Table 3 illustrates these parameters for each classifier for all four embedding rates tested. It can be observed that ANN Classifier gives the best performance of detecting PQ. Overall we conclude that the proposed model detects PQ of lower embedding rate with a better accuracy, the reason being that lesser changeable coefficients in the lower embedding rates of PQ cause a prominent change in the HBCL statistics as fewer changes in the source symbols show up a significant variation in the number of particular bit Huffman codes. However, the performance of the Random Forests (RF) classifier with these features show an exception to this behavior, as RF builds a separate classifier for each label during which it would not consider the original HBCL statistics.

Table 3. Performance Evaluation Parameters for the Identified Classifiers

Classifier/ Parameters		0.05 bpac	0.1 bpac	0.2 bpac	0.4 bpac	Classifier/ Parameters		0.05 bpac	0.1 bpac	0.2 bpac	0.4 bpac
ANN	P	0.014	0.015	0.019	0.020	RF	P	0.025	0.016	0.012	0.011
	AUR	0.975	0.969	0.962	0.958		AUR	0.951	0.967	0.975	0.977
k-NN	P	0.017	0.018	0.020	0.021	C-SVM	P	0.014	0.015	0.020	0.022
	AUR	0.965	0.962	0.958	0.956		AUR	0.972	0.968	0.956	0.954

For performance evaluation, the experimental results obtained in this work are compared with four other previously published attacks of which SVBS [21] is a targeted steganalysis for PQ and the others use blind steganalytic features to detect PQ [5, 6, 15].

Table 4. Comparison of Detection Accuaracy with [21] and [6]; Error Probability (*P*) with [5]

Payload (bpac)	SVBS	CBS	HFS	Fridrich	HFS
0.05	NA	NA	97.25%	6.73%	1.375%
0.1	NA	81%	96.99%	4.49%	1.505%
0.2	67%	83%	96.16%	3.19%	1.880%
0.4	76%	94%	95.80%	NA	2.000%

Table 4 (column 2) compares the detection accuracy of the proposed model with that proposed in [21] that uses Singular Value Decomposition (SVD) features for steganalysis (SVBS). The proposed model is also compared with Contourlet Based Steganalysis (CBS) (column 3) in which contourlet transforms are used for feature extraction and non linear SVM classifier to evaluate steganalysis of PQ [6].

Columns 5 and 6 gives a comparison of the error probabilities (*P*) for different payloads using the proposed model with the work of Fridrich et al, where in a 274-dimensional feature vector consisting of 193 extended DCT features and 81 Markov features is used for blind steganalysis using a Gaussian kernel SVM classifier [5].

The area under ROC (AUR) is compared with the work of Mehdi et al, [7] in which several steganalysis techniques are compared and shown that Feature Based Steganalysis (FBS) proposed by [22] detects PQ more accurately compared to two other methods; Binary Similarity Measures (BSM) [23] and Wavelet Based Steganalysis

(WBS) [24]. The comparative results are as shown in Table 5. The proposed HFS algorithm can effectively detect PQ even when the model was evaluated against a image database consisting stego images embedded with LSB encoding, JPEG Hide & Seek and Model Based Steganography [14]. Fig. 2 shows the confusion matrix which illustrates that PQ can be detected with 100% accuracy in a blind steganalytic environment too.

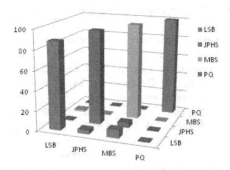

Table 5. Comparison of AUR with [15]

Payload (bpac)	FBS	HFS
0.05	0.8509	0.9725
0.1	0.8555	0.9699
0.2	0.8579	0.9624
0.4	0.8696	0.9582

Fig. 2. Confusion matrix in a blind steganalytic approach

7   Conclusions

This paper proposes a new steganalysis approach for Perturbed Quantization (PQ) namely, the HFS algorithm using a combination of HBCL statistics and FR Index.The proposed HFS algorithm and the features can be effectively used for steganalysis of PQ of several payloads with a high accuracy.

The experiments also prove that the detection accuracy of PQ by our proposed model is superior compared to some of the existing techniques. The response of several classifiers (excepting RF) with the HFS algorithm is almost similar which proves that these steganalytic features are accurate, consistent and monotonic in nature. The proposed model gives a reliable detection of PQ using five feature vectors compared to a huge number of feature vectors used for steganalysis by other prediction models. The HFS algorithm can also be used in a blind steganalytic environment to detect PQ with 100% accuracy.

The proposed HFS algorithm can be further fine-tuned to estimate the size of the embedding payload of PQ. These features extracted can be further tested on more recent steganographic methods like Yet Another Steganographic Scheme (YASS).

References

1. Birgit, P.: Information hiding terminology-results of an informal plenary meeting and additional proposals. In: The Proceedings of the First International Workshop on Information Hiding, pp. 347–350 (1996)
2. Miche, Y., Bas, P., Lendasse, A., Jutten, C., Simula, O.: Reliable steganalysis using a minimum set of samples and features. EURASIP Journal of Information Security, 350–354 (2009)

3. Fridrich, J., Goljan, M., Soukal, D.: Perturbed quantization steganography. ACM Multimedia and Security Journal 11(2), 98–107 (2005)
4. Fridrich, J., Goljan, M., Soukal, D.: Perturbed quantization steganography using wet paper codes. In: The Proceedings of ACM MM&S Workshop, Germany, pp. 4–15 (2004)
5. Fridrich, J., Pevny, T., Kodovsky, J.: Statistically undetectable JPEG steganography: dead ends. Challenges, and Opportunities. In: The Proceedings of ACM MM&S Workshop, Dallas, TX, pp. 3–14 (2007)
6. Hedieh, S., Mansour, J.: CBS: contourlet-based steganalysis method. Journal of Signal Processing Systems (2010)
7. Kharrazi, M., Sencar, H.T., Memon, N.: Benchmarking steganographic and steganalysis techniques. Journal of Electronic Imaging 15(4) (2006)
8. Zuzana, O., Jiri, H., Ivan, Z., Roman, S.: Steganography detection by means of neural network. In: The Ninteenth International Conference on Database and Expert Systems Application, pp. 571–574. IEEE Computer Society, Los Alamitos (2008)
9. Joachims, T.: Text categorization with support vector machines: learning with many relevant features. In: Nédellec, C., Rouveirol, C. (eds.) ECML 1998. LNCS, vol. 1398, pp. 137–142. Springer, Heidelberg (1998)
10. Chiew, K.L., Pieprzyk, J.: Blind steganalysis: a countermeasure for binary image steganography. In: The Proceedings of the International Conference on Availability, Reliability and Security, pp. 653–658 (2010)
11. Dautrich, J.: Multi-class steganalysis. Machine Learning Course Research Project Distinguishing Images Embedded using Reversible Steganographic Schemes (2009)
12. Yoan, M., Benoit, R., Amaury, L., Patrick, B.: A feature selection methodology for steganalysis. In: Gunsel, B., Jain, A.K., Tekalp, A.M., Sankur, B. (eds.) MRCS 2006. LNCS, vol. 4105, pp. 49–56. Springer, Heidelberg (2006)
13. Bosch, A., Zisserman, A., Munoz, X.: Image classification using random forests and ferns. In: The IEEE Eleventh International Conference on Computer Vision, pp. 1–8 (2007)
14. Bhat, V.H., Krishna, S., Shenoy, P.D., Venugopal, K.R., Patnaik, L.M.: HUBFIRE - A multi-class SVM based JPEG steganalysis using HBCL statistics and FR index. In: SECRYPT (in press, 2010)
15. Kharrazi., M., Sencar., H.T., Memon, N.: A performance study of common image steganography and steganalysis techniques. Journal of Electronic Imaging 15(4) (2006)
16. Fawcett, T.: ROC graphs: notes and practical considerations for data mining researchers. Technical Report HPL-2003–4. HP Laboratories, Palo Alto (2003)
17. Bishop, C.: Neural networks for pattern recognition. Oxford, Oxford University, UK (1995)
18. Ho, T.: Random decision forest. In: The Third International Conference on Document Analysis and Recognition, pp. 278–282 (1995)
19. http://debian.mc.vanderbilt.edu/R/CRAN/web/packages/randomForest/randomForest.pdf
20. Burges, C.: A tutorial on support vector machines for pattern recognition. Data Mining and Knowledge Discovery 2, 121–167 (1998)
21. Gül, G., Dirik, A.E., Avcıbas, I.: Steganalytic features for JPEG compression-based perturbed quantization. IEEE Signal Processing Letters 14(3), 205–208 (2007)
22. Fridrich, J.: Feature-based steganalysis for JPEG images and its implications for future design of steganographic schemes. In: Fridrich, J. (ed.) IH 2004. LNCS, vol. 3200, pp. 67–81. Springer, Heidelberg (2004)
23. Avcibas, I., Kharrazi, M., Memon, N., Sankur, B.: Image steganalysis with binary similarity measures. EURASIP Journal on Applied Signal Processing, 2749–2757 (2005)
24. Lyu., S., Farid, H.: Steganalysis using color wavelet statistics and one-class support vector machines. In: SPIE Symposium on Electronic Imagining, San Jose, CA (2004)

A View Recommendation Greedy Algorithm for Materialized Views Selection

T.V. Vijay Kumar[1], Mohammad Haider[1,2], and Santosh Kumar[1,3]

[1] School of Computer and Systems Sciences, Jawaharlal Nehru University,
New Delhi-110067, India
[2] Mahatma Gandhi Mission's College of Engineering and Technology,
Noida, UP-201301, India
[3] Krishna Institute of Engineering and Technology,
Ghaziabad, UP-201206, India
tvvijaykumar@hotmail.com,
{haider.mohammad,santoshg25}@gmail.com

Abstract. View selection is one of the key problems in view materialization. Several algorithms exist in literature for view selection, most of them are greedy based. The greedy algorithms, in each iteration, select the most beneficial view for materialization. Most of these algorithms are focused around algorithm HRUA. HRUA exhibits high run time complexity. As a result, it becomes infeasible to select views for higher dimensions. This scalability problem is addressed by the greedy algorithm VRGA proposed in this paper. Unlike HRUA, VRGA selects views from a smaller search space, comprising of recommended views, instead of all the views in the lattice. This enables VRGA to select views efficiently for higher dimensional data sets. Further, experimental results show that VRGA, in comparison to HRUA, requires significantly lesser benefit computations, view evaluation time and memory. Alternatively, HRUA has a slight edge over VRGA as regards to the total cost of evaluating all the views.

Keywords: Materialized view, view selection, greedy algorithm.

1 Introduction

In today's scenario, organizations are increasingly exploiting historical data, accumulated over a period of time, to design business strategies for remaining competitive in the market. The organization stores such historical data in a data warehouse in order to support their strategic decision making queries [11]. Since the information in the data warehouse grows continuously over time, the query response time for decision making queries, which are usually analytical, long and complex, is high. This time needs to be reduced so that decision making can become more efficient. This can be addressed by materializing views over a data warehouse [28]. Materialized views are pre-computed and summarized information stored in a data warehouse [15] for the purpose of reducing the response time for analytical queries [11].

S. Dua, S. Sahni, and D.P. Goyal (Eds.): ICISTM 2011, CCIS 141, pp. 61–70, 2011.

All views cannot be materialized, as the number of possible views is exponential in the number of dimensions and therefore would not conform to the storage space constraints [5, 10]. Further, selecting an optimal subset of views is shown to be an NP-Complete problem [10]. Thus, there is a need to select a subset of views that contains the relevant and the required information for answering queries, while fitting within the available space for materialization [5]. The selection of such views is referred to as the view selection problem in literature [5]. Views can be selected based on past query patterns [12, 25], or heuristically, using algorithms that are greedy [10], evolutionary [29, 30], etc. This paper focuses on the greedy based view selection.

In greedy based view selection, at each step, the view having maximum benefit is selected for materialization. This selection of views at each step continues till a pre-defined number of views are selected or available space for view materialization is exhausted. Several greedy view selection algorithms have been proposed in literature [1, 2, 4, 6, 7, 8, 9, 10, 13, 14, 16, 17, 18, 20, 21, 22, 23, 24, 26, 27]. Most of these algorithms select views based on size of the view. Among these, the greedy algorithm proposed in [10], which hereafter, in this paper, will be referred to as HRUA, is considered the most fundamental greedy algorithm. HRUA, in each iteration, selects the most beneficial view from a multidimensional lattice. In each iteration, HRUA computes benefit of each view using the cost of the views, defined in terms of its size, and then selects the most beneficial view for materialization. The benefit of a view is computed as

$$Benefit(V) = \sum \{(size(SMA(W)) - size(V)) \mid V \in A(W) \wedge (size(SMA(W)) - size(V)) > 0\}$$

where size(V) is size of view V, size(SMA(W)) is size of smallest materialized ancestor view of W and A(W) is the ancestor view of W.

The view having maximum value for Benefit(V) is selected for materialization. The selection of the most beneficial view, in each iteration, continues till a pre-defined number of views are selected for materialization. HRUA is shown to achieve atleast 63% of the optimal solution [10].

HRUA exhibits high run time complexity as the number of views is exponential in the number of dimensions. As the dimensions of the data set increase, the number of views requiring computation of benefit values, i.e. Total Benefit Computations (TBC), would also increase and for higher dimensional data sets, it would become almost infeasible for HRUA to selects views for materialization. One way to address this problem is by reducing TBC without compromising much on the quality of views selected for materialization. This would enable view selection for higher dimensional data sets.

The proposed View Recommendation Greedy Algorithm (VRGA) aims to address this problem by selecting views in two phases. In the first phase, it recommends views, among all possible views in the lattice, for materialization. This is followed by greedily selecting the most beneficial view from these recommended views. That is, VRGA selects views from a smaller set of recommended views, instead of all possible views as considered in HRUA. This would result in lesser number of benefit computations enabling view selection to become more efficient and feasible for higher dimensional data sets.

The paper is organized as follows: The proposed algorithm VRGA is given in Section 2 followed by examples based on it in Section 3. The experimental results are given in Section 4. Section 5 is the conclusion.

2   VRGA

As discussed above, one of the key limitations with HRUA is that it exhibits high run time complexity. As a result, it would become infeasible to select views for higher dimensional data sets. This scalability problem is addressed by the proposed algorithm VRGA, which selects views from a smaller set of recommended views instead of from all the views in the lattice. This would reduce the number of views requiring computation and thereby would make view selection feasible for higher dimensional data sets. VRGA recommends views based on the following heuristic:

"At each level, the view with smallest size is recommended for selection."

The heuristic is based on the rationale that the minimum sized view at each level in the lattice is likely to have maximum benefit with respect to any views above it in the lattice. As a result, it is more likely to be selected for materialization. To understand the basis of this heuristic, consider the four dimensional lattice shown in Fig. 1.

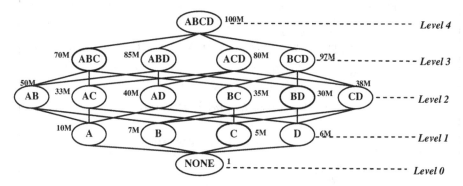

Fig. 1. 4-Dimensional Lattice along with Size of Each View

HRUA assumes view ABCD to be materialized, as any query on it cannot be answered by any other view in the lattice. At level 0, there is only one view NONE and thus it is selected. At level 1 there are four views A, B, C and D and their benefits, as computed using HRUA, with respect to root view ABCD is 180M, 186M, 190M and 188M respectively. The view C, which has the smallest size at level 1 has maximum benefit and therefore has maximum likelihood, among the views in level 1 of being selected for materialization. The views at level 2 are AB, AC, AD, BC, BD and CD having benefits 200M, 268M, 240M, 260M, 280M and 248M respectively. View BD, which is the smallest sized view amongst all the views at this level, has maximum benefit and therefore is most likely to be selected for materialization. Similarly, at level 3 view ABC is the smallest in size and has benefit 240M, which is more than benefit 120M, 160M and 24M of views ABD, ACD and BCD respectively. Thus ABC is most likely to be selected for materialization. In this manner, VRGA recommends views for selection in each iteration. The most beneficial view is then selected from amongst these recommended views. That is, from amongst views NONE, C, BD, and ABC, view BD is selected for materialization. The algorithm VRGA, based on the above heuristic, is given in Fig. 2. VRGA takes lattice of views, along with the size of each view, as input and produces the top-T views as output.

```
INPUT: Lattice of views with size of each view
OUTPUT: top-T views V_T
METHOD:
L = {L_max, L_max -1,....,0} where L is the set of Levels in the lattice
L_max is the root level and 0 is the bottommost level of the lattice
VR is the View Recommended in each iteration
View Recommended Set (VRS) is a set of views recommended for
materialization
               VRS = Φ
V_T is the set of top-T views
V_T = Φ
N_VS is the number of views selected
               N_VS=0
WHILE (N_VS ≤ T)
 BEGIN
        // Phase-1: Compute the view recommended set VRS
        Level L=0
        VR = Φ
        WHILE (L < L_max - 1)
        BEGIN
            V'={views at level L}
            Find the set of views V_L in V' such that V_L ∉ VRS ∧ V_L
∉ V_T i.e.
                V_L = V' - VRS - V_T
                Find minimum sized view V_M in V_L
```

$$V_M = \left\{ v \mid v \in V_L \wedge size(v) = \underset{v' \in V_L}{Min}\left(size(v')\right) \right\}$$

```
            IF |V_M| > 1 then
                        Select any one element from V_M say V"
                        V_M = {V"}
            END IF
            VR = VR ∪ V_M
            L=L+1
        END WHILE
        VRS=VRS ∪ VR
        // Phase-2: Greedy selection of a view from VRS using HRUA
        B_M=0
        FOR V ∈VRS
            Benefit(V) = 0
            FOR W ∈ Desc(V) // Desc(V): Descendent views of V
                    IF (size(SMA(W)) - size(V)) > 0  // SMA(W):
Smallest Materialized Ancestor of W
                        Benefit(V) = Benefit(V) + (size(SMA(W)) -
size(V))
                    END IF
            END FOR
            IF (Benefit(V) > B_M)
        V_b=V
                    B_M=Benefit(V)
            END IF
        END FOR
        V_T = V_T ∪ V_b
        VRS=VRS - V_b
        N_VS = N_VS + 1
END WHILE
RETURN top-T views V_T
```

Fig. 2. Algorithm VRGA

VRGA comprises of two phases namely View Recommended Set (VRS) generation followed by greedy view selection. In the VRS generation phase, views are recommended for selection, based on the heuristic defined above. The recommendation of views starts at the bottom of the lattice. In the first iteration, the view at the bottommost level in the lattice is added to VRS. Then the views at a level above it are considered for recommendation. Among these views, as per the heuristic, the view having the smallest size is recommended and added to VRS. Then the views in the level above it are considered and from amongst them the view having smallest size is recommended and added to VRS. In this manner views are recommended and added to VRS till the level of the root view is reached. In the greedy selection phase, the benefit of each view in VRS is computed, as in HRUA, and the most beneficial view is selected for materialization. This most beneficial view is then removed from VRS. In subsequent iterations, view recommendation starts from the bottom of the lattice and the smallest sized view at each level, which has not yet been recommended, is recommended and added to VRS. This is followed by selecting the most beneficial view from among the views in VRS. This process continues till top-T views are selected. It can be observed that VRGA recommends views in a greedy manner, where at each step, starting from the bottom of the lattice, the view having the smallest size at each level is recommended for selection. This enables VRGA to select views from a smaller set of recommended views in VRS. This helps in overcoming the scalability limitation of HRUA as lesser number of views would require benefit computation resulting in view selection becoming feasible for higher dimensional data sets. Examples based on VRGA are illustrated in the next section.

3 Examples

Consider the three dimensional lattices shown in Fig. 3. Suppose top-3 views are to be selected, using HRUA and VRGA, from each of these lattices. The size of each view (in million rows (M)) is given alongside the node in the lattice. VRGA, like HRUA, assumes the view ABC as materialized, as queries on ABC cannot be answered by any other view in the lattice. Considering only ABC as materialized, queries on each view would be evaluated using ABC thereby incurring a cost of ABC i.e. 100 million rows. This would result in the total cost of evaluating all the views, referred to as Total View Evaluation Cost (TVEC), as 800 million rows.

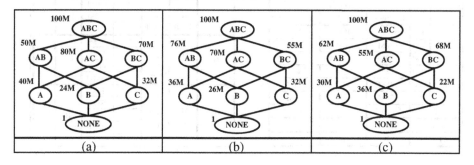

Fig. 3. 3-Dimensional Lattice Along with Size of Each View

HRUA		Benefit		
View	Size	Iteration-I	Iteration-II	Iteration-III
AB	50M	200M		
AC	80M	80M	40M	20M
BC	70M	120M	60M	30M
A	40M	120M	20M	10M
B	24M	152M	52M	34M
C	32M	136M	86M	
NONE	1	100M-1	50M-1	32M-1
View Selected		AB	C	B
TBC		7	6	5

VRGA	Iteration	VR	VRS	Benefit View	Benefit	V_b	TBC
	I	AB, B, NONE	AB, B, NONE	AB	200M	AB	3
				B	152M		
				NONE	100M-1		
	II	BC, C	BC, B, C, NONE	BC	60M	C	4
				B	52M		
				C	86M		
				NONE	50M-1		
	III	AC, A	AC, BC, A, B, NONE	AC	20M	B	5
				BC	30M		
				A	10M		
				B	34M		
				NONE	32M-1		

Fig. 4. Top-3 Views Selection from the Lattice of Fig. 3(a)

First let us consider the top-3 views selected using HRUA and VRGA from the 3-dimensional lattice shown in Fig. 3(a). These selections are shown in Fig. 4.

HRUA and VRGA select the same top-3 views AB, C and B. These views if materialized would reduce the TVEC from 800 million rows to 480 million rows. Though the TVEC due to views selected using HRUA and VRGA are same, the TBC value of VRGA is 12, which is less than the TBC value of 18 due to views selected using HRUA. This implies that VRGA is able to select views having same quality, as those selected using HRUA, against comparatively lesser number of computations of benefit value.

HRUA and VRGA need not always select views with same TVEC. Let us consider top-3 views selection from the 3-dimensinal lattice shown in Fig. 3(b). The top-3 views selection using HRUA and VRGA is shown in Fig. 5.

HRUA selects BC, A and B as the top-3 views whereas VRGA select BC, AC and B as the top-3 views. The TVEC of views selected using HRUA is 498 million rows whereas for those selected using VRGA is 502 million rows.

HRUA		Benefit		
View	Size	Iteration-I	Iteration-II	Iteration-III
AB	76M	96M	48M	24M
AC	70M	120M	60M	30M
BC	55M	180M		
A	36M	128M	83M	
B	26M	148M	58M	39M
C	32M	136M	46M	27M
NONE	1	100M-1	55M-1	36M-1
View Selected		BC	A	B
TBC		7	6	5

VRGA	Iteration	VR	VRS	Benefit View	Benefit	V_b	TBC
	I	BC, B, NONE	BC, B, NONE	BC	180M	BC	3
				B	148M		
				NONE	100M-1		
	II	AC, C	AC, B, C, NONE	AC	60M	AC	4
				B	58M		
				C	46M		
				NONE	55M-1		
	III	AB, A	AB, A, B, C, NONE	AB	24M	B	5
				A	53M		
				B	58M		
				C	46M		
				NONE	55M-1		

Fig. 5. Top-3 Views Selection from the Lattice of Fig. 3(b)

That is, the views selected using VRGA incur a slightly inferior TVEC as compared to those selected using HRUA. The reason behind this difference is that HRUA, in each iteration, selects the most beneficial view from the entire search space whereas VRGA selects the most beneficial view from a very limited search space, over only the recommended views. VRGA gains over HRUA on the TBC value.

VRGA need not always select views with inferior TVEC. Let us consider the top-3 views selection using HRUA and VRGA from a 3-dimensional lattice shown in Fig. 3(c). These selections are shown in Fig. 6.

The top-3 views selected using HRUA is AC, B, and C as against AC, AB, and C selected by VRGA. The TVEC due to views selected using VRGA (478) is less than TVEC (490) due to views selected using HRUA. This implies that, VRGA not only requires lesser number benefit computations to select views but can also select views having lesser total cost of evaluating all the views

From the above, it can be inferred that VRGA is able to select fairly good quality views efficiently. Further, in order to substantiate this claim, experiment based comparison of HRUA and VRGA were carried out and their results are given next.

4 Experimental Results

The HRUA and VRGA algorithms were implemented using JDK 1.6 in Windows-XP environment. The two algorithms were compared by conducting experiments on an Intel based 2 GHz PC having 1 GB RAM. The comparisons were carried out on parameters like Total View Evaluation Time (TVET), TBC, TVEC and Total Memory Usage (TMU). The experiments were performed for selecting top-10 views for materialization for dimensions 5 to 10. These graphs are shown in Fig. 7.

HRUA

View	Size	Benefit		
		Iteration-I	Iteration-II	Iteration-III
AB	62M	152M	76M	38M
AC	55M	180M		
BC	68M	128M	64M	32M
A	30M	140M	50M	31M
B	36M	128M	83M	
C	22M	156M	66M	47M
NONE	1	100M-1	55M-1	36M-1
View Selected		AC	B	C
TBC		7	6	5

VRGA

Iteration	VR	VRS	Benefit		V_b	TBC
			View	Benefit		
I	AC, C, NONE	AC, C, NONE	AC	180M	AC	3
			C	156M		
			NONE	100M-1		
II	AB, A	AB, A, C, NONE	AB	76M	AB	4
			A	50M		
			C	66M		
			NONE	55M-1		
III	BC, B	BC, A, B, C, NONE	BC	32M	C	5
			A	50M		
			B	45M		
			C	66M		
			NONE	55M-1		

Fig. 6. Top-3 Views Selection from the Lattice of Fig. 3(c)

First, the two algorithms were compared on TBC and the corresponding graph is shown in Fig. 7(a). The graph shows that VRGA, in comparison to HRUA, requires lesser TBC. The difference in the TBC value becomes significant for higher dimensions. Further, a graph (Fig. 7(b)) was plotted to determine the impact of this

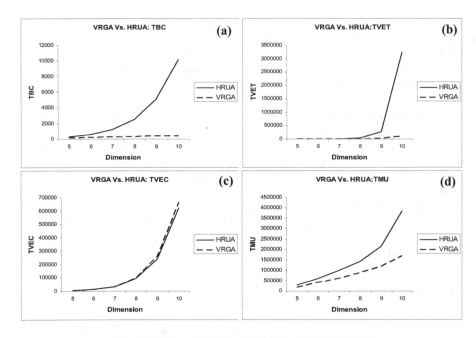

Fig. 7. HTUA vs. VRGA – TBC, TVET, TVEC, TMU

difference in the TBC value on TVET. The graph shows that TVET(in milliseconds) of VRGA is less than that of HRUA and, for dimensions beyond 8, the difference becomes significant. This implies that VRGA is able to select views efficiently for higher dimensional data sets.

Further, in order to ascertain the impact of lower TBC on TVEC, a graph (Fig. 7(c)) was plotted. It is observed from the graph that the views selected using VRGA incur a slightly higher TVEC than those selected using HRUA. Thus, it can be inferred that VRGA selects views efficiently at the cost of a slight drop in TVEC.

Furthermore, graph (Fig. 7(d)) was plotted to compare VRGA and HRUA in terms of the TMU (in bytes) to select top-10 views for materialization. The graph shows that the VRGA has a lower TMU than HRUA and this difference becomes significant for higher dimensions. This may be due to the fact that for higher dimensions significantly lesser number of views requires evaluation of benefit value in case of VRGA resulting in lesser memory requirement.

It can be inferred from the above results that VRGA, in comparison to HRUA, performs better on TBC, TVET and TMU. HRUA has a slight edge over VRGA on TVEC.

5   Conclusions

In this paper, an algorithm VRGA is proposed that selects top-T views greedily from a multidimensional lattice. This algorithm attempts to overcome the scalability limitation of the most fundamental algorithm HRUA. VRGA, in each iteration, recommends a set of views from among all the views in the lattice. This is followed by greedily selecting the most beneficial view from these recommended views. Unlike HRUA, which selects the most beneficial view from among all the views in the lattice, VRGA selects it from a smaller search space comprising of set of recommended views. As a result, VRGA would be able to select views efficiently and make view selection feasible for higher dimensional data sets.

Further, experiment based comparison of VRGA and HRUA showed that VRGA was found to achieve significant improvement in TBC, TVET and TMU at the cost of a slightly inferior TVEC value, of the views selected for materialization. This shows that VRGA is able to select fairly good quality views in a more efficient manner.

References

1. Agrawal, S., Chaudhuri, S., Narasayya, V.: Automated selection of materialized views and indexes in SQL databases. In: Proceedings of VLDB 2000, pp. 496–505. Morgan Kaufmann Publishers, San Francisco (2000)
2. Aouiche, K., Jouve, P.-E., Darmont, J.: Clustering-based materialized view selection in data warehouses. In: Manolopoulos, Y., Pokorný, J., Sellis, T.K. (eds.) ADBIS 2006. LNCS, vol. 4152, pp. 81–95. Springer, Heidelberg (2006)
3. Aouiche, K., Darmont, J.: Data mining-based materialized view and index selection in data warehouse. Journal of Intelligent Information Systems, 65–93 (2009)
4. Baralis, E., Paraboschi, S., Teniente, E.: Materialized view selection in a multidimensional database. In: Proceedings of VLDB 1997, pp. 156–165. Morgan Kaufmann Publishers, San Francisco (1997)
5. Chirkova, R., Halevy, A., Suciu, D.: A formal perspective on the view selection problem. The VLDB Journal 11(3), 216–237 (2002)
6. Gupta, H.: Selection of views to materialize in a data warehouse. In: Proceedings of ICDT, Delphi, Greece, pp. 98–112 (1997)
7. Gupta, H., Harinarayan, V., Rajaraman, A., Ullman, J.: Index selection in OLAP. In: The Proceedings of the ICDE 1997, Washington, DC, USA, pp. 208–219. IEEE Computer Society, Los Alamitos (1997)
8. Gupta, H., Mumick, I.: Selection of views to materialize under a maintenance cost constraint. In: Beeri, C., Bruneman, P. (eds.) ICDT 1999. LNCS, vol. 1540, pp. 453–470. Springer, Heidelberg (1998)
9. Gupta, H., Mumick, I.: Selection of views to materialize in a data warehouse. IEEE Transactions on Knowledge and Data Engineering 17(1), 24–43 (2005)
10. Harinarayan, V., Rajaraman, A., Ullman, J.: Implementing data cubes efficiently. In: The Proceedings of SIGMOD, pp. 205–216. ACM Press, New York (1996)
11. Inmon, W.H.: Building the data warehouse, 3rd edn. Wiley Dreamtech (2003)
12. Lehner, R., Ruf, T., Teschke, M.: Improving query response time in scientific databases using data aggregation. In: Proceedings of Seventh International Conference and Workshop on Databases and Expert System Applications, pp. 9–13 (1996)
13. Liang, W., Wang, H., Orlowska, M.: Materialized view selection under the maintenance time constraint. Journal of Data and Knowledge Engineering 37(2), 203–216 (2001)
14. Nadeau, T.P., Teorey, T.J.: Achieving scalability in OLAP materialized view selection. In: Proceedings of DOLAP, pp. 28–34. ACM Press, New York (2002)
15. Roussopoulos, N.: Materialized views and data warehouse. In: The Fourth Workshop KRDB 1997, Athens, Greece (1997)

16. Serna-Encinas, M.T., Hoya-Montano, J.A.: Algorithm for selection of materialized views: based on a costs model. In: The Proceedings of the Eighth International Conference on Current Trends in Computer Science, pp. 18–24 (2007)
17. Shah, A., Ramachandran, K., Raghavan, V.: A hybrid approach for data warehouse view selection. International Journal of Data Warehousing and Mining 2(2), 1–37 (2006)
18. Shukla, A., Deshpande, P., Naughton, J.: Materialized view selection for multidimensional datasets. In: The Proceedings of the VLDB, pp. 488–499. Morgan Kaufmann Publishers, San Francisco (1998)
19. Teschke, M., Ulbrich, A.: Using materialized views to speed up data warehousing. Technical Report IMMD 6, Universität Erlangen-Nümberg (1997)
20. Uchiyama, H., Ranapongsa, K., Teorey, T.J.: A progressive view materialization algorithm. In: The Proceedings of the Second ACM International Workshop on Data Warehousing and OLAP, Kansas City Missouri, USA, pp. 36–41 (1999)
21. Valluri, S.R., Vadapalli, S., Karlapalem, K.: View relevance driven materialized view selection in data warehousing environment. In: The Proceedings of CRPITS, Darlinghurst, Australia, Australia, pp. 187–196. Australian Computer Society (2002)
22. Vijay Kumar, T.V., Ghoshal, A.: A reduced lattice greedy algorithm for selecting materialized views. In: Prasad, S.K., Routray, S., Khurana, R., Sahni, S. (eds.) ICISTM 2009. Communications in Computer and Information Science, vol. 31, pp. 6–18. Springer, Heidelberg (2009)
23. Vijay Kumar, T.V., Ghoshal, A.: Greedy selection of materialized views. International Journal of Computer and Communication Technology (IJCCT) 1, 47–58 (2009)
24. Vijay Kumar, T.V., Haider, M., Kumar, S.: Proposing candidate views for materialization. In: Prasad, S.K., Vin, H.M., Sahni, S., Jaiswal, M.P., Thipakorn, B. (eds.) ICISTM 2010. Communications in Computer and Information Science, vol. 54, pp. 89–98. Springer, Heidelberg (2010)
25. Vijay Kumar, T.V., Goel, A., Jain, N.: Mining information for constructing materialised views. International Journal of Information and Communication Technology 2(4), 386–405 (2010)
26. Vijay Kumar, T.V., Haider, M.: Materialized views selection for answering queries. LNCS, vol. 6411
27. Vijay Kumar, T.V., Haider, M.: A query answering greedy algorithm for selecting materialized views. In: Pan, J.-S., Chen, S.-M., Nguyen, N.T. (eds.) ICCCI 2010. LNCS(LNAI), vol. 6422, pp. 153–162. Springer, Heidelberg (2010)
28. Widom, J.: Research problems in data warehousing. In: The Fourth International Conference on Information and Knowledge Management, Baltimore, Maryland, pp. 25–30 (1995)
29. Zhang, C., Yao, X., Yang, J.: An evolutionary approach to materialized views selection in a data warehouse environment. IEEE Transactions on Systems, Man and Cybernetics, 282–294 (2001)
30. Zhang, C., Yao, X., Yang, J.: Evolving materialized views in data warehouse. In: Proceedings of the Congress on Evolutionary Computation, vol. 2, pp. 823–829 (1999)

Experience Based Software Process Improvement: Have We Found the Silver Bullet?

Neeraj Sharma[1], Kawaljeet Singh[2], and D.P. Goyal[3]

[1] Department of Computer Science, Punjabi University, Patiala - 147002, Punjab, India
[2] University Computer Centre, Punjabi University, Patiala - 147002, Punjab, India
[3] Management Development Institute, Gurgaon - 122007, Haryana, India
sharma_neeraj@hotmail.com, director@pbi.ac.in,
dpgoyal@mdi.ac.in

Abstract. Software product quality is colossally dependent upon the software development process. Software engineers have always been experimenting with various models in an endeavor to improve upon their software processes. Recently a new area of software process improvement through experience management has got attention of the software engineering community and developers are experimenting with experience management to improve software process. Though literature is abound with studies of problems faced by software organizations and strategies adopted by these firms but there is dearth of literature suggesting how experience and its management can support SPI efforts in an organization. This paper explores the role of experience management in mitigating the effects of software crisis and investigates the current state of EM in Indian software engineering environment.

Keywords: Experience management, SPI, Experience Bases, Software crisis.

1 Introduction

Software has become '*woven into the threads of our daily lives*' [18]. It has become an inherent constituent of survival for almost all organizations as well as for individuals in today's world of competition. May it be telecommunications, transportation, medical services or the defense of a nation, software is a critical component. With Internet and mobile technologies becoming omnipresent, and growing use of embedded software in consumer products, individuals are increasingly becoming dependent on software. For this ever increasing dependence over software, problems in developing software can have ravaging effects not only at individual and organizational levels but also at national and international levels. As a result, the improvement of software development processes has become the pressing area of concern for software engineering (SE) professionals and researchers.

For a long time the general public view of the software engineering profession has been the delayed projects and poor quality software and the field is overwhelmed with reports of software projects missing their deadlines, surpassing their budgets, delivery of poor quality software and dissatisfied clients. Learning in software development teams is often from scratch and each team has to relearn the mistakes of its

S. Dua, S. Sahni, and D.P. Goyal (Eds.): ICISTM 2011, CCIS 141, pp. 71–80, 2011.

predecessors. Reuse of small-scale code components has been practiced in some situations but systematic learning from organization's own products, processes and experience is still uncommon. Both software development professionals as well as research community in the domain are seeking ways and means to overcome these problems. Recently a whole new subfield under the notion of Software Process Improvement (SPI) within SE has gained popularity and has become one of the dominant approaches to improve quality and productivity in software engineering [1].

2   Present Study

The major objective of the paper is to study the use of experience management (EM) systems in software engineering environments. It is especially interesting to study this in a setting with people who are very skilled in using computers, like people developing software. The paper also investigates the current EM practices followed by Indian software engineering organizations. For the purpose of this study, an exhaustive literature search has been carried out. For the empirical evidence of EM in Indian SE environment, five major SE companies in India were selected randomly. Primary data was collected through interviews and secondary data was collected from internal documentation and official websites of the companies.

3   Literature Review

A lot of research has been reported about KM in software engineering e.g. [29] and [37]. An infrastructure to deal with KM in software engineering environments is presented in [26]. There are studies which investigate the need for experience bases in software projects [6]. Talking of experience management, much research exists on many aspects of EM including approaches to EM (e.g. [6], [30]), how to collect experience (e.g. [2], [22]), how to structure experience (e.g. [21], [23]), tools for EM (e.g. [20], [25]) and applications of EM in practice (e.g. [6], [16], [31]). However, the literature on the use of EM for SPI is limited though pivotal e.g. [24]. Commonly two approaches to SPI are found in literature. The first approach tries to improve the process through standardization. Another approach, the Quality Improvement Program (QIP), is more common in software engineering environments [9]. An attempt at establishing an overview of the SPI field is described in [19]. An empirical investigation of the critical success factors of SPI is given by [32].

4   Experience and Experience Management

Knowledge and experience are closely associated and the latter can be conceived as a special form or instance of the former. Experience is defined as previous knowledge or skill one obtained in everyday life [33]. Commonly experience is understood as a type of knowledge that you have gained from practice. In this sense, experience can be regarded as a specialization of knowledge consisting of the problems encountered and successfully tackled in the past. The salient distinguishing characteristic of experience is that it resides in humans and moving it from one person to another is almost impossible as one has to experience it to call it an experience.

The area of EM is increasingly gaining importance. EM can be defined as a special kind of KM concerned with discovering, capturing, storing & modeling and re-/using knowledge of the members of an organisation, who acquired this knowledge through learning from experience [3], [10]. Therefore, methodologies, tools and techniques for KM can be directly reused for EM. However management of experience warrants different methods as experience has some special features different from knowledge. A unique differentiating feature of EM is the additional stage where experience is transformed into knowledge [10]. Since the evolution of the human race, all invaluable experience is transformed into knowledge and this knowledge is shared among humans in varied forms. Creation of knowledge is an indispensable stage in the KM process (exemplified by data mining and knowledge discovery techniques); generation of experience from heaps of knowledge is still a challenging task and is not addressed in EM research [27], [10], [33].

5 Improving Software Process through Managing Experience

The major problem with the existing SPI approaches is that none of these explicitly value the knowledge and experience of software engineers in improving the software processes in an organization and nor do they provide for any system or method to capture, manage and use this accumulated knowledge to avoid repeating the mistakes and to enable the software engineers learn from past experience. There is no denying the fact that accumulating and managing software development experiences plays a very significant role in improving the software process. Reusing experience in the form of processes, products, and other forms of knowledge is essential for improvement, that is, reuse of knowledge is the basis of improvement [5]. The fundamental idea is to improve the software process on the basis of the accumulated knowledge and experiences of the software engineers working in the organization. Insights from the field of knowledge and experience management are therefore potentially useful in software process improvement efforts so as to facilitate the creation, modification, and sharing of software processes in an organization. Management of knowledge and experience are key means by which systematic software development and process improvement occur [37]. They further assert that although remedies such as fourth generation programming languages, structured techniques and object-oriented technology, software agents, component based development, agile software practices, etc. have been promoted; a *silver bullet*[1] has yet to be found (*ibid.*). EM provides for the systems and mechanisms for managing the knowledge and experience of software engineers in an organisation that becomes the basis of process improvement. SPI in practice is about managing software knowledge.

---

[1] A term rooted in folklore of the American Civil War, supposedly became popular from the practice of encouraging a patient who was to undergo field surgery to bite down hard on a lead bullet to divert the mind from pain [35]. According to another folklore, one seeks bullets of silver to magically lay to rest the werewolves who could transform unexpectedly from the familiar into horrors. The software project is capable of becoming a monster of missed schedules, blown budgets, and flawed products, calling desperately for a silver bullet [11].

It was in 1968 NATO conference that the problems of project overruns, over budgeting and poor functionality of the software were discussed and eventually identified as *software crisis* or *software gap* to expose the gravity of the problems in the discipline. But even after over 40 years, not much seems to have changed; late delivery of the systems, ill-functionality and cost overruns are commonly reported by companies [13].

Following text discusses the common SE problems and explains how EM can address these issues to a great extent, thus promising a silver bullet of some sort.

5.1 Reinventing the Wheel

SE is an extremely experimental discipline [8]. With every software development project, new experience is gained but this experience is seldom captured and stored systematically so as to learn from it in future projects to avoid project delays and cost overruns.

Often software development teams carry out similar kinds of projects without understanding that they could have achieved the results more easily by following the practices adopted in previous projects [7], [12]. In every project, software engineers acquire new knowledge and experience and this new knowledge can be used to better the future projects but much of this knowledge and experience goes unnoticed and is hardly shared among team members in an organisation. The underlying fact is that SEs do not benefit from existing experience and instead repeat mistakes over and over again by reinventing the wheel [12].

EM in general and Software Experience Bases in particular can help SE organisations capture the experience from across the projects and store it in a central repository so as to make it available to the whole organisation. An experience management system encourages knowledge growth, communication, preservation and sharing. EM in SE projects can be used to capture the knowledge and experience generated during the software development process and these captured experiences can be used to better the performance of the developers in similar kinds of projects in future. Reusing experience forbids the repetition of past failures and steers the solution of recurrent problems.

5.2 Understanding the Problem Is the Problem

SE is also highly interdisciplinary in approach. A software engineer is required to have knowledge not only about its own domain but must also possess sufficient knowledge about the domain for which software is being developed. Many a times writing software for a new domain demands learning a new technique or tool or a programming language or application of a new project management technique. Even when the organisation has been working on a particular domain extensively, the deep application-specific knowledge required to successfully build complex software is thinly spread over many software development engineers [14]. Although individual development engineers understand different components of the application domain, building large and complex software requires integration of different pieces of domain knowledge. Software developers working in these complex domain environments admit that *"writing code is not the problem, understanding the problem is the problem"* (ibid.).

EM serves two goals in this direction – (a) identifying expertise or competence already existing in the organization and helping leverage and package this knowledge in a reusable form and (b) helping organization in acquiring new domain(s) knowledge by identifying the knowledge gaps that exist in the organization.

5.3 Here Today, Gone Tomorrow Technology

Emergence of new technologies makes software more powerful, but at the same time new technologies is every project manager's worst nightmare [12]. The fast pace of technology change makes the already complex software development business a highly demanding one, necessitating software developers to acquire knowledge of emerging technologies while at the same time adhering to organizational processes and methodologies. Though these problems are ubiquitous across industries, but the software industry is probably worse than other industries due to the fact that the pace of change is faster here. Lack of time causes a lack of experience, constantly pushing the boundaries of an organization's development set of skills and competencies [20]. When a project team uses a new technology, the software engineers all too often resort to the 'learning by doing' approach, often leading to serious project delays [12].

EM provides an opportunity to SE organizations to create a common language of understanding among software developers, so that they can interact, negotiate and share knowledge and experiences [4]. Furthermore, an EM system compliments the organizational ability to systematically manage innovative knowledge in the SE domain. Experience management nurtures a knowledge sharing culture within the company and eases sharing of knowledge about new technologies. EM also provides for the systems that help in actively searching for knowledge not only within organization but outside as well. Knowledge sharing among communities of practice is the best example of such a system.

5.4 Knowledge Walks Home Everyday

The success of any organization in performing its core business processes depends heavily upon the employees' relevant knowledge about these processes. But the problem is that this knowledge is within the control of its employees and not with the organization as such. Serious knowledge gaps are created when the employees who possess this core knowledge leave the organization. The problem becomes graver if the outgoing employee is the only expert in that field of activity. Still serious is the case when no one else in the organization knows what knowledge he possesses [7]. Also in usual working culture of SE organizations, people from different departments with different sets of skills and knowledge are taken together to form a team for a specific project. These teams dissolve after the completion of the project, taking the cross-functional experience along, without storing these experiences in organizational repository of some sort. These experiences remain with the individuals who will take it with them when they leave the organization.

EM helps organizations in systematically capturing, storing (in a repository called experience base) and disseminating knowledge and experience at all levels of the organization, by creating a knowledge-sharing culture. EM also facilitates employees by providing who knows what.

EM helps organizations create systems and frameworks for capturing core competency that can help retain some knowledge when employees leave. This core competency would at least help in getting insights as to what the employee who left knew and what profile his successor needs to have to fill the position. This is even more relevant in the current scenario of economic slowdown.

5.5 Trust in Virtual Teams

Another important trend being witnessed by SE organizations is the development of the systems through, what is called, the virtual teams. A virtual team consists of its members who are not co-located but separated by geographic distances and national cultures. Increase in globalization has led to this trend [28], [36]. Other reasons behind the popularity of virtual teams in SE are availability of experts at cheap costs in other parts of the world, enhanced telecommunication infrastructure, and competitiveness. Members of the virtual teams are required to complete interdependent tasks, requiring different sets of technical skills. Collaboration among members, having different domain expertise and cultural backgrounds, requires knowledge transfer among them so as to create a sense of mutuality. Knowledge transfer across space and time has problems in virtual teams. The root-cause of this problem, among other things, is the concept of 'localness of knowledge'. People usually get knowledge from their organizational neighbors. The knowledge market depends on trust, and individuals usually trust people they know. Face-to-face meetings are often the best way to get knowledge. Reliable information about more distant knowledge sources is not available. Also, mechanisms for getting access to distant knowledge tend to be weak or nonexistent [15].

EM can redress the problem of localness of knowledge as it acknowledges the value of knowledge transfer and communication. Collaboration among members of virtual teams is related to mutual sharing of knowledge, which is facilitated if the software artifacts and their status are stored and made part of an organizational repository.

5.6 Organizational Culture

It is a well known fact that every software development organization fosters a specific software development culture and creates (and thus promotes) policies and conventions in the light of this culture. Often this type of culture exists as folklore and is informally maintained within the brains of the experienced developers [29]. This knowledge is passed on to new and/or inexperienced developers through informal meetings with the effect that not everyone gets the knowledge they need [34]. 'Water-cooler meetings' are the common sight in SE environments where software engineers share their knowledge informally but this shared knowledge never becomes available to everyone who needs it in the organization as it is not formalized in any repository.

EM provides for both formal as well as informal ways of sharing SE knowledge and effective means of communication accessible to every member of the organization. Lightweight knowledge management approaches attempt to capture the informal knowledge that is shared on a daily basis so that it can be disseminated on a larger scale.

6 The Indian Scenario

We now identify the trends and EM practices adopted by Indian SE companies. Infosys Technologies founded in 1981, has been a pioneer in the Indian software industry for implementing a comprehensive KM program. A SEI-CMMI Level 5 certification company, Infosys has won four Global Most Admired Knowledge Enterprise (MAKE) awards and has thus become the first Indian company to be inducted into the Global MAKE Hall of Fame. The KM portal, K-Shop supports process management as the repository of over 30,000 knowledge components, best practices and experiences across all projects managed for clients across domains and technologies. KShop, a central repository, provides access to knowledge assets across Infosys. "People-Knowledge Map," an application, locates experts within Infosys who have volunteered to be knowledge resources for others. Knowledge Currency Units (KCU) has been designed to reward and encourage experts to submit papers and volunteer on People-Knowledge Map. High KCU scorers are awarded with incentives. Infosys has pioneered a five-stage KM Maturity Model.

Wipro Technologies, the first PCMM Level 5 and SEI CMM Level certified IT Services Company globally, has been awarded KMWorld's prestigious KM reality award for best practices in KM. The company has devised a KNet Framework to convert the explicit knowledge of documents and other reusable components into tacit knowledge and store it into a knowledge repository. The main components of this repository are DocKNet (The repository of documents), KoNnEcT (yellow pages with an associated database of experts, queries, responses and ratings), War Rooms (a virtual workspace for time-bound and task-oriented jobs) and the repository of software reusable components and tools developed in-house. Konnect helps share the tacit knowledge within the organization. War Rooms facilitate document sharing, exchange of information, real-time online discussions, sharing work plans and online updates, monitoring the progress of activities.

Satyam Computers Services Ltd. started the Knowledge Initiative to address the pain-areas of its project managers by filling the current knowledge gaps. Satyam also has the knowledge networking portal called K-Window. Some of the highly successful ideas implemented in Satyam's KI programme include Satyam Pathshala, Communities of Excellence (CoE), K-Radio and the K-Mobile. Satyam Pathshala is a tacit knowledge sharing program facilitated by employees volunteering to share their experiences with fellow colleagues. CoE are virtual communities of professionals who volunteer to share their experience with fellow employees across Satyam. K-Radio is a query broadcast and response service provided under K-Window, wherein employees can post their queries online and can receive responses-all pooled at one place-from the members of the targeted CoE. K-Mobile is an application of SMS in the field of KM, with the aim of facilitating knowledge request through mobile by employees who are on the move.

Tata Consultancy Services (TCS) is the leading Indian software solutions provider and has been providing KM solutions to a wide spectrum of organizations. The KM philosophy of TCS is manifested through its motto of managing mindsets. The KM initiative of TCS is known for its 5iKM3 Knowledge Management Maturity Model.

Patni Computer Systems, India's software services giant is also one of the few organizations that make extensive use of experience bases in software engineering.

The company has created a knowledge centre, which allows its employees to learn about new technologies, have discussions and get technical queries answered. The knowledge portal of the company has a searchable experience base of large reusable software components.

7 Conclusions

For the continuous improvement of software process, the knowledge and experience of its employees cannot be overemphasized in an organization. Large amounts of knowledge in the form of project data, lessons learnt, software artifacts, code libraries etc. could be accumulated for a software organization but to make this knowledge usable, it needs to be structured, organized, modeled and stored in a generalized and reusable form in an organizational repository or 'experience base.'

However, learning and reuse of knowledge usually occur only because of individual efforts or by accident. This necessarily leads to a loss of the experience and knowledge after the completion of the project and therefore a reuse-oriented software development process in which learning and feedback are regarded as integrated components, and experiences are stored in an experience base, is potentially the best solution. Reuse of experience in SE environments requires that processes and products from software development projects are systematically collected, packaged and stored in an experience base. Managing experiences through experience bases will potentially provide SE organizations with some remedies against the software crisis and help them not only in improving their software processes but will also lead them to a learning organization that constantly improves its work by letting employees share experience with each other, thus proving to be a silver bullet of some sort. This enlightenment seems to be following major software engineering organizations in India.

References

1. Aaen, I., Arent, J., Mathiassen, L., Ngwenyama, O.: A conceptual MAP of software process improvement. Scandinavian Journal of Information Systems 13, 81–101 (2001)
2. Althoff, K., Birk, A., Hartkopf, S., Muller, W., Nick, M., Surmann, D., Tautz, C.: Systematic population, utilization, and maintenance of a repository for comprehensive reuse. In: Ruhe, G., Bomarius, F. (eds.) SEKE 1999. LNCS, vol. 1756, pp. 25–50. Springer, Heidelberg (2000)
3. Althoff, K., Decker, B., Hartkopf, S., Jedlitschka, A., Nick, M., Rech, J.: Experience management: the Fraunhofer IESE experience factory. In: Perner, P. (ed.) Industrial Conference on Data Mining. Institute for Computer Vision and applied Computer Sciences, Leipzig, Germany (2001)
4. Aurum, A., Jeffery, R., Wohlin, C., Handzic, M.: Managing software engineering knowledge. Springer, Heidelberg (2003)
5. Basili, V., Caldiera, G., Rombach, H.: The experience factory. In: Marciniak, J. (ed.) Encyclopedia of Software Engineering, pp. 468–476. John Wiley & Sons, NJ (1994)
6. Basili, V., Caldiera, G., Mcgarry, F., Pajerski, R., Page, G., Waligora, S.: The software engineering laboratory - an operational software experience factory. In: The Fourteenth International Conference on Software Engineering, pp. 370–381 (1992)

7. Basili, V., Costa, P., Lindvall, M., Mendonca, M., Seaman, C.: An experience management system for a software engineering research organization. In: Twenty-Sixth Annual NASA Goddard Software Engineering Workshop, pp. 29–35 (2001)
8. Basili, V., Rombach, H.D.: Support for comprehensive reuse. IEEE Software Engineering Journal 22(4), 303–316 (1991)
9. Basili, V., Schneider, K., Hunnius, J.-P.V.: Experience in implementing a learning software organization. IEEE Software, 46–49 (May/June 2002)
10. Bergmann, R.: Experience management: foundations, Development Methodology and Internet-Based Applications. LNCS (LNAI), vol. 2432, p. 25. Springer, Heidelberg (2002)
11. Brooks, F.P.: No silver bullet: essence and accidents of software engineering. Computer 20(4), 10–19 (1987)
12. Brossler, P.: Knowledge management at a software engineering company - an experience report. In: The Workshop on Learning Software Organizations, Kaiserslautern, Germany, pp. 163–170 (1999)
13. Parnas, D.L.: Which is riskier: OS diversity or OS monopoly? Inside risks. Communications of the ACM 50(8) (2007),
http://www.csl.sri.com/users/neumann/insiderisks.html
14. Curtis, B., Krasner, H., Iscoe, N.: A field study of the software design process for large systems. Communications of the ACM 31(11), 1268–1289 (1988)
15. Davenport, T.H., Prusak, L.: Working knowledge: how organizations manage what they know. Harvard Business School Press, Boston (1998)
16. Diaz, M., Sligo, J.: How software process improvement helped Motorola. IEEE Software 14, 75–81 (1997)
17. Disterer, G.: Management of project knowledge and experiences. Journal of Knowledge Management 6(5), 512–520 (2002)
18. Glass, R.L.: The relationship between theory and practice in software Engineering. Communications of the ACM 39(11), 11–13 (1996)
19. Hansen, B., Rose, J., Tjornhoj, G.: Prescription, description, reflection: the shape of the software process improvement field. International Journal of Information Management 24(6), 457–472 (2004)
20. Henninger, S., Schlabach, J.: A tool for managing software development knowledge. In: Bomarius, F., Komi-Sirviö, S. (eds.) PROFES 2001. LNCS, vol. 2188, pp. 182–195. Springer, Heidelberg (2001)
21. Houdek, F., Schneider, K., Wieser, E.: Establishing experience factories at Daimler-Benz: an experience report. In: The Twentieth International Conference on Software Engineering, pp. 443–447 (1998)
22. Land, L., Aurum, A., Handzic, M.: Capturing implicit software engineering knowledge. In: The 2001 Australian Software Engineering Conference, pp. 108–114 (2001)
23. Lindvall, M., Frey, M., Costa, P., Tesoriero, R.: Lessons learned about structuring and describing experience for three experience bases. In: The Third International Workshop on Advances in Learning Software Organizations, pp. 106–119 (2001)
24. Martinez, P., Amescua, A., Garcia, J., Cuadra, D., Llorens, J., Fuentes, J.M., Martín, D., Cuevas, G., Calvo-Manzano, J.A., Feliu, T.S.: Requirements for a knowledge management framework to be used in software intensive organizations. IEEE Software, 554–559 (2005)
25. Mendonca, M., Seaman, C., Basili, V., Kim, Y.: A prototype experience management system for a software consulting organization. In: The International Conference on Software Engineering and Knowledge Engineering, pp. 29–36 (2001)
26. Natali, A.C.C., Falbo, R.A.: Knowledge management in software engineering environments. In: The Fourteenth International Conference on Software Engineering and Knowledge Engineering, Ischia, Italy (2002)

27. Nilsson, N.J.: Artificial intelligence: a new synthesis. Morgan Kaufmann Inc., San Francisco (1998)
28. Nonaka, I.: A dynamic theory of organizational knowledge creation. Organization Science 5(1), 14–37 (1994)
29. Rus, I., Lindvall, M.: Knowledge management in software engineering. IEEE Software 19(3), 26–38 (2002)
30. Schneider, K.: LIDs: a light-weight approach to experience elicitation and reuse. In: Bomarius, F., Oivo, M. (eds.) PROFES 2000. LNCS, vol. 1840, pp. 407–424. Springer, Heidelberg (2000)
31. Sharma, N., Singh, K., Goyal, D.P.: Knowledge management in software engineering environment: empirical evidence from Indian software engineering firms. Atti Della Fondazione Giorgio Ronchi 3, 397–406 (2009)
32. Sharma, N., Singh, K., Goyal, D.P.: Software process improvement through experience management: an empirical analysis of critical success factors. In: Prasad, S.K., Vin, H.M., Sahni, S., Jaiswal, M.P., Thipakorn, B. (eds.) ICISTM 2010. Communications in Computer and Information Science, vol. 54, pp. 386–391. Springer, Heidelberg (2010)
33. Sun, Z., Finnie, G.: Intelligent techniques in e-commerce: a case-based reasoning perspective. Springer, Heidelberg (2004)
34. Terveen, L.G., Sefridge, P.G., Long, M.D.: From 'folklore' to 'living design memory'. In: The ACM Conference on Human Factors in Computing Systems, pp. 15–22 (1993)
35. Tiwana, A.: The knowledge management toolkit: practical techniques for building knowledge management systems. Prentice Hall PTR, Englewood Cliffs (1999)
36. von Krogh, G., Ichijo, K., Nonaka, I.: Enabling knowledge creation. Oxford University Press, New York (2000)
37. Ward, J., Aurum, A.: Knowledge management in software engineering- describing the process. In: The 2004 Australian Software Engineering Conference. IEEE Computer Society, Los Alamitos (2004)

Framework for Spatial Query Resolution for Decision Support Using Geospatial Service Chaining and Fuzzy Reasoning

Jayeeta Mukherjee, Indira Mukherjee, and Soumya Kanti Ghosh

School of Information Technology
Indian Institute of Technology, Kharagpur-721302, India
{itsmejayeeta,indira.mukherjee.23}@gmail.com,
skg@iitkgp.ac.in

Abstract. Geospatial data play a vital role in various decision making systems. Technological advancements have enabled users to access geospatial functionalities as services over web. In many decision support systems, it is required that more than one service is to be involved to help decision makers, calling for a service chaining. Chaining distributed geospatial services requires dealing with several heterogeneity issues such as semantic, syntactic issues. In addition, properties of geospatial data are fuzzy by nature. The fuzziness may exist in thematic definition and in spatial properties. This may lead to inflexible, less accurate decision making. Thus, to avail more accurate decision making, the uncertainty associated with the spatial information should be captured. In this paper, an approach for service chaining in decision support systems has been taken, along with fuzzy logic to resolve user queries and process imprecise information. Service chaining has been used for integrating distributed geospatial services, and fuzzy logic has been incorporated with services to capture fuzziness and uncertainty that are intrinsic to the data.

1 Introduction

Geospatial data have been playing a central role in various socio-economic fields, due to its inherent capability of holding geospatial references. Geospatial data are unique in the way that they have got both spatial and non-spatial attributes. Geospatial Information Systems (GIS) is an information system which captures, collects, stores, manages and analyzes geospatial data to get geospatial information. Earlier times GIS services were only accessible to corresponding domain experts. Advancements in computing systems and technologies have enabled computers throughout the world to be connected, thus improving geospatial information sharing [1]. As a consequence, GIS has transformed to a service based model from stand-alone systems, and geospatial functionalities can be consumed as services over the web.

Although now a days one can use GIS services over web, there can be situations where a user requirement can not be accomplished by a single geospatial service. It needs involvement of more than one geospatial services to resolve

S. Dua, S. Sahni, and D.P. Goyal (Eds.): ICISTM 2011, CCIS 141, pp. 81–90, 2011.
© Springer-Verlag Berlin Heidelberg 2011

client requirement. Such situations call for integrating or chaining geospatial services, forming a service chain. For example, suppose fire has broken out in a place and people are required to be rescued from that place and shifted to some safe shelters. Now, this needs to identify how many people to evacuate, where to place them, what will be a safe shelter etc. Resolving this query requires integrating or chaining more than one geospatial service, may be from different E-GIS units. Pipelining of services where input of one service is the output from another service, is called service chaining. Although integration of these services can resolve the spatial queries and facilitate decision support system, but, are unable to produce more accurate result, which is necessary in real world application. In this work, a fuzzy logic concept has been incorporated with the spatial services to process the imprecise information.

Chaining services from different geospatial communities requires to deal with several fuzziness and heterogeneity issues. The heterogeneity issue mainly related to syntactic and semantic interoperability[2]. Semantic interoperability is a major, mostly unresolved issue when integrating services from several distributed sources. The fuzziness are mainly associated with improper definition of data like, buffer are of fire effected zone, nearest shelter to a disaster zone. Consider a user query that wants to retrieve buffer zone around fire affected location; classical service processing mechanism just returns a area around the fire effected location that have equal impact from center to outer boundary of the buffer area. However, in reality, intensity of the fire decreases gradually from center to outer boundary. In this paper, an approach for service chaining with fuzzy reasoning for supporting decision support systems has been proposed. Service chaining approach solves situations where one single service is incapable of serving decision maker with adequate information, while fuzzy reasoning has been used to address fuzziness and ambiguity among data.

A motivating example has been considered throughout the work to show working of the proposed approach. Suppose in a place fire has broken out, and rescue team is required to sent there for evacuating people from there and relocate them to some safe shelter. Now, a decision support system will help a decision maker by providing the danger zone regarding the fire break-out, number of affected people to be evacuated, where to relocate them etc. Geospatial services will be of help here and no single geospatial service is capable of serving with all those data. Hence a service chaining approach can be helpful for decision making, while fuzzy engine helps to identify danger zone and shelters along with their suitability.

Rest of the paper in organized as follows. Section 2 discusses the previous works. Concepts of service chaining has been discussed in Section 3. Section 4 discusses about basic concept of fuzzy set theory. Section 5 provides overview of the proposed framework. Methodology of service chaining has been discussed in Section 6. In Section 7 description about fuzzy engine has been given. A case study has been discussed next to show working of the framework in Section 8. Finally conclusion has been drawn in Section 9.

2   Related Works

In this section background works related to spatial service chaining and fuzziness in spatial data have been discussed. Three types of service chaining has been proposed by ISO, followed by Alameh [3][4], where different geospatial services have been considered for chaining. Three types of service chaining are client-coordinated service chaining, aggregate service chaining and work-flow managed service chaining. However semantic aspect of service chaining has not been considered here. Service chaining approach in context of Spatial Data Infrastructures (SDI) facilitates resolving geospatial problems for disaster management, and incorporating commercial services in daily activities [1]. However both syntactic and semantic is crucial for chaining distributed services [5][6], while service chaining methodology has been decomposed into few steps, separate steps for taking care of different heterogeneity issues. A work-flow managed service chaining has been attempted here with considering ontology for capturing semantics of data. Semantic interoperability is a major, mostly unresolved issue when integrating distributed services [7][8] as semantic heterogeneity between services may lead to inappropriate result, even improper interpretation. Ontology has always been considered an efficient tool to resolve semantic heterogeneity. However, automatic formation of service chaining based on demand can get benefited from rule based systems, where services has been considered to exist as a rule [9].

In real world application, however, information are defined abstractly or fuzzily [10]. Different kind of imperfect information adheres to the geospatial information have been studied in literature [11]. The current standards models the geospatial information crisply, and can address the issue of interoperability; however, it is unable to model the fuzziness that are intrinsic to most geospatial features. Attempts have been made in [12][13] where the uncertainty in the objects has been mapped in the UML model using fuzzy class and fuzzy association.

3   Concepts of Service Chaining

Over the past years, GIS technologies have come a long way from stand-alone systems where data were tightly coupled with the system, to a distributed model [3][4] owing to the inventions of web service. Organizations are highly using geo information and related services and there is an increasing need for integrating the services, called as service chaining. Three types of service chaining are there: user-controlled, aggregate, and workflow managed. In user controlled service chaining, the whole control of the chain is in the hand of the user. However, to accomplish such a service chain, the user should have prior information of all the services. On the contrary in aggregated service chaining, one aggregate service looks up the chaining of services. Here the service chain is fully opaque to the user and appears as a single service to the user. Work-flow managed Service chaining approach combines the facilities of the previous two approaches. Here a mediating service is maintained to monitor the service chaining.

4   Fuzzy Sets

Fuzzy set was introduced by Zadeh in 1963 [14]. Nowadays, fuzzy sets theory and fuzzy logics are widely used for capturing the inherent vagueness that exists in real life applications. In classical set theory an element either belong to a set or not; while in fuzzy set theory elements belong to a set with a certain degree. Formally, let U be the collection of elements called universe of discourse. A crisp subset A of U is collection elements U that is defined by a *characteristic function* $U_a(u)$ that assigns any element $u \in U$ to a value 1 if the element belongs to set A and 0 if it does not belong to set A. On other hand, for a fuzzy subset A of U is defined by a *membership function* $\mu_A(u)$ for $\forall u \in U$. This membership function assigns any $u \in U$ to a value between 0 and 1 that represents degree to which an element of U belong to subset A of U.

5   Overview of the Fuzzy Geospatial Reasoning Framework

This section gives an overview of the proposed fuzzy logic based geospatial service chaining for decision making process. Fig 1 shows the integration of spatial services for supporting decision making in decision making systems. Spatial services are located in several locations, these services are need to be chained for resolving decision making queries. The *Service Chaining Engine* facilitates the integration of spatial web services and resolve user's or client's queries. The clients may vary from human users to other systems and services. The service chaining engine searches for the required services pertaining to the given query and chain the services.

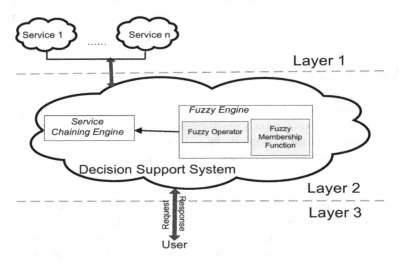

Fig. 1. Fuzzy Geospatial Information Integrating Framework

In real world scenario the query processing mechanism are complex, like, "*Generation of Danger zone around a fire effected location*". The danger zone indicates area which is prone to get affected by the fire breakout, and it can be generated by a buffer generating procedure by generating a buffer around the center of fire breakout. For crisp processing services, the buffer zone generated is crisp area, but in reality whole area do not have the same impact as the center. Again, the effect of the fire reduces gradually over the buffer zone, the effect does not fall certainly. Therefore, reflection of these gradual changes are required to be captured for achieving more accuracy in decision support system. In this framework, a fuzzy engine has been used by spatial web services to process the information. The fuzzy engine attempts to address the fuzziness associated with the information. The fuzzy engine has predefined fuzzy operators and fuzzy membership function, which are used to process the information in order to resolve decision support queries. Components of the framework are described below:

- **Decision Support System:** A decision support system is one which allows users to access and process data for making some decision to resolve problems. In geospatial decision support systems, service chaining approach can be beneficial when a single service is not adequate. The decision making can be done by chaining more than one distributed geospatial services.
- **Chaining Engine:** The chaining engine manages chaining of services by identifying which services to chain, searching the repository for services, chain the services and execute the service chain. Working of the chaining engine depends on the chaining approach taken in the system.
- **Fuzzy Engine:** A fuzzy engine deals with ambiguity and imprecision in data, helping in flexible decision making.
- **Geospatial Web Services:** These are web services pertaining to geospatial data and functionalities. These services can be accessed over web, and allows to retrieve information from geospatial data, process the data, and to view those data in form of a map. All of those services are compliant with OGC standard and they may be from different service providers.

6   Chaining Methodology of Geospatial Services

The proposed framework accomplishes chaining of geospatial services in a finite number of steps. The steps are : User query decomposition, Service Discovery, Service Chain composition and Execution of Service Chain. In all the previous works of service chaining, main thrust has been put into service discovery and service composition, while in order to search for services one needs to know which services to search in service repository. Here decomposition of user query is proposed to identify required services to chain. The service chaining steps are elaborated below:

6.1   User Query Decomposition

More than one service may required to resolve suer queries. However, in order to construct a chain of services, the user requirement must be decomposed to identify which services are at all required. Here it is proposed to decompose the user requirement into more detailed subtasks and the decomposition should be repetitive in nature and will continue until the task can be executed by a series of atomic GIS services. By decomposing user requirement, processing of the services and further match making of services become easier. Hence after decomposition of user requirement, set of atomic services is available to chain. Once services are identified, they can be further searched in service repository. Services are represented in some formal method before they are searched in service repository. Formal description of services owes to better searching of services and better match-making between services. Here Description Logic (DL) has been used for formal description of services. Description Logic is a subset of First Order Logic (FOL). This formalizing technique is used in several domain, especially in knowledge representation systems to capture and represent domain knowledge in a structured way. Here concepts of geospatial domains and their relations is captured by description logic.

Suppose there is a service which is provided by service provider X and provides image with a blue border. Using description Logic, this service can be represented as below:

$$S \equiv service(hasProvider.X) \sqcap (hasOutput.Image \sqcup hasBorder.Blue)$$

6.2   Service Discovery

Service descriptions are stored by a service catalog, that are searched for services. Once formal representation of services are done, services are searched with their corresponding descriptions. Services description plays an important part in services discovery, as an improper service description may result in inefficient service discovery. Discovery of services includes matching service description from service catalog with service request. Web Service Description Language (WSDL), a W3C standard for describing web services, has been used for storing service description.

Service discovery requires that services are to be stored in some sort of service repository. OGC compliant Catalog Service Web (CSW) has been used as service repository. Service providers have to register their services with their corresponding service description in the catalog service. Catalog service also allows to publish their service description over in a standard interface.

Match-making between services are done concerning service description from catalog service and service request. The service request pertains to the atomic geospatial services obtained in the user query decomposition phase. Once a service match is found , the service selected for chaining.

6.3   Service Chain Composition and Execution

Service discovery is followed by chaining those services in a competent service chain. As there are different service chaining mechanism is available, depending upon user efficiency different mechanism can be chosen. An aggregate service chaining approach is suitable for novice users as total chaining mechanism is opaque to the user in it, while for geospatial experts and an user coordinated service chaining is most suited. A work-flow managed service chaining approach combines both the facilities of previous approaches. In case of a aggregate service chaining, service discovery, chain composition all these work-flows are kept hidden from the user.

Once services are discovered from service repository and these are chained to form a service chain, the chaining engine will execute the service chain for resolving user requirement. It invokes the starting service and controls the data flow among services while invoking the services one after another. Every composite process is executed by invoking all the component atomic processes.

7   Fuzzy Engine

According to given standards, several geospatial services are chained for resolving user query. Services have some predefined logics based on which the spatial queries are resolved. The predefined logics are framed assuming that the information related to the spatial and attribute data are crisp or well defined in nature. However, in real world applications the properties of the spatial and attribute data are not well defined. In this work a fuzzy logic concepts has been applied in the predefined logics to address the issue of fuzziness.   In the proposed framework, a fuzzy engine has been appended to the predefined procedure of the spatial services. Fuzzy engine have some fuzzy membership functions defined by domain experts and fuzzy operators. The fuzzy membership function helps to capture the smooth transition of one state of information to another. The fuzzy operators are used to carry out several fuzzy operations. The case study described in the following section explains the concept and its applicability in decision making process.

8   Case Study

This section provides a case study to show efficacy of the proposed approach of service chaining along with fuzzy reasoning for a decision support system in a disaster management. As described in motivating example, a fire has broken out in a place, people from danger zone is to be evacuated and relocated to safe shelters. For this purpose, several geospatial services are required; for example one processing service to identify danger zone around the fire break out place, one feature service for providing shelter information and finally one map service to display all those data. As no single service is capable of processing all those, hence a service chaining approach will be of use here to help the decision makers while taking decisions about rescuing people.

Identifying danger zone requires finding a buffer zone around the center of fire breakout, depending upon the severity of fire. It can be done using a processing service. Following that one feature service will be of use to retrieve number of people affected. After that output from another feature service providing information regarding possible shelters can be clipped with the population

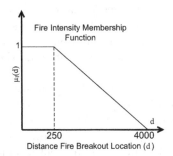

Fig. 2. Fire Intensity Membership Function

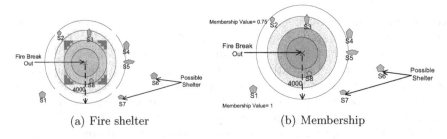

(a) Fire shelter   (b) Membership

Fig. 3. Intermediate Boundary and Shelter for Fire Effected Zone along with their membership value

information in order to find suitable shelters for the rescued people. Finally, one map service is required to overlay all the data in a map to visualize all the information. Here the required services are a processing service. Once services are discovered, they are chained by the chaining engine and it executed the service chain when required.

The determination of a suitable shelter near a fire breakout zone has been carried out in this work. This may help in indexing the shelters with respect to their suitability values. Let us consider that there are 8 shelters available near the fire breakout zone (refer figure 3(a)). In order to find the suitability of shelter the various spatial services are used for processing the gathered data. The fuzzy engine facilitates in processing the imprecise information in following manner-

1. Determine the core, an alpha-cut region of the fire impact zone (values provided by domain expert). Here the core the region where effect of the fire is maximum is considered to be 250m and the alpha-cut regions are considered up to 4000m
2. Next, the fuzzy membership functions is determined according to Fig. 2. The universe of discourse of *Distance* in *Fire intensity* membership function $\mu_I(d)$ w.r.t *Distance*(R_d) is 0 to 4000 meters; i.e. the intensity of the fire is 0 beyond 4000m. However, according to domain expert maximum impact of

the fire is within 250m of the effected zone. Therefore, the membership value of fire intensity is 1.0 within 250m. The membership function $\mu_I(d)$ w.r.t is d can be given as follows.

- $\mu_I(d) = 1$ for $0 \le d \le 250$
- $\mu_I(d) = [0,1]$ for $250 \le d \le 4000$

The core zone is region under the circle with radius 250m and fire break out zone as the center. Therefore, this zone has the highest value and zone beyond 250m has impact within 0.0 and 1.0.

3. Determination of the spatial relationship of the locations with respect to other spatial objects. Here the possible locations (building, schools) where the people from the fire effected zone can be evacuated.

4. Finally, suitability of the shelter has been determined based on spatial web services and fuzzy reasoning framework.

The *Fire Intensity* ($\mu_I(d)$) membership function with respect to *Distance*(d) is given by

$$\mu_I(d) = \begin{cases} 1 & \text{if } 0 \le d \le 250, \\ (4000-d)/3750 & \text{if } 250 \le d \le 4000, \end{cases} \tag{1}$$

Spatial relation of fire breakout zone with the shelter

- S1 is located at distance of 5000m from fire breakout location.
- S2 is located at distance of 3050m from fire breakout location. . .
- S8 is located at distance of 350m from fire breakout location.

Now, considering the membership function of each factor $\mu_I(R_d)$ and the spatial relationships, the membership value of S1 can be determined as

- $d = 1000$ according to Distance membership function $\mu_I(d) = 0$, $d \ge 4000$, therefore, $\mu_I(d) = 0.0$. The suitability of the shelter is inverse of fire intensity. therefore suitability of shelter S1 $= 1-\mu_I(d)=1-0=1.0$.
- Similarly, $\mu_I(d)$ for S2 $=1-0.25= 0.75$

Thus, the degree of suitability of shelter S1 $=\mathbf{1.0}$. Similarly, the degree of suitability for other shelters can also be determined. Thus this in turn helps the decision maker to accommodate peoples on those zone which overlap within the danger zone of fire breakout but effect of the fire is much lesser there.

9   Conclusion

This paper presents an service chaining approach in a decision support systems, along with fuzzy engine to avail more accuracy in decision making. The proposed approach uses service chaining for integrating distributed services required for resolving user queries to support decision making. Service chaining has been done with a finite number of steps, while service description plays an important role in it. Since uncertainties exist in spatial information, fuzzy engine will help in processing such information more accurately. A detailed case study has also been presented to demonstrate the efficacy of the system, where danger zone around a fire affected location is generated with multiple alpha-cut. Each alpha-cut is associated with fuzzy membership values that indicate intensity of fire in the particular alpha-cut region. It in turn determines the suitability of different possible shelters.

References

1. Aditya, T., Lemmens, R.: Chaining distributed gis services. In: Prosiding Pertemuan Ilmiah Tahunan XII (2003)
2. Mark, D.: Toward a Theoretical Framework for Geographic Entity Types. In: Spatial Information Theory. Springer, Heidelberg (1993)
3. Alameh, N.: Service chaining of interoperable geographic information web services (2010), http://web.mit.edu/nadinesa/www/paper2.pdf (accessed on June 2)
4. Alameh, N.: Chaining geographic information web services. In: Internet Computing. IEEE, Los Alamitos (2003)
5. Lemmens, R., Wytzisk, A., de By, R., Granell, C., Gould, M., van Oosterom, P.: Integrating semantic and syntactic descriptions to chain geographic services. In: Internet Computing, vol. 10, pp. 42–52. IEEE, Los Alamitos (2006)
6. Lemmens, R., de By, R., Gould, M., Wytzisk, A., Granell, C., van Oosterom, P.: Enhancing geo-service chaining through deep service descriptions. Transactions in GIS 11(6), 849–871 (2007)
7. Visser, U., Stuckenschmidt, H.: Interoperability in gis – enabling technologies. In: Ruiz, M., Gould M., Ramon, J. (eds.) 5th AGILE Conference on Geographic Information Science, pp. 291–297 (2002)
8. Lutz, M., Sprado, J., Klien, E., Schubert, C., Christ, I.: Overcoming semantic heterogeneity in spatial data infrastructures. Computers & Geosciences 35(4), 732–752 (2009)
9. Di, L., Yue, P., Yang, W., Yu, G., Zhao, P., Wei, Y.: Ontology-supported automatic service chaining for geospatial knowledge discovery. In: American Society of Photogrammetry and Remote Sensing (2007)
10. Sozer, A., Yazici, A., Oguztuzun, H., Tas, O.: Modeling and querying fuzzy spatiotemporal databases. Information Sciences 178(19), 3665–3682 (2008)
11. Worboys, M.: Imprecision in finite resolution spatial data. GeoInformatica 2(3), 257–279 (1998)
12. Sicilia, M., Mastorakis, N.: Extending uml 1.5 for fuzzy conceptual modeling: A strictly additive approach. WSEAS Transaction on Systems 3, 2234–2239 (2004)
13. Ma, Z.M., Yan, L.: Fuzzy xml data modeling with the uml and relational data models. Journal of Data & Knowledge Engineering 63, 972–996 (2007)
14. Zadeh, L.: Fuzzy sets. Information and Control 8, 338–353 (1963)

An Ontology Based Approach for Test Scenario Management

P.G. Sapna and Hrushikesha Mohanty

University of Hyderabad,
Hyderabad, India
sapna_pg@yahoo.com, mohanty.hcu@gmail.com

Abstract. Management of test scenarios involves ordering and selecting a set of test scenarios for testing with the objective of fulfilling criteria like maximizing coverage or discovering defects as early as possible. To do this, knowledge about the main activities in the domain and the interactions between them need to be maintained. This knowledge helps in categorizing scenarios as per the approaches used for testing. Current approaches to test management involve generating scenarios related to use cases. However, the use case based approach does not capture knowledge about the dependencies that exist among activities in the domain and therefore cannot provide complete information for scenario management. Ontologies provide a mechanism to represent this knowledge. They also provide a means to share and reason on knowledge that is captured. The objective of this work is to use ontologies to aid test management.

Keywords: UML, use case, scenario, ontology, test scenario management.

1 Introduction

One of the challenges faced in software engineering is building software with quality at reduced cost, time and effort [2]. This requires a clear understanding of the elements used to build a software and the relations among them. In this direction ontologies can aid in performing inference on knowledge gathered from requirements captured using UML diagrams. Ontologies also enable communication among different people involved in building a software and hence can be used as a common mechanism for requirements analysis.

Test management is a way of organizing artifacts of the system like requirements, scenarios and test cases such that it enables accessibility and usability with the objective of delivering quality software within constraints of cost and time. Requirements and scenarios are specified by UML models. Given the large number of use cases and scenarios for even a medium sized system, there is a need for effective scenario management. For this purpose, not only domain but also the process of testing should be specified comprehensively as well as understandably so that the stakeholders in testing can participate effectively. With the background of the wide use of ontology in software engineering, it is proposed to use ontologies for test management in this work. The use of analyzing the

S. Dua, S. Sahni, and D.P. Goyal (Eds.): ICISTM 2011, CCIS 141, pp. 91–100, 2011.

requirements of a domain through an ontology of software testing is to allow reasoning about the information represented [1]. For instance :

– Consistency. Suppose it is declared a and b are instances of the concept U. Also, it is defined that a and b are different individuals and the relation 'includes' is irreflexive and asymmetric. Then, if it is defined that a includes b and b includes a, then there is inconsistency.. This error can be detected on reasoning of test ontology.
– Inferred relationships. If concept X is equivalent to concept Y, and concept Y is equivalent to concept Z, then X is equivalent to Z.
– Membership. If a is an instance of concept U and U is a subconcept of Y, then it can be inferred that a is an instance of Y.
– Classification. Suppose it is declared that certain property-value pairs are a sufficient condition for membership in a concept A. Then, if an individual x satisfies the condition, it can be said that x is an instance of A.

This kind of inference and reasoning based on specific relationships can be better done on ontology repository than a use case based approach. In this work, an ontology is introduced to facilitate querying requirements for test management. The contribution of this work includes :

– Development of an ontology for software testing
– Querying the ontology

The ontology is implemented in OWL using Protege as the editor. Built ontology shows the relations among the testing concepts outlined in SWEBOK[1]. Section 1.1 provides a motivating example for using ontology. Section 1.2 discusses the use of an ontology as a repository for test management. The use of ontologies in different areas and particularly in testing is discussed in Section 2. Application of the steps to build an ontology for testing based on the testing concepts extracted from SWEBOK and the primitives of relevant UML diagrams is discussed in Section 3. The section includes design, implementation and transformation steps involved in building an ontology. Reasoning on the ontology is discussed in Section 4. Section 5 summarizes the work.

1.1 Motivating Example

This section discusses an example that motivates work on test management using an ontology. Fig. 1 shows the relationship between use case, scenario and classes. A requirement shown in a use case diagram and detailed by scenarios is implemented by classes. For querying on test process with respect to a requirement, there has to be traceability for the relations existing among the artifacts. e.g. 'project management' has scenario 'wage calculation' implemented by classes, say, 'pay-calculate', 'award-calculate', 'utility-award', etc. The relations among these artifacts can be shown using an ontology. Using an E-R diagram for the

---

[1] Software Engineering Book of Knowledge provides a consistent view of software engineering, its boundary and contents. Available at www.swebok.org/

same example does not show the impact on testing. That is, we need to show that for every use case that is to be tested, corresponding scenarios must be tested. Also, unit testing of all classes that make up the scenario must be tested. This requires a transitive relationship to be defined such that for each use case, all corresponding scenarios as well as related classes have to be tested. Usually class repositories are stored in an RDBMS, but still has limitations. Fig. 1. Entity Relationship diagram for the relation 'Use case has scenarios, and each scenario is made of classes'.

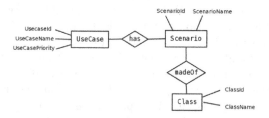

Fig. 1. Entity Relationship diagram for the relation 'Use case has scenarios, and each scenario is made of classes'

1.2   Ontology as a Repository

The common practice is to use a database to store the requirements of a system and query the same. However, it is not possible to capture all constraints related to a system using the representation provided by relational databases. Examples showing situations where a relational database is insufficient to represent the conceptual relations are discussed in Section 1.1.

For the following disabilities [7] found with RDBMS, ontology is considered as useful repository.

– Lack of hierarchy. Relational models have no notion of a concept/query hierarchy.
– Representing m:n relationships. Relation between entities are represented using 1:n relationship in RDBMS. m:n relationships are represented as a set of m, 1:n relationship.
– Access to data. There is need to have access to the database to query the database.
– Knowledge of a query language. User must have knowledge of a query language supported by the specific database.

Representing requirements using ontologies is found more useful than RDBMS for being expressive and flexible to manage. Querying on requirements stored in a relational repository requires prior knowledge on entity relations(i.e. domain structure) whereas in case of ontology, repository queries can be exploratory for traversing links among concepts that make the ontology. The use of ontologies in different areas and particularly in testing is discussed in Section 2.

2   Survey

A survey of current work on ontologies for software reveals that ontologies have been built with varied objectives in mind. One of the objectives of building ontologies is to facilitate collaboration of remote teams in multi-site distributed software development. SWEBOK has been used as the point of reference to build ontologies of software engineering [11, 12]. The ability of ontologies to query necessary and relevant information regarding the domain concerned is exploited. Dillon et al [3] have focused on the same issue by developing a software engineering ontology that defines common sharable software engineering knowledge as well as information about particular projects. The software engineering ontology consists of five sub-ontologies for software requirements, design, construction, testing and tools and methods. The software testing ontology in particular, consists of subontologies, namely: test issues sub-ontology, test targets sub-ontology,test objectives sub-ontology, test techniques sub-ontology and test activities sub-ontology. Both [11, 3] provide the advantage of development of a consistent understanding of the meaning of issues(terminology and agreements) related to a project by different people distributed geographically across locations.

A second objective for building ontologies is to access ontology mediated information [3]. Rules can be written about relationships between concepts in an ontology and the same can be used for query processing. The advantage over databases is that new facts can be inferred or reasoned with asserted facts. An example of this objective is the Protein Ontology built to integrate protein knowledge and provide a structured and unified vocabulary to represent concepts related to protein synthesis.

Thirdly, ontologies are used in the area of semantic web services [3]. Several issues need to be addressed in the area of web services that include, selection of architecture, discovery of service, selection of service and composition and coordination of services to meet requirements. Web services are semantically annotated to assist in the process of discovering and selection for which a combination of ontologies and Web 2.0 is used.

A fourth area in which ontologies are used is multi-agent systems [3, 9]. Multi-agent systems involve multiple autonomous agents that collaborate with one another to fulfil goals of the system. Agents have a knowledge base that provides some intelligence. Also, the system is distributed and decentralized where agents are geographically distributed and communicate mainly through message passing. To maintain coherence and consistency of knowledge among agents, an ontology is used as a common knowledge base that is shared by all agents. This facilitates communication and coordination among agents.

Other ontologies that have been built include the disease ontology, manufacturing ontology and different financial system ontologies [3]. A software test ontology(SwTO), that deals with the software testing domain has been built by [2]. The ontology is used along with a test sequence generator to generate test sequences to test the operating system domain, Linux. Zhu et al [13] in their work present an ontology of software testing. They discuss use of an ontology in a multi-agent software environment context to support the evolutionary development and maintenance of web-based applications. Software agents use the ontology as the content language to register into a system and for test engineers and agents to make test requests and report results. The paper describes how the concepts of the ontology and the relations between them are defined in UML.

Thus, ontologies have found use in different areas like software engineering, semantic web services and multi-agent systems. The next section details the steps involved in building an ontology with focus on testing.

3   Building an Ontology for Testing

One of the well-known methods to build an ontology is the Methontology strategy [6]. A generally applicable method to construct domain knowledge model and validate the same is proposed. The ontology development process is composed of the following steps: planning, understanding, knowledge elicitation, conceptualization, formalization, integration, implementation, evaluation, documentation and maintenance. Noy et al [8] follow an iterative design process to build an ontology. Steps to build an ontology is proposed. This work adapts ideas from both works. The set of steps to build an ontology is given in the subsection below.

3.1   Design of Ontology

Steps described in building an ontology are applied in the context of software testing and are explained below:

a. Determine domain and scope of ontology : The objective of this work is to provide a framework for testing, specifically specification based testing wherein specification is captured using UML. Representation and management of use cases and scenarios are in focus while building an ontology. The objective is to use this ontology for enumerating test scenarios to form a test suite for testing.

In the ontology, concepts related to requirements and scenarios like, use cases,related users(actors), scenarios related to each use case are included. At the same time concepts like type of testing and resource are referred from SWE-BOK. In this work, the objective of building an ontology is to provide a means to manage scenarios. Some of the questions that the ontology should answer includes:

- Which requirements involve use of activity 'x'?
- What are the scenarios required to test use case 'UC'?
- What are the use cases that include use case 'UC' ?
- What are the use cases that extend use case 'UC' ?
- List all scenarios of the use case 'UC' that include use case 'UC1' having priority 'p'.

b. Defining concepts in the ontology : To define an ontology for testing, a list of terms and the related properties are listed. For example, important terms related to this work include: use case, actor, scenario, activity diagram, class, priority, testtype, testsuite, etc; different types of priority include customer assigned priority and technical priority. Fig. 3 shows the concepts in the ontology.

c. Create a class hierarchy : First, terms that independently describe objects are considered. The terms form classes in the ontology. For example, use case, actor, scenario, priority form classes in the ontology. The next step involves organizing classes into hierarchical taxonomy. For example, customer assigned priority and technical priority are subclasses of the class priority. i.e.customer assigned priority is a type of priority. Therefore, customer priority is a subclass of the priority class.

d. Defining properties and constraints : Concepts themselves are incapable of answering questions such as those enumerated in Step 1. There is need to describe internal structure of concepts. For example, consider the concept use case. Properties related to the concept includes haspriority, hasscenario, hasactor. For each property, the class it describes must be determined. The properties are attached to concepts. With reference to Fig. 1, the properties include has between concepts UseCase and Scenario and property madeOf between concepts Scenario and Class.

e. Creating instances : Instances are individuals belonging to a concept. Instances of concepts in Fig. 1 are shown in Fig. 2.

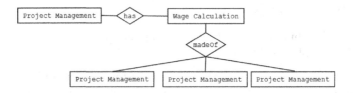

Fig. 2. Instances of concepts in Figure 1

3.2   Defining Relations between Concepts

Classes and relations between the concepts are shown in Fig. 3. For each concept, conditions are defined as to how the concepts interact for realizing an objective. An example of sufficient and necessary condition for some of the concepts is elaborated below:

Use case - Condition. The concept UseCase is a sub concept of System. The hasActor property is used to relate individuals of the UseCase concept with the individuals of the Actor concept. Also, a UseCase must be related to atleast one actor of the Actor concept through the hasActor property.

Use case - Properties. hasActor is an inverse object property of isActorOf, whose domain is UseCase and range is Actor. hasUseCasePrecondition is a functional object property, whose domain is UseCase and range is UCPrecondition. hasUseCasePostcondition is a functional object property, whose domain is

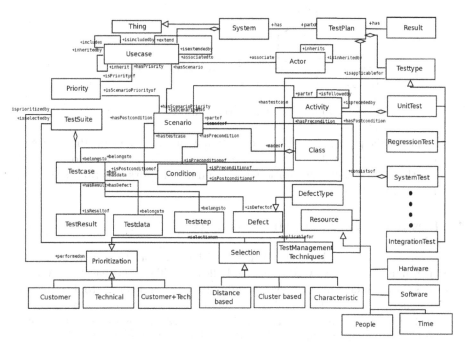

Fig. 3. Class diagram showing relation among concepts of the ontology

UseCase and range is UCPostcondition. hasUseCasePriority is a functional object property, whose domain is UseCase and range is UCPriority. hasUseCase-Name is an object property.

3.3   Implementation Using OWL

OWL has been chosen to implement the ontology due to its knowledge representation capabilities(concepts, individuals, properties, relationships and axioms) and the possibility to reason about the concepts and individuals[5]. Other ontology languages that can be used include RDF, DAML and DAML+OIL. Given use case and activity diagrams, an XSLT(eXtensible Stylesheet Language Transformation) document is written that defines the transformation rules to covert an XML file to OWL file[4]. An XSLT processor aids in performing the transformation based on the rules. Following are the transformation steps from XML to OWL using XSLT as shown in Fig. 4.

- The use case and activity diagrams serve as input to the transformation process. In this work, scenarios have been generated from activity diagrams.
- Run the XSLT processor. The XSLT file provides the template for transformation. The input file is read and if it finds the predefined pattern, it replaces the pattern with another according to the rules.

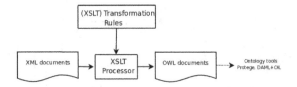

Fig. 4. The transformation process

– Output of the process is OWL specification. Ontology tools like Protege[2] can be used to verify syntax and validate semantics captured. It is to be noted here that test scenario generation algorithm in this work generates scenarios for system testing.

4   Results - Querying the Ontology

As shown in Fig. 5 the Protege OWL editor has support to edit and execute rules through the Query window using Pellet/FACT++ as reasoner. A user expresses requirements as a query and submits it to the Protege tool to obtain results. Queries include determining the number of tests that passed based on some criteria, tests based on a criteria that failed as well as displaying results. A sample of the queries that can be given to the ontology include:

1. What are the use cases the include use case 'x' ?
2. List all scenarios have test result = fail.
3. List all actors related to use cases and scenarios having priority 'y'.
4. List all scenarios for use case 'uc'. (This lists all scenarios for use case 'uc' as well as for use cases that are included by use case 'uc').

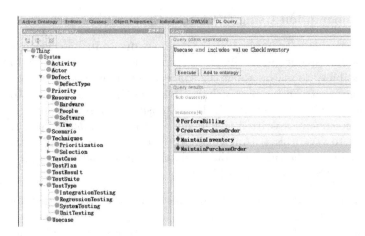

Fig. 5. Querying the ontology - list all use cases that include use case 'Check Inventory'

---

[2] Protege is an open source ontology editor and knowledge-base framework. Available at http://protege.stanford.edu

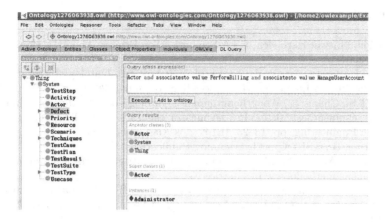

Fig. 6. Querying the ontology - list all actors associated with both use cases 'Perform Billing' and 'Manage User Account'

5. List of all unit tests for the Software under Test.
6. List all scenarios used by an actor.

Two examples querying the ontology is shown. In the first example(Fig. 5), use cases, 'PerformBilling', 'CreatePurchaseOrder' and 'MaintainInventory' include use case 'CheckInventory'. The constraint applied on relation 'include' is that it is transitive. Hence, the query 'Usecase and includes value CheckInventory' gives the above three use cases as well as the use case 'MaintainPurchaseOrder' which includes use case 'CreatePurchaseOrder'. This inference is useful in testing the use case 'Maintain Purchase Order' where the scenarios belonging to 'Check Inventory' also have to be tested. A second example requires that all actors related to both use cases 'PerformBilling' and 'Manage User Account' be given. Fig. 6 show the results of the query.

5   Conclusion

In this work, an approach for test scenario management is proposed. The increasing size of software systems makes it necessary to manage the test scenarios in an efficient way. In this regard, ontologies help to analyze the relationship between requirements captured using UML diagrams. Scenarios generated from activity diagrams(i.e. each use case has atleast one activity diagram) along with activities, are linked to use cases. Further, software testing concepts adopted from SWEBOK, are used to build an ontology for test management. Protege, an OWL editor is used for querying and reasoning. Thus, ontologies have a common reference point for architects, developers and test engineers engaged in software development. In particular, the use of ontologies for test scenario management to achieve desired quality in terms of test coverage is shown.

References

1. Antoniou, G., van Harmelen, F.: A Semantic Web Primer. The MIT Press, Cambridge (2004)
2. Bezerra, D., Costa, A., Okada, K.: swtoI (Software Test Ontology Integrated) and its Application in Linux Test. In: Proceedings of Ontose 2009, pp. 25–36 (2009)
3. Dillon, T.S., Chang, E., Wongthongtham, P.: Ontology-based software engineering-software engineering 2.0. In: Proceedings of the 19th Australian Conference on Software Engineering, pp. 13–23. IEEE, Los Alamitos (2008)
4. DuCharme, B.: XSLT Quickly. Manning Publications
5. Mendes, O., Abran, A., Montral Qubec, H.K.: Software engineering ontology: A development methodology. In: Metrics News
6. Juristo, N., Ferndandez, M., Gomez-Perez, A.: METHONTOLOGY: From Ontological Art Towards Ontological Engineering. AAAI Technical Report SS-97-06, 15(2) (1997)
7. Mabotuwana, T., Warren, J.: An ontology based approach to enhance querying capabilities of general practice medicine for better management of hypertension. Artificial Intelligence in Medicine (2009)
8. Noy, N.F., McGuinness, D.L.: Ontology Development 101: A Guide to Creating Your First Ontology. Stanford Knowledge Systems Laboratory Technical Report KSL-01-05 and Stanford Medical Informatics Technical Report SMI-2001- 0880, 15(2) (2001)
9. Nguyen, C.D., Perini, A., Tonella, P.: Ontology-based Test Generation for Multi-agent Systems. In: AAMAS 2008: Proceedings of the 7th International Joint Conference on Autonomous Agents and Multiagent Systems. International Foundation for Autonomous Agents and Multiagent Systems, pp. 1315–1320 (2008)
10. Andrea Rodriguez, M., Egenhofer, M.J.: Determining semantic similarity among entity classes from different ontologies. IEEE Transactions on Knowledge and Data Engineering 15(2), 442–456 (2003)
11. Chang, E., Wongthongtham, P., Cheah, C.: Software Engineering Sub Ontology for Specific Software Development. In: Proceedings of 29th Annual EEE/NASA Software Engineering Workshop (SEW 2005), pp. 27–33. IEEE, Los Alamitos (2005)
12. Wongthongtham, P., Chang, E., Dillon, T.: Towards Ontology-based Software Engineering for Multi-site Software Development. In: 3rd IEEE International Conference on Industrial Informatics (INDIN), pp. 362–365. IEEE, Los Alamitos (2005)
13. Zhu, H., Huo, Q.: Developing software testing ontology in UML for a software growth environment of web-based applications. Software Evolution with UML and XML, 263–295 (2005)

Creating Value for Businesses by Providing CRM Solutions: A Case of Oracle CRM on Demand

Swati Singh[1] and Sanjeev Kr. Singh[2]

[1] Institute of Technology & Science, Ghaziabad, U.P. 201007, India
[2] IIIT Hyderabad, Gachibowli, Hyderabad, Andhra Pradesh - 500 032, India
{swatisingh2620,sanjeev3094}@gmail.com

Abstract. Information technology integrations are playing great role in increasing business efficiency. IT based solutions have brought wonders to the enterprises. Organizations are looking for the better technologies for improving their performance at low cost. On the Demand CRM solutions are one of the major tools of increasing business efficiency by providing customer knowledge. This research attempts to study the role of On Demand CRM solutions in creating value for the organizations. The research establishes the importance of on Demand CRM solutions in profitable business implementation with the help of the case of Oracle CRM on Demand solutions. The research focuses at discussing the role of Oracle CRM on Demand solution in improving organizations' performance by analyzing related business and technological issues.

Keywords: CRM, On Demand Solutions, Technology, Business performance, Profitability, and Customization.

1 Introduction

The information technology revolution and, in particular, the World Wide Web has created an opportunity to build better , long term and profitable relationship with customers than has been previously possible in the traditional offline world. Leveraging Business models with updated technology has changed the way of marketing. By combining the abilities to respond directly and immediately to customer requests and complaints and to provide the customer with a highly interactive, customized experience, companies have a greater ability today to establish, nurture, and sustain long-term profitable customer relationships than ever before. The eventual objective is to transform these relationships into greater profitability by increasing repeat purchase, cross purchases and up purchase rates and reducing customer acquisition costs. Indeed, this revolution in customer relationship management or CRM as it is called has been referred to as the new "mantra" of marketing. Companies like Siebel, E.piphany, Oracle, Broadvision, Net Perceptions, Kana and others have filled this CRM space with products that do everything from tracking customer behaviour on the Web to predicting their future behaviours to sending direct e-mail communications. This has created a worldwide market for CRM products and services of $34 billion in 1999 and which is forecasted by IDC to grow to 12 million by the end of 2010. Companies rely on CRM and Data Integration Solutions for having complete

S. Dua, S. Sahni, and D.P. Goyal (Eds.): ICISTM 2011, CCIS 141, pp. 101–109, 2011.

knowledge about the customers which they use for building long term customer relationships. Customer Relationship Management (CRM) is a combination of information technology and strategic management processes, focused at providing better customer service [3]. CRM facilitates collaborative relations amongst specific functional areas of the enterprise [13]. The areas of the organization which are most commonly affected by CRM are sales, marketing and customer service [11]. The ultimate goal of CRM is to align departmental capabilities - the core competencies of all departments affected by CRM [2] - in order to increase customer retention and loyalty, as well as revenues [2]. Technology plays a key role in effective CRM implementation. Key role being played by technology creates the space for integration of business and information technology.

2 Research Purpose

The purpose of this study was to discuss the role of CRM solutions in creating value for the businesses. The study focuses on Oracle CRM on Demand and attempts to analyze the various aspects related to this application. The study aimed at establishing the role of CRM on Demand Solution in creating a long-term profitable relationships with the customers with the help of Oracle application.

3 Literature Review

Customer relationship management (CRM) is emerging as the core marketing activity for the businesses operating in dynamic and competitive business environment. Scholars and researchers agree that CRM as a system and process has potential to bring results for the organization. CRM in the literature has been defined as "comprehensive strategy and process of acquiring, retaining and partnering with selective customers to create superior value for the company and the customer" [19]. Past researchers have attempted to provide more conceptual clarity of CRM by synthesizing the relevant marketing, management, and IT literature [29]. There are four key areas which has been identified as necessary for successful CRM implementation: (1) *strategy,* (2) *people,* (3) *processes,* and (4) *technology* [7, 8, 22]. Literature suggest that there are four dimensions of CRM implementation (1) focusing on key customers [24, 26] , (2) organizing around CRM [10, 12], (3) managing knowledge [25], and (4) incorporating CRM based technology [1, 4]. The role of technology is inevitable in the success of CRM initiatives. This study focuses on the technology and incorporating CRM based technology. The technology component of a CRM implementation is often centred on a customer relational database management system (RDBMS) [27]. A RDBMS is a method of managing data, such as customer data, by linking records via the use of indices, and by allowing structured searches or queries, to be performed on the data [20]. The RDBMS is shared among select functional areas of the enterprise, or in some cases the whole enterprise, facilitating sales force automation [5] and customer relationship management [23]. Such software systems are often called Contact Management Software (CMS), or can also be referred to as simply CRM [23].The CRM technology enables efficiencies that ultimately provide increased customer value, including a sales force that is more reactive to customer needs and more specialized by product line, more intuitive interaction methods (between the company and the client), as well as improved market segmentation [30]. Technology has resulted in success of the companies. The past researches reveal that CRM helps in successful business implementations [9]. CRM success lies in its ability to establish and effectively use enterprise customer care information systems [2], CRM implementations are considered successful when they have achieved a predetermined level of return on investment (ROI) [6]. CRM implementations are considered failed when a certain percentage of end users boycotted or ignored the CRM, thus making ROI impossible to achieve [28]. The analysis of past literature put forth the fact that technology plays a great role in improving business performance. The future of CRM looks very interesting as social networking becomes integrated into CRM functionality. Sales force automation and contact management are common everyday needs in large organisations, and some of the less sophisticated suppliers can offer more than adequate functionality for a modest cost [14]. It is up to the individual organisation to match needs with available solutions – and the suppliers analysed in this short report cover most of the needs any organisation will have. The growth of on demand, or Software as a Service (SaaS), CRM solutions has significantly increased, this growth is fuelled by adoption among organizations of all types and sizes seeking ways to quickly and cost-effectively deploy CRM to meet specific business objectives [18]. Lower initial and ongoing costs, faster time to market, and less reliance on technical expertise are just a few of the drivers behind the acceptance of on demand as a viable CRM deployment option. Now these advantages are attracting the attention of larger enterprises. The enterprises are looking for On Demand business solutions in order to achieve better efficiency and profitability at low cost. Despite of the fact that On Demand CRM solutions are new business mantra for organizations, this area has yet not received considerable attention of academic researcher. This research is an attempt to explore the research scope related to CRM on Demand solutions. Oracle CRM on Demand is one of the On Demand CRM solutions, and this study unravels the role of this software in creating business value for organizations.

4   Research Methodology

The research method used for the study included case study and Delphi method. The focus of case study was the detail analysis of Oracle CRM on Demand solution. The information was collected from research reports of Research Firms like Forrester, Martin Butler, and Nucleus Research and Gartner RAS Core Research. The detailed analysis of Oracle website was carried out in order to gather updated information about the application. The case study was developed in two stages, in first stage the data was collected from research reports, journals, oracle website, and analysis of related new. This was followed by five step Delphi method. In first stage all the issues and facts related to the Oracle CRM on Demand was laid down and discussed with a group of experts. The expert group included five experts from the field of Information Technology Education, Management Education and Software Industry. In the second step the scope and applicability of Oracle CRM on demand were discussed and debated. Based on discussion key features related to the solution were filtered out. In the fourth stage, critical evaluation of the solution was carried out. The fifth stage comprised of comprehensive discussion on customer's experience related to Oracle CRM on Demand. The step Delphi method helped and gave the direction to the research. The case was developed with help of analyzed information and the results of the discussion of Delphi method.

5 Oracle CRM on Demand

Oracle Customer Relationship Management (CRM) On Demand is a Software-as-a-Service (SaaS) solution that has tremendous business potential to simplify an organization's IT, reduce costs and increase productivity. Oracle Customer Relationship Management (CRM) On Demand brings clarity to sales processes; build loyal, long-term customer relationships; accelerate productivity; boost call center efficiency; and derive customer intelligence. The solution helps in getting real-time, actionable business intelligence through interactive dashboards, custom reports, and historical trending to uncover new opportunities and identify issues, which impact business and business strategies.

5.1 Three Level of Customization by Oracle CRM on Demand

Oracle CRM on Demand enables customization at all levels of the application – the user interface (UI) level, the business process level, and the data level. This customization flexibility combined with Oracle CRM on Demand's rich, prebuilt functionality and embedded best practices enables deployment of CRM that is a perfect fit for your business, without having to build it yourself from the ground up. It provides high personalized application. The display of sales data, analytics reports may be customized by the user as per the need of sales representative and executives. The solution also creates customization at data level. It allows custom data relationships enabling organizations to relate custom objects to prebuilt objects. The customization is even extended to Business level, with features like time based workflow, scripting and advanced field management. Custom workflows allow deep insight of the events and the customers a prerequisite of long term profitable customer relationship.

Data Level. The most basic form of integration involves the integration at data layer via Web services, batch integration, and data import and export. The solution helps in critical analysis of key information, such as customer and product data, which needs to be appropriately synchronized in real-time or accessed regardless of where the data originally resides. This ensures access of the latest information by the user from any source without fear of duplication. Oracle CRM on Demand provides real-time flow for data such as accounts, contacts and products, where information frequently changes. For instance, customer orders can be delayed or even lost if account shipping information in CRM and ERP systems are not kept up-to-date, severely impacting customer satisfaction.

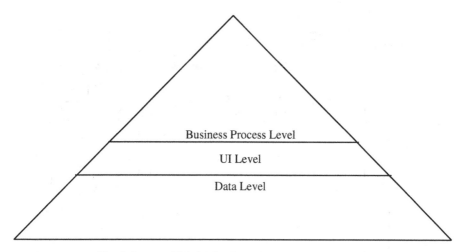

Fig. 1. Proposed Model for Understanding of the application

User Interface Level. A more highly developed integration occurs at the user interface layer where advanced integration extensions enable usability features like mashups and web links to embed custom HTML and third party content within the Oracle CRM On Demand user interface. As a result, users can view information from other applications within a single user interface, vastly improving usability and user productivity. It provides the facility by which users can interact with content through one application rather than needing to navigate across multiple applications. The solution also extends the scope of access to back office information – such as products and quotes – from the front office application, time and money needed to be spent on training salespeople on multiple applications is saved. And sales users have more information available at the point of interaction with their customers, enabling a better service experience.

Business Processes Level. The highest level of integration involves business processes that span across applications. In addition to business process orchestration, business logic, such as data validation rules, need coordination as well to ensure consistency throughout the entire process. Business processes based on BPEL are the key enabler for adopting a Service Oriented Architecture (SOA), enabling steps in a business process to be stitched together across applications to form a complete, end-to-end flow. By leveraging business process integration, organizations can build processes around the customer that span application boundaries rather than conforming to integration constraints. With business process integration, a complete business flow such as lead conversion can reconcile account information from a lead in a CRM application with similar information in an ERP system, eliminate redundant and duplicate data, and automatically convert the lead to an opportunity. Rather than change the application when business needs change – which can be a lengthy process – organizations can simply alter the business process in real time to adjust to changing dynamics. Oracle CRM on Demand integrates with other applications like Oracle e Business and Oracle Siebel CRM on Demand. The prebuilt integration spans both desktop and back office applications. It integrates seamlessly with desktop

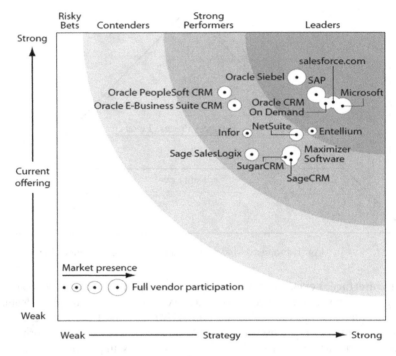

Fig. 2. Source: Forrester Research, Inc.

applications like Microsoft Office, MS outlook and IBM Lotus Notes this helps and improves users' efficiency. It has built a bidirectional data synchronization with oracle E business Suite and JD Edwards Enterprise One which streamlines the 360 degree view of the customer.

Research reports states that Oracle CRM on Demand has huge business potential with key features like competitive price point, high customization, and high integration with other applications [21]. Research reports also discuss that Oracle CRM on demand is the leader in the customer- record- centric product [15]. The solution is high on dimensions of current offering as well as in the market presence; refers Fig. 2. Current offering included parameters like customer service, field service, Internationalization, Industry business process support, Architecture and platform, Usability and cost. Market presence included parameters like customer base, employees and financial performance. The report also states that Oracle CRM On Demand "competes with other SaaS vendors, providing faster time-to-value, good usability and lower upfront costs. Oracle CRM on Demand is innovating with new capabilities such as Social CRM (Fusion Edge applications) to increase collaboration, innovation and adoption of CRM within organizations.

5.2 The Customer Experience

Oracle discusses about the Customer Experience on its website. Nucleus Research also discussed the customer experience of Oracle CRM on Demand in form of case studies. The in depth analysis reveals that the solution has brought fruitful results in short span of time. As discussed in Nucleus research case study – ROI case study: CRM on Demand Rackable Systems deployed Oracle CRM on Demand to get full view of customer data, enabling pipeline forecasting, improving management staff productivity and sales [17]. Another case study written by Nucleus research – ROI case study: Oracle & Antenna Software states that Direct TV used Oracle CRM on Demand to deliver key customers and partner data and metrics to sales managers, which improved their productivity and enabled them to sell more intelligently [16]. A customer response displayed on the Oracle website proposes that Equifax Inc. increased revenues and profit margins to achieve a 392% ROI and payback in only 10 months with the help of Oracle CRM on Demand. The customer's experience put forth the fact that Oracle CRM on demand has helped the companies to run their business profitably by providing them timely and accurate information. Business intelligence developed by Oracle CRM on Demand ensures customization with CRM initiatives focusing company's specific requirements.

6 Conclusions

CRM solutions have successful leveraged the business models of the organizations. This study indicates the fact that information technology is integrating with business more effectively and is impacting the efficiency of the business. Oracle CRM on Demand case suggests that deploying a CRM solution can bring drastic change in the business performance in the short span of time. However, it is also worth mentioning that CRM solution is not panacea for all business problems. The study creates the scope for further research in this context to find out the role of technology in business performance.

References

1. Bhaskar, R.: A Customer Relationship Management System to Target Customers at Cisco. Journal of Electronic Commerce in Organizations 2(4), 63–73 (2004)
2. Brown, S.A.: Strategic customer care: An evolutionary approach to increasing customer value and profitability. Wiley, Ontario (1999)
3. Chan, J.O.: Toward a unified view of customer relationship management. Journal of American Academy of Business 6, 32–38 (2005) (retrieved September 9, 2005, from Business Source Premier Database)
4. Chen, J.-S., Ching, R.K.H.: An Empirical Study of the Relationship of IT Intensity and Organizational Absorptive Capability on CRM Performance. Journal of Global Information Management 12(1), 1–17 (2004)
5. Coker, D.M., Del Gaizo, E.R., Murray, K.A., Edwards, S.L.: High performance sales organizations: Creating competitive advantage in the global marketplace. McGraw-Hill, New York (2000)
6. Corner, C., Rogers, B.: Monitoring qualitative aspects of CRM implementation: The essential dimension of management responsibility for employee involvement and acceptance. Journal of Targeting, Measurement & Analysis for Marketing 13, 267–274 (2004) (retrieved September 9, 2005, from Business Source Premier database)

7. Crosby, L.A., Johnson, S.L.: Technology: Friend or Foe to Customer Relationships? Marketing Management 10(4), 10–13 (2001)
8. Fox, T., Steve Stead, L.: Customer Relationship Management-Delivering the Benefits', White Paper of Secor Consulting (2001)
9. Gordon, I.: Relationship marketing. John Wiley and Sons Canada, Ltd., Ontario (1998)
10. Homburg, C., Workman Jr., J.P., Jensen, O.: Fundamental changes in marketing organization: the movement toward a customer-focused organizational structure. Journal of the Academy of Marketing Science 28(4), 459–478 (2000)
11. Itkis, D.: The buzz about CRM. Broker Magazine 7, 46–48 (2005) (retrieved September 8, 2005, from Business Source Premier Database)
12. Langerak, F., Verhoef, P.C.: Strategically embedding CRM. Business Strategy Review 14(4), 73–80 (2003)
13. Mack, O., Mayo, M.C., Khare, A.: A Strategic approach for successful CRM: A European perspective. Problems & Perspectives in Management 2, 98–106 (2005) (retrieved September 9, 2005, from Business Source Premier Database)
14. Butler, M.: A Comparison of CRM-on-Demand Solutions' (2009),
    http://www.martinbutlerresearch.com/index.php/
    free-reports/comparison-reports/crm-ondemand/
15. Petouhoff, N.L.: The Forrester Wave: Customer Service Software Solutions, Q4' (2008),
    http://www.forrester.com/rb/Research/
    wave%26trade%3B_customer_service_software_solutions%2C_q4_2008/
    q/id/45543/t/2
16. Nucleus Research: ROI Case Study: Oracle and Antenna Software (2008),
    http://nucleusresearch.com/research/roi-case-studies/
    roi-case-study-oracle-and-antennasoftware-directv-group/
17. Nucleus Research: ROI Case Study: Oracle CRM On Demand' (2008),
    http://nucleusresearch.com/research/roi-case-studies/
    roi-case-study-oracle-crm-ondemand-rackable-systems/
18. Oracle Taking On Demand CRM Integration to the Next Level, Oracle Corporation (2009)
19. Parvatiyar, A., Shet, J.: Customer Relationship Management: Emerging Practice, Process & Discipline'. Journal of Economic & Social Research 3, 1–34 (2001)
20. PC Magazine Online,
    http://www.pcmag.com/encyclopedia_term/
    0,2542,t=back+office&i=38342,00.asp (retrieved October 1, 2005)
21. Desisto, R.P.: Magic Quadrant for Sales Force Automation' Gartner RAS Core Research Note G00168995, RA3. 307302010 (2009),
    http://www.gartner.com/technology/mediaproducts/reprints/
    oracle/article86/article86.html
22. Ryals, L., Knox, S.: Cross-Functional Issues in the Implementation of Relationship Marketing Through Customer Relationship Management. European Management Journal 19(5), 534–542 (2001)
23. Simpkins, R.A.: The secrets of great sales management. AMACOM, New York (2004)
24. Srivastava, R.K., Shervani, T.A., Fahey, L.: Marketing, Business Processes, and Shareholder Value: An Organizationally Embedded View of Marketing Activities and the Discipline of Marketing. Journal of Marketing 63(4), 168–179 (1999)
25. Stefanou, C.J., Sarmaniotis, C., Stafyla, A.: CRM and customer-centric knowledge management: An empirical research. Business Process Management Journal 9(5), 617–634 (2003)

26. Vandermerwe, S.: Achieving Deep Customer Focus. MIT Sloan Management Review 45(3), 26–34 (2004)
27. Webster Jr., F.E.: Market-driven management: How to define, develop, and deliver customer value. Wiley, New Jersey (2002)
28. Wu, I.L., Wu, K.W.: A hybrid technology acceptance approach for exploring e-CRM adoption in organizations. Behavior & Information Technology 24(4), 303–316 (2005)
29. Hong-kit Yim, F., Anderson, R., Swaminathan, S.: Customer Relationship Management: Its Dimensions And Effect On Customer Outcomes'. Journal Of Personal Selling & Sales Management 24, 263–278 (2004)
30. Zoltners, A.A., Sinha, P., Zoltners, G.A.: The complete guide to accelerating sales force performance. AMACOM, New York (2001)

Societal Responsibility of Business and Information Systems

Hugues Poissonnier[1], Claudio Vitari[1], Raffi Duymedjian[1], and Renaud Cornu-Emieux[2]

[1] Grenoble Ecole De Management
[2] Ecole de Management Des Systems D'Information
12, rue Pierre Sémard - BP 127 - 38003 Grenoble Cedex 01
hugues.poissonnier@grenoble-em.com,
renaud.cornu-emieux@emsi-grenoble.com

Abstract. In this chapter, we present the development of the (CSR) corporate social responsibility as a factor of development of the use of the IS. First, we examine in detail how companies can seek to optimize the (positive or negative) externalities associated with ICTs in their lifecycle. Then we insist on the fact that businesses appear increasingly forced to act in order to communicate. Finally, we discuss the potential offered by ICT and SID in terms of preserving the environment and social progress. The IS are then used in a resolutely proactive approach towards CSR.

Keywords: Societal responsibility, environment, social progress, development.

1 Introduction

Two types of representation of corporate social responsibility coexist today. The first one considers the social or environmental objectives as objectives "out of business". The second representation incorporates these objectives to the business strategy. A good news is linked to the fact that this second representation is more and more dominant.

Sustainable development and CSR are still largely in the public space incantatory speeches. It's even more real during a period of economic crisis.

This perception is growing not only in the citizen public space, but also in smaller spaces and theoretically under best "control", i.e. the Organization internal businesses. Information systems, strategic alignment and activity control fundamental instruments are thus first online when it comes to a company to maintain a strategic course integrating CSR and sustainable development concerns.

In this chapter, we propose a presentation of the roles of IS towards the implementation of CSR principles.

2 Optimizing Externalities Associated with ICTs

The first link that we discuss in this part, between SI and CSR is linked to the minimization of the negative impacts of the SI. We present, as a first step, externalities associated with various phases of the lifecycle of the SI, to highlight the most recent and future solutions.

S. Dua, S. Sahni, and D.P. Goyal (Eds.): ICISTM 2011, CCIS 141, pp. 110–117, 2011.

2.1 The Lifecycle of ICTs and the Externalities Associated with Each Step

ICTs generate new opportunities but also new risks in terms of sustainable development.

ICTs are quickly changing the world. However, alongside the dematerialization, globalization of the economy allowed among others by ICTs, stimulates the production of hardware products. Finally, ongoing innovation in ICT accelerates the renewal of skills, knowledge and existing products.

The life cycle theory con be used to explain the nature of trade flows or marketing and the developments in demand for a product. It can also explain the phenomena of emergence and renewal of ICTs. Identification of the different stages of the lifecycle of ICTs can serve as basis for the definition of externalities associated with each step. Three major steps can be taken: design, use and recycling. These externalities can be done on society and the environment, be positive (case should take advantage) or negative (it will be then to minimize).

2.2 Towards a Better Use of the Capacity of ICT to Generate Positive Externalities

The use of ICT can first generate positive externalities for society and the environment. Boudreau, Chen and Huber (2007) [1] attributed to ICT the capacity to contribute to:

- Reduction of costs of transport across the fleet vehicle routing and enterprise management systems.
- Limitation of travelling by air through collaboration and teleconferencing systems.
- Traceability of environmental information to the tower building products, components and services, through environmental information management systems.
- Increasing efficiency in the management of emissions and waste through environmental coaching systems

The provision of information to customers in order to increase transparency on the processes and the resources in place by the company through environmental reporting systems.

Next to these positive externalities, a number of efforts should be taken to reduce the negative externalities associated with ICTs in their lifecycle.

2.3 To Reduce the Negative Externalities of ICT Solutions

Organizations, like all social actors, contribute in their choice concerning ICT, to developments in the world.

ICTs can facilitate:

- The development of processors and disks that use less energy design
- The replacement of standard computers by energy efficient ones
- The use of the software to operate several software operating on the same server virtualization
- The reduction of energy consumption in the data center
- The reduction of electronic waste from obsolete it equipment

All these initiatives allow:

1. The reduction of energy consumption to operate ICT business and therefore prevent pollution caused by the production of this otherwise necessary energy.

2. The development of the sorting and recycling of ICT business and therefore to prevent pollution by ICT production and mining resources are exhausted.
3. Changes in the habits of work with ICT by replacing physical transport telecommunication and therefore prevent pollution caused by transport and travel.

Companies that voluntarily start one or several initiatives are committed on a path of accountability towards society and the environment. Jevons paradox (Jevons, 1865) sets out that technological improvements increase the efficiency with which a resource is used, but the total consumption of that resource can increase rather than decrease. Therefore any substitution or optimization by ICTs might be compensated by quantitative increase which prevents to achieve sustainable development.

The following section presents some keys to identify positive and negative externalities.

2.4   Identification of Positive and Negative Externalities: Practical Tips

If today the priority is the achievement of the goals of the Millennium Development Goals (MDGs) for 2015 deadline; companies can use this list as an operational framework to analyze their contribution to SD.

First, to make products and services that are consistent with the MDGs, companies must establish consistent economic objectives with the goals of society (the Millennium goals for development). To do so, companies must first identify the priorities for improving the sustainability of their products and services, based on emerging trends and the actions of competitors. Then companies must develop these options for possible improvements, including a plan of action for these improvements and a system of continuous measurements of the new sustainable performance. A critical element of this layout consistency is the identification of the value drivers and their potential. These generators can be diverse and varied as customer satisfaction, customer loyalty, leadership, superior value brand, innovative product or service, expansion of business opportunities or improving cost structure.

In this first part, we insisted on initiatives to reduce the environmental and social negative impact of ICT in their lifecycle. Development of ICTs, and the IS, more generally contributes to the formation of new constraints for businesses. In effect also, offered communication opportunities are constraints in the sense that as competitor it is important to communicate.

3   Act to Communicate: To New Constraints

The development of ICTs probably generates as many constraints as new opportunities. It is now easier to communicate to the general public on how to conduct business and its commitments in terms of CSR; the development of this new type of communication gives rise to an additional differentiation factor. Communication on commitments in terms of CSR is therefore not only linked to new obligations established by law (including the 2001 France NRE Law) and to new standards of communication resulting from the practices of competitors.

3.1 Risks and Opportunities Generated by ICT in Terms of Communication

Access to information that extends the development of ICT and the IS is originally linked to new demands from society. The relative loss of control of Governments in the development of world trade and the weakening trade unions as regulatory instruments are often presented being responsible for abuses concerning the working conditions of workers. In this context, the international dimension and the flexibility of nongovernmental organizations (NGOs) contribute to the development of these last effective ramparts against mentioned social problems.

It is therefore now possible to report on recent developments in the area of governance underlining on the one hand strengthening expectations of shareholders and other societal pressures emerging with the NGOs.

Carrol (1970) defines three major dimensions of the overall performance : compliance with legal and economic responsibilities, the adoption of an ethical behavior with respect to commercial and financial partners and finally the development of philanthropic actions (support to associations for example). Today, it is generally accepted that the overall performance incorporates three major dimensions: economic, social and environmental. This recognition is indeed valid in the academic world (Capron and Quairel, 2004) [2] as well as among practitioners (triple bottom line in force for many years at Danone, pioneer in this area, for example).

3.2 The Strengthening of the Societal Requirements: Focus on the Problem of Traceability

Among the new societal demands, those concerning the problem of traceability seem very important. We are back in the lines that follow on the origins of expectations regarding traceability to submit the answers provided by ICT.

3.2.1 The Origins of Expectations Concerning Traceability

Among the various requirements emanating from society, traceability deserves special attention. The new traceability requirements relate to information on the chain of production and distribution of a product. Initially, these new requirements appeared in the sector of health due to awareness of potential problems arising from dangerous products (contaminated blood). Very quickly, the food industry, due to various health problems (BSE, dioxin, chicken...), has taken the relay, making this requirement shared by consumers. Today, traceability requirements are essential to measure the social content of the latter. Textile is an interesting example of integration of social content products in the requirements of consumers in terms of traceability.

Today, three large objects of traceability are retained: (probably the oldest object tracking) people, things and processes identify three key issues associated with traceability:

- safety (especially in the agri-food sector),
- counterfeiting (primarily in the manufacturing sector),
- customer relationship (to meet the demand of information by clients).

3.2.2 The Answers Provided by Businesses Thanks to the SI

Recently, traceability has benefited from progress related to computing and miniaturization:

- bar codes allow identification of products as electronic chips information storage capacity increase;
- radio-labels allow product tracking
- Finally, the capacity of databases allows following a growing list of references.

Development of security and authentication technologies therefore provides answers to questions related to the three major current issues in terms of traceability.

However, any other issue emerges with the dissemination of these technologies, because they can provide the right to respect for privacy, which is protected by articles 9 of the civil code and 8 of the European Convention of human rights violations. E.g. radiolabels embedded in objects can enable geo locate individuals against their willingness.

4 Use the SI as SD Media

Information systems can also be used as genuine media of sustainable development. It is this proactive use of the IS vision that we propose to explain and illustrate in this part. First, we will discuss the role of IS as a media of environmental policies, and then as a support for social policies.

4.1 Environmental Policies Media SI

The geographic information systems (GIS) are computer tools aimed at organizing and presenting spatially referenced alphanumeric data, and at producing plans and maps. If their application domains are quite important (especially in marketing to customers, or urban planning for the land, and even locate biology and study of the movement of animal populations...), they can contribute to a better management of natural resources. Applications of GIS contributing to the implementation of sustainable development policies open new perspectives in water resources management or implementation of monitoring stations to avoid forest fires (used by the National Office for the environment of Madagascar).

4.2 The Media of Social Policies IS

At a time when information and knowledge are the new levers for value creation, information systems must play an important role in certain regions disenclose. It is what explains Mamadou Decroix Diop, Minister of information, communication and promotion of technology information (note the title of this Department) of Senegal: "we missed the first two revolutions of the modern era: the agricultural and the industrial." "Africa cannot be in rest in the information society".

In addition to the necessary efforts of developing countries, the responsibility of the developed countries is to participate in these efforts. SIST (System of scientific information and technical) project seems a good illustrative approach. It is a project of the French Ministry of Foreign Affairs, whose aim is precisely to help the African research cooperation. It is based on support for the development of the IS at institutions of higher education.

Initiative "Resafad-Tice" (network of support French for the adaptation and the development of technologies of information and communication in Education), which dates back to 1997, recalls the role of education in development and the potential of e-learning. It's based on the idea that access to knowledge through access to technology is essential.

Many other examples deserve to be mentioned here. Thus, in St. Louis in Senegal, an internet service has recently been developed for fishermen to improve their safety, their incomes and the management of the resources of the sea. A weather service at the Cafe of the port helps in reducing injuries related to adverse weather conditions, often causing of drowning. Furthermore, a database was created to learn more about transport with trucks passing port day filling rate capacity. This served to reduce rate Sin fish not finding not preneur (a third previously). Tygerberg in Cape Town South Africa children's Hospital gives another example of the impact of ICT and the SI on improving health. With telemedicine service, these hospital specialists can communicate directly with three hospitals in disadvantaged neighborhoods and provide their expertise on difficult operations.

4.3 Social Responsibility Issues Associated with the Use of the IS in Customer-Supplier Relations

The new communication, control and information exchange opportunities offered by the IS changes the nature of the customer-vendor relationships. Development of e-commerce in B-to-B context promotes acceleration level supplier involvement since the latter appear particularly much easier to find, without prospecting field.

In parallel to the development of potentially less stable relationships between customers and suppliers, the IS share data that can contribute to the implementation of partnership relations. Accounting in open-book provides a good example.

Use of the IS in inter-organizational context can therefore assign customer-supplier relations in different meanings. In any case, their role in the dynamics of development of these relations is established and deserves to be taken into account when one asks the interactions between SI and CSR. In this paragraph, we are back on the impacts for suppliers of the sustainability developed relationships and the richness of their customer. We then show how the inter-organizational IS contributes to the weakening of relations, then, when certain conditions are met, to their sustainability.

4.3.1 The Role of Inter-organizational Relations Development Providers

The Green Paper of the European Commission defines CSR as the 'voluntary social concerns throughout the chain of production and supply, combined to a dialogue with all stakeholders 'integration'. This definition has the merit of drawing attention to the inter-organizational nature of CSR. Several studies show how the nature of the relationship developed between customer and supplier are crucial for the latter. The current crisis gives us every day near examples: payment periods, attempting to blend purchases.

Client companies therefore have an important role to play in the development of their suppliers. More generally, it is possible to distinguish three types of learning's from inter-organizational relations for suppliers in developing countries:

- Learning the trade, or the acquisition of new skills (we refer here to the famous "learning by doing");
- Organizational learning, linked to a new (and better) management structure and the organization itself (this is especially true when organizational asymmetry between customer and supplier is important, supplier then tends to adopt certain principles of functioning of its client);
- Learning relational behavior that allows the supplier to develop a genuine "customer orientation".

Of course, these different learning's can only emerge if the customer-supplier relationship is sufficiently stable, future-proof, that does not always favor the use of the IS.

4.3.2 The IS at the Origins of the Weakening of Inter-organizational Relations Managers?

By enabling the development of new relations very quickly, the inter-organizational IS contributes to increasing instability of customer-supplier relations. While sharing culture and shared values played an important role in the strengthening of relationships there are still a few years (these could be described as "claniques"), the distance requires radically different management modes. "Management by the senses", it remove by the "management by numbers" (Torres, 2004) [10] that allow the new IS.

In this context, the fragility of relations and the formalism of the latter contribute to reduce the possibilities of transfer to suppliers (among three types of learning's mentioned in the preceding point, essential elements pass through a completely informal).

4.3.3 The Open-Book: When the SI Contribute to the Development of Suppliers

Accounting in open-book, sometimes regarded as a form of control of suppliers, is much more than that. It consists for business partners to share information related to the cost of products or components sold. This method of operation comes from Japan. Transmission of such information by suppliers to their customers in strategic nature is a widespread practice. The objective of this practice is to build trust between trading partners and facilitate the negotiation. A benefit may come for both firms in savings due to reflection joint costs if this reflection develops effectively. Another form of income is the result of the fact that, by giving access to their accounts, suppliers certify their good faith and the price they offer is less prone to attack. In this sense, the formalism introduced the development of the open-book in customer-supplier relations is likely to strengthen mutual trust. However it should be kept in mind that perverse effects can arise when the information provided is used as simple arguments negotiation assistance.

Nevertheless, the inter-organizational IS undoubtedly has an essential role to play in the development of these practices to strengthen the partnership and joint progress.

References

1. Boudreau, M.-C., Chen, A., Huber, M.: Green IS: Building Sustainable Business Practices. In: Watson, R.T. (ed.) Information Systems: A Global Text (2008), http://globaltext.terry.uga.edu/node/21
2. Capron, M., Quairel, F.: Myths and Realities of Responsible Business. Paris, La Découverte edn. (2004)
3. Fabbe-Costes, N.: Transformative Role of SIC and ICTs on the Distribution and Logistics. In: Fabbe-Costes, N., Colin, J., Paché, G., Vuibert, R. (eds.) Doing Research in Logistics and Distribution, Multi-Stakeholder Interfaces, pp. 171–194 (2000)
4. Kuhndt, M., Likinc, B., Stöcker, T.: Indetifying Value Drivers for Sustainable Business Development. Wuppertal Institute, Wuppertal (in preparation)
5. ORSE: How to Develop a Summary of Sustainable Development, Meetings of Working Group ORSE-EPA 2002, p. 34 (2003)
6. Viruega, A., Pellaton, J.-L.: Use of Traceability for Safety: Analysis by the Theory of Translation. In: 1st day of Research Relations between Industry and Grande Distribution Food, Avignon, March 29 (2007)
7. Fishmonger, H.: Client Relational Dynamics - Provider and Learning Opportunities: An Analysis of the Situation Suppliers French Distributors Clothing. In: CIFEPME (Congrès International Francophone in entrepreneurship and small business), Montpellier (October 2004)
8. Quairel, F., Auberger, M.: Newfoundland. Responsible for Management and Small Business: A Review of the Concept of "Corporate Social Responsibility". La Revue des Sciences de Gestion, Direction and Management 211-212, 111–126 (2005)
9. Salge, F.: The Sharing of Geographic Information for Sustainable Development, Seminar Sustainable Development, Clermont-Ferrand, pp. 24–25 (September 2008), http://www.agroparistech.fr/Colloque/ dd_ingenierie_territoriale/IMG/PDF/Atelier10.PDF
10. Torres, O.: Theorization of SME Management Test: Globalisation in the Proxémie Authorization at Leading Research in Management Sciences, University of Caen Basse (December 2004)

Application of Vector Quantization in Emotion Recognition from Human Speech

Preeti Khanna[1] and M. Sasi Kumar[2]

[1] SBM, SVKM's NMIMS, Vile Parle, Mumbai, India
[2] CDAC, Kharghar, Navi Mumbai, India
preeti.khanna@nmims.edu, sasi@cdacmumbai.in

Abstract. Recognition of emotions from speech is a complex task that is furthermore complicated by the fact that there is no unambiguous answer to what the "correct" emotion is for a given speech sample. In this paper, we discuss emotion classification of a well known German database consisting of 6 basic emotions: sadness, boredom, neutral, fear, happiness, and anger using Mel frequency Cepstral Coefficients (MFCCs). A concern with MFCC is the large number of features. We discuss the use of LBG-VQ algorithm to minimize the amount of data to be handled. At last, emotion classification is done using Euclidean distance, Manhattan distance and Chebyshev distance of the codebooks between neutral state and other emotional states for the same sample.

Keywords: Emotion recognition, Mel frequency cepstral coefficient, vector quantization, German database.

1 Introduction

Many researchers in the area of speech technology during the last decade have worked on different aspects of emotions in speech. Recently acoustic investigation of emotions expressed in speech has gained increased attention partly due to the potential value of emotion recognition for spoken dialogue management ([1, 2, 3]). For instance, anger due to unsatisfied services about client's requests could be dealt with smoothly by transferring the user to human operator. However, in order to reach such a level of performance we need to extract a reliable feature set that is largely immune to inter- and intra-speaker variability in emotion expression. That is one of the reasons in speech processing area, emotion recognition is usually considered a difficult task. Despite the progress in understanding the mechanisms of emotions in human speech, progress in the development and design of emotion recognition systems for practical applications is still in its infancy. The reasons behind limited progress in developing an emotion recognition system include: (1) challenges in identifying what input features are suitable and optimal to achieve reliable recognition; (2) variability in emotion resulting from language and culture differences and from inter-speaker differences. Therefore, it is often difficult to

S. Dua, S. Sahni, and D.P. Goyal (Eds.): ICISTM 2011, CCIS 141, pp. 118–125, 2011.

extend and generalize the results obtained from a particular domain and database to other cases; (3) lack of reliable databases of emotions across acted or forced emotion versus natural emotions arising in day to day activity. The consensus among researchers has, however, shown that primary emotional states such as the "Big Six" - anger, happiness, sadness, disgust, fear, and surprise - share similarity across culture and language [2] and provide at least a good starting point to explore specific scientific questions in automatic emotion recognition.

In recent years, a lot of approaches have been performed with different set of features and different models like Linear Discriminant, K-Nearest Neighbors, Neural Network, Support Vector machine etc (as mentioned below in much detail). Our proposal consists in studying the performance of Mel Frequency Cepstral Coefficient (MFCC) features for emotion recognition. The aim of this paper is to use spectral features such as MFCC and the technique like vector quantization to handle input data and to identify the emotions from the input databases. For our experiments we are considering six classes of emotions – happy, angry, sad, bore, neutral and fear. The study has been done on the standard database (Danish Emotional Speech) for both genders, male as well as female.

2   Related Work

Many researchers have done work on emotion recognition from speech. Most of them use prosodic information as their feature parameters ([4, 5, 6, 7, 8]). Nevertheless, the performance of the reported techniques is hard to compare because of the lack of a common database though researchers report being able to improve the classification performance in emotion classification by determining a subset of the extracted features instead of employing all of them in the classification. Many classification algorithms exist and have been proposed, such as neural network (NN) ([4, 5, 6]), decision tree [6], k-nearest neighbour (K-NN) [6], support vector machine (SVM) ([5, 6, 7]), discriminant analysis [8], and hidden markov model (HMM) [8].

Most commonly, prosodic features are used in emotion recognition from speech. The prosodic features are made of fundamental frequency and energy component which carries useful information for discriminating emotions. However, there exists another set of phonetic features of speech, like Mel Frequency Cepstral Coefficients (MFCC) which could also be used to discriminate emotions. Even if a small amount of useful information is available in phonetic features it will enhance the accuracy of emotion recognition by increasing the number of independent phonetic features. Many algorithms of emotion recognition using MFCC have been proposed. Kwon et al. 2003 [8] proposed the algorithm based on utterance-level features such as maximum, average, variance etc. Each of these is made from MFCC and they have omitted the detailed structure of the frame-level features. We explore the effectiveness of MFCC as the feature for emotion recognitions and vector quantization techniques to minimize the amount of data to be handled.

3 Mel Frequency Cepstral Coefficients

MFCC's are based on the known variation of the human ear's critical bandwidths with frequency; filters spaced linearly at low frequency and logarithmically at high frequencies have been used to capture the phonetically important characteristics of speech. This is expressed in the mel-frequency scale, which is linear frequency spacing below 1000 Hz and a logarithmic spacing above 1000 Hz. The block diagram for the computation of MFCC approach is given in Fig. 1. The various blocks are discussed below in brief.

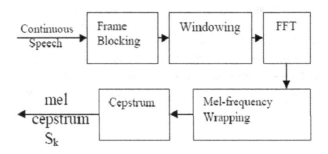

Fig. 1. Block Diagram of the Computation Steps of MFCC

3.1 Frame Blocking

This step involves the blocking of speech signal into frames. Input signal is blocked into N samples, with adjacent frames being separated by M (M<N). The first frame consisting of first N samples while the second frame begins M sample after the first frame, and overlaps it by N-M samples and this process continues.

3.2 Windowing

This concept is used to minimize the signal distortion by using the window to taper the signal to zero at the beginning and end of each frame. Typically the popular Hamming window as a time window is used, given by

$$w(n) = 0.54 - 0.46 * \cos(2\pi n/N\text{-}1), \quad 0 \leq n \leq (N\text{-}1) \tag{1}$$

3.3 Fast Fourier Transform (FFT)

The next step is Fast Fourier Transform, which converts each frame of N samples from the time domain into frequency domain.

3.4 Mel-Frequency Wrapping

A psychophysical study shows that human perception of the frequency contents of sound for speech signals does not follow a linear scale. The speech signal consists of tones with different frequencies. For each tone with an actual Frequency, f, measured in Hz, a subjective pitch is measured on the 'Mel' scale. The mel-frequency scale is linear frequency spacing below 1000Hz and a logarithmic spacing above 1000Hz. As a reference point, the pitch of a 1 kHz tone, 40dB above the perceptual hearing threshold, is defined as 1000 mels. Therefore we can use the following formula to compute the mels for a given frequency f in Hz [10].

$$mel(f) = 2595*log10(1+f/700) \qquad (2)$$

The filter bank has a triangular band pass frequency response, and the spacing as well as the bandwidth is determined by a constant mel frequency interval.

3.5 Cepstrum

In the final step, the log mel spectrum has to be converted back to time. The result is called the mel frequency cepstrum coefficients (MFCCs). The cepstral representation of the speech spectrum provides a good representation of the local spectral properties of the signal for the given frame analysis. Because the mel spectrum coefficients are real numbers (and so are their logarithms), they may be converted to the time domain using the Discrete Cosine Transform (DCT). The MFCCs may be calculated using this equation ([10, 11]):

$$c_n = \sum_{k=1}^{K} (\log S_k) \cos[n(k-0.5)\frac{\pi}{K}] \qquad (3)$$

where n = 1, 2 ...K

K denotes the number of mel cepstrum coefficients. The number of mel cepstrum coefficients, K, is typically chosen as 13. This set of coefficients is called an acoustic vector.

4   Vector Quantization

In 1980, Linde, Buzo, and Gray (LBG) [12] proposed a VQ design algorithm based on a training sequence. The use of a training sequence bypasses the need for multi-dimensional integration. A VQ that is designed using this algorithm are referred to in the literature as an LBG-VQ.

The algorithm requires an initial codebook. The initial codebook is obtained by the splitting method. In this method, an initial code vector is set as the average of the entire training sequence. This code vector is then split into two. The iterative algorithm is run with these two vectors as the initial codebook. The final two codevectors are split into four and the process is repeated until the desired number of codevectors is obtained. The algorithm is summarized in the flowchart of Fig. 2.

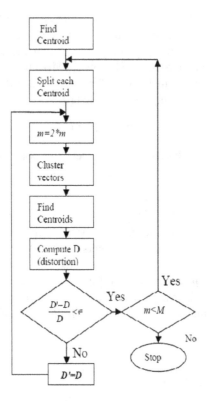

Fig. 2. The Flowchart of LBG-VQ Algorithm

The distance from a vector to the closest codeword of a codebook is called a VQ distortion. It shows how close the approximation is.

5 Our Experiment

For our experiments we have used Danish Emotional Speech (DES) database [9] that covers six emotional states. The design of this database is especially oriented toward speech synthesis purpose, but it can also provide a first approximation to emotional speech analysis and emotion recognition. A high quality microphone was used, which did not influence the spectral amplitude or phase characteristics of the speech signal. Ten actors (5 female and 5 male) simulated the emotions producing 10 German utterances (5 short and 5 longer sentences) which are used in everyday communication across various emotional states. From this corpus, for our experiment, we have selected total of 260 emotional utterances for female and 217 emotional utterances for male across the six emotional states as anger, happy, fear, sad, neutral and bore.

It is believed that acoustic features are the primary indicator of speakers' emotional states. Research to analyze emotional speech indicates that fundamental frequency, energy and formant frequencies are potentially effective parameters to distinguish certain emotional states. In this paper we will be discussing one of the acoustic features - MFCC (Mel-scale Frequency Cepstral Coefficient). The feature extraction process was performed on a frame by frame basis. A Hamming window was used to segment each speech utterance into frames. The sampling rate is 22 KHz. The frame length is set to 20 ms with an overlap of 10ms. First 13 coefficients of MFCC are computed by applying the procedure described above. This set of coefficients is called an acoustic vector. These acoustic vectors can be used to represent and recognize the voice characteristic of the speaker. Therefore each input utterance is transformed into a sequence of acoustic vectors. The next section describes how these acoustic vectors can be used to represent and recognize the voice characteristic of a speaker. The system has been implemented in Matlab7.1 on Windows Vista platform. VQ-LBG algorithm has been applied using a codebook size of 16. For example, for one of the input signal for 135 columns (input signal is for about 1.35 sec) each 13 coefficient of MFCC has been calculated. Hence 13*135 vectors have been transformed to 13*16 vectors by using VQ-LBG method. This exercise has been done for all the sample data (5 male and 5 female) across six emotional states. Then we used three different distance method to classify emotional states. These are Chebyshev distance, Euclidean distance and Manhattan distance. Each of these distances is calculated with respect to neutral state of the respective sample.

6 Results

We ran our experiments for male as well as for female sample. All the distances have been computed with reference to the neutral state of the respective sample. Fig. 3, 4 and 5 shows the variation of Euclidean, Manhattan and Chebyshev distances for the emotional state-anger, bore, happy, fear, and sad.

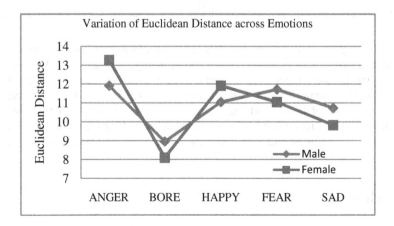

Fig. 3. Results using Euclidean Distance with Reference to Neutral State for Male and Female

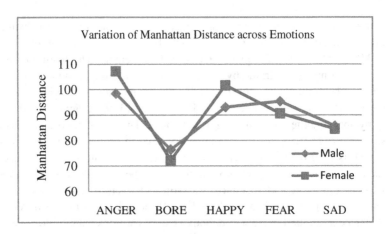

Fig. 4. Results using Manhattan Distance with Reference to Neutral State for Male and Female

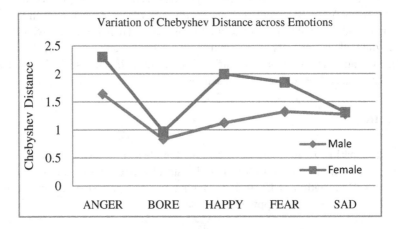

Fig. 5. Results using Chebyshev Distance with Reference to Neutral State for Male and Female

The above figures shows the average variation of all subjects both for male as well as for female across five emotions (anger, bore, happy, fear and sad. We found the both male and female are showing the same trend across five emotions for Euclidean distance and Manhattan distance. Similar trend is for Chebyshev distance except for the emotional state of happy and fear. Anger and bore are the two emotional state which shows the best recognized as compared to others. Anger has the highest values while bore is at the lowest and in between these two emotional states happy, fear and sad is lying.

7 Conclusions

In this paper, we proposed an emotion recognition system using spectral features such as MFCC. We apply the LBG-VQ algorithm to get the codebooks of the input signal.

The experiment has been done for three distances method such as Euclidean, Manhattan and Chebyshev distances. As a whole, this proposed system is simple in design and could work in defining three categories of emotions. The first category consists of anger, the second category consisting of bore and third is of fear and sad. It has been found that "anger" and "bore" are clearly distinguished from other emotional categories and more visible so best to recognized. The other two emotions, "fear" and "sad" lies in between these two extreme categories which is "anger" and "bore". This trend is seen from the Figs. 4, 5, and 6.The present study is still ongoing. In the future work we will compare across the size of codebook (e.g. 8, 16, and 32) for each of these distance method.

References

1. Cowie, R., Douglas-Cowie, E., Tsapatsoulis, N., Votsis, G., Kollias, S., Fellenz, W., Taylor, J.: Emotion recognition in human-computer interactions. IEEE Signal Proceedings 18(1), 32–80 (2001)
2. Litman, D., Forbes, K.: Recognizing emotions from student speech in tutoring dialogues. In: The Proceedings of the ASRU 2003 (2003)
3. Lee, C.M., Narayanan, S.: Towards detecting emotion in spoken dialogs. IEEE Trans. on Speech and Audio Processing 13(2) (2005)
4. Tato, R., Santos, R., Kompe, R., Pardo, J.: Emotional space improves emotion recognition. In: The Proceedings of the Seventh International Conference on Spoken Language Processing, vol. 3, pp. 2029–2032 (2002)
5. Yacoub, S., Simske, S., Lin, X., Burns, J.: Recognition of emotions in interactive voice response systems. In: The Proceedings of the Eighth European Conference on Speech Communication and Technology, pp. 729–732 (2003)
6. Oudeyer, P.Y.: The production and recognition of emotions in speech: features and algorithms. International Journal of Human Computer Interaction 59(1-2), 157–183 (2003)
7. Yu, F., Chang, E., Xu, Y.Q., Shum, H.Y.: Emotion detection from speech to enrich multimedia content. In: The Proceedings of the Second IEEE Pacific Rim Conference on Multimedia, pp. 550–557 (2001)
8. Kwon, O.W., Chan, K., Hao, J., Lee, T.W.: Emotion recognition by speech signals. In: The Proceedings of the Eighth European Conference on Speech Communication and Technology (EUROSPEECH), pp. 125–128 (2003)
9. German Emotional Speech Database, http://emotion-research.net/biblio/tuDatabase
10. Deller, J., Hansen, J., Proakis, J.: Discrete-time processing of speech signals, 2nd edn. IEEE Press, New York (2000)
11. Soong, F., Rosenberg, E., Juang, B., Rabiner, L.: A vector quantization approach to speaker recognition. AT&T Technical Journal 66, 14–26 (1987)
12. Linde, Y., Buzo, A., Gray, R.: An algorithm for vector quantizer design. IEEE Transactions on Communications 28, 84–95 (1980)

Fuzzy Time Series Prediction Model

Bindu Garg[1], M.M. Sufyan Beg[1], A.Q. Ansari[2], and B.M. Imran[1]

[1] Department of Computer Engineering, Jamia Millia Islamia,
New Delhi-110025, India
[2] Department of Electrical Engineering, Jamia Millia Islamia,
New Delhi-110025, India
bindugarg80@gmail.com, mmsbeg@hotmail.com,
aqansari@ieee.org, abbabeta@gmail.com

Abstract. The main objective to design this proposed model is to overcome the drawbacks of the exiting approaches and derive more robust & accurate methodology to forecast data. This innovative soft computing time series model is designed by joint consideration of three key points (1) Event discretization of time series data (2 Frequency density based partitioning (3) Optimizing fuzzy relationship in inventive way. As with most of cited papers, historical enrollment of university of Alabama is used in this study to illustrate the new forecasting process. Subsequently, the performance of the proposed model is demonstrated by making comparison with some of the pre-existing forecasting methods. In general, the findings of the study are interesting and superior in terms of least Average Forecasting Error Rate (AFER) and Mean Square Error (MSE) values.

Keywords: Time Series, Soft Computing, Fuzzy Logic, Average Forecasting Error Rate, Mean Square Error.

1 Introduction

In past few decades of research and development, many methodologies & tools emerged to deal with the forecasting processes. Forecasting using conventional time series methodology is one of the oldest and most reliable techniques to foresee future events. Various approaches have been developed on conventional time series forecasting. Among them ARMA models and Box-Jenkins are highly famous. These traditional statistical time series models can predict problem arising from new trends, but fail to forecast the data with linguistic values. Such models also do not attempt to represent nonlinear dynamics. Consequently, other models came into existence. In recent years, soft computing techniques are being used to handle prediction problems. A vast and broad logical survey of forecasting models using artificial neural network has been accomplished [1]. Neural Network models are better in terms of their ability to handle non-linear problem but inefficient because of large training time and predicted values are also not too accurate due to inability to handle non-stationary nature of data. Another soft computing technique which has recently received attention is the fuzzy time series based approach. Primary reason for fuzzy time series

S. Dua, S. Sahni, and D.P. Goyal (Eds.): ICISTM 2011, CCIS 141, pp. 126–137, 2011.

popularity is, it can relate trend or cyclic component in time series observation and hence can utilize historical data more effectively. So further in this paper, fuzzy logical relation in time series data are established rather than random and non random functions in usual time series analysis. This paper can broadly be divided into seven sections. Section 1 is current section, briefly elaborating on the evolution of various fuzzy time series models. Section 2 lists all the definitions used in this paper. Section 3 describes the new dynamic computational algorithm for forecasting in detail. Section 4 is to exercise the proposed method on enrollment data of University of Alabama. Section 6 evaluates and compares the result of proposed forecasting model with previous forecasting models. Section 7 has conclusion.

2 Related Work

The initial work of Zadeh [2] concerning fuzzy set theory has been applied in several diversified areas. Immense work has been done on forecasting problems too using fuzzy time series. Song and Chissom [3, 4, 5, 6] used fuzzy set theory to develop models for fuzzy time series and applied them on the time series data of University of Alabama to forecast enrollments. Song and Chissom used an average auto correlation function as a measure of dependency. Chen [7, 8, 9] presented simplified arithmetic operations and considered high order fuzzy time series model. Hunrag [10, 11], Hsu and Chen [12], Hwang and Chen [13], Lee Wang and Chen [14], Li and Kozma [15], Melike and Degtiarev [16]; all developed number of fuzzy forecasting methods with some variations. Jilani, Burney and Ardi [17, 18] partitioned the universe of discourse into equal length interval and developed method based on frequency density partitioning. Singh [19] developed forecasting models using computational algorithm. Stevson and Porter [20] used concept of percentage change of data to define universe of discourse followed by usage of straightforward defuzzification function to defuzzify value in their forecasting model. However, above methods ignored the impact of previous values on next forecasted value and how these could be further helpful in forecasting process.

Aforementioned research on fuzzy time series for forecasting problems focused on obvious linguistic values; while ignoring slight, but potentially crucial clues behind those obvious ones. In this paper an innovative forecasting model is proposed to rectify these imperfections. Proposed method capitalizes on available information with different perspectives. Model endeavor the issue of improving forecasting accuracy by introducing the concept of dynamic event discretization function and novel approach of weighted frequency density based distributions for length of intervals. This new concept eliminates the inadequacies due to the way of defining the universe of discourse on static historical data. Event discretization function controlled the forecasting error by considering percentage change of historical time series data for universe of discourse. This is quite encouraging as it highlights impact of trend & seasonal components by yielding dynamic change of values from time t to t+1. In short proposed dynamic computational time series based forecasting model presents and integrates concept of fuzzy logic relationship, event discretization function and a dynamic weighted frequency based distribution function in an inventive way to maintain consistency and robustness in forecasting accuracy. This has been demonstrated by comparing results of the proposed method with already existing methods on the same enrollment data of University of Alabama.

3 Proposed Dynamic Computational Algorithm

In this section, we present the new dynamic computational algorithm for forecasting of time series data. Strength of algorithm lies in integrated usage of event discretization function, dynamic weighted frequency based distribution function and fuzzy logic relationship in an inventive way. Primarily it calculates event discretization function for time series in terms of RoCs and defines universe of discourse on these RoCs. Thereafter, it applies the concept of weighted frequency density based distributions on intervals so that intervals get weightage on the basis of maximum rate of change along with frequency of RoCs. As a next step, the fuzzy logical relationships is formulated in such a way that impact of previous and next value is also accounted on current time series fuzzy value F_i besides the influence of fuzzy intervals F_{i-1} and F_{i+1} on F_i. Subsequently, to utilize derived information effectively, some arithmetic operations are performed in a creative way. Finally, forecasted value is generated using predicted ROC.

3.1 Key Concepts of Algorithm

Event discretization function. The discretization operation causes a large reduction in the complexity of the data. This process is usually carried out as a first step toward making data suitable for numerical evaluation. In our problem, event discretization function can be defined in a way, so that its value at time t index correlates with the occurrence of the event at particular specified time in the future.

RoC(t) = (X(t+1) - X(t)) \ X(t), where X(t+1) is value at time t+1 index and X(t) is actual value at time t index. ROC is the rate of change of value from time t to t+1.

Frequency density based partitioning procedure
- Determine number of RoC fall in each fuzzy interval
- Identify the interval with maximum frequency of RoC.
- Search for existence of similar interval (sign ignored) having frequency of RoC with a difference of less than or equal to one.eg [-1 -2] and [2,1] are same interval (sign Ignored)
- In case any such interval does not exist, continue with normal procedure of Frequency distribution
- Otherwise select the interval having maximum rate of change among these two and divide it into four sub intervals.
- Repeat the same process for next two intervals of highest frequency. Further divide these intervals into three and two sub intervals respectively. Let all subsequent intervals remain unchanged in length.frequency density procedure is demonstrated in Table 3.

Optimization of fuzzy relationship procedure. Using this procedure, more optimized F_i at particular time t would be generated. Obtain the fuzzy logical relation for $F_i \rightarrow F_j$

and $F_j \rightarrow F_k$, where F_j is fuzzy value at time y. F_i is fuzzy value at time y-1. F_k is fuzzy value at time y+1. Get their corresponding RoCs. D is time variant parameter which is calculated as:

$$\text{Calculate } D = \| (RoCj - RoCi) \| - | (RoCk - RoCj) \| \qquad (1)$$

To generate nearest and optimized value, some simple arithmetic operations are performed [19]. F_{val} is defuzzified value obtained in step 6 of proposed algorithm in section 3.3. X_i, Y_i, Z_i and W_i are positive added fraction in F_{val} of D, D/2, D/4 and D/8 respectively. XX_i, YY_i, ZZ_i and WW_i are positive subtracted fraction in F_{val} of D, D/2, D/4 and D/8 respectively.

Notations used in the procedure

[*Fj] is corresponding interval uj for which membership in Fj is Supremum (i.e. 1).

L[*Fj] is the lower bound of interval uj.

U[*Fj] is the upper bound of interval uj.

l[*Fj] is the length of the interval uj whose membership in Fj is Supremum (i.e. 1).

M[*Fj] is the midvalue of the interval uj having Supremum value in Fj

Initialize Count=1 and Target = F_{val}

---

i. $X_i = F_{val} + D$, if $X_i \leq L$ [*Fj] and $X_i \geq U$ [*Fj] then Target= X_i and Count=Count+1

ii. $XX_i = F_{val} - D$, if $XX_i \leq L$ [*Fj] and $XX_i \geq U$ [*Fj] then Target=Target + XX_i and Count=Count+1

iii. $Y_i = F_{val} + D/2$, if $Y_i \leq L$ [*Fj] and $Y_i \geq U$ [*Fj] then Target= Target + Y_i and Count=Count+1

iv. $YY_i = F_{val} - D/2$, if $YY_i \leq L$ [*Fj] and $YY_i \geq U$ [*Fj] then Target=Target + YY_i and Count=Count+1

v. $Z_i = F_{val} + D/4$, if $Z_i \leq L$ [*Fj] and $Z_i \geq U$ [*Fj] then Target= Target + Z_i and Count=Count+1

vi. $ZZ_i = F_{val} - D/4$, if $ZZ_i \leq L$ [*Fj] and $ZZ_i \geq U$ [*Fj] then Target= Target + ZZ_i and Count=Count+1

vii. $W_i = F_{val} + D/8$, if $W_i \leq L$ [*Fj] and $W_i \geq U$ [*Fj] then Target= Target + W_i and Count=Count+1

viii. $WW_i = F_{val} - D/8$, if $WW_i \leq L[*Fj]$ and $WW_i \geq U[*Fj]$ then Target= Target+WW_i and Count=Count+1

ix. $FRoC_j = Target/Count$ (FRoC is optimized forecasted RoC at time y)

x. Return $FRoC_i$

3.2 Basic Definitions Used in Algorithm

Fuzzy Set: A fuzzy set is a pair (A, m) where A is a set and m:A$\rightarrow$ |0,1|

For a finite set A={x_1, ..., x_n}, the fuzzy set (A, m) is often denoted by {($m(x_1)/x_1$), ..., ($m(x_n) / x_n$)}. For each, x $\in$ A, m(x) is called the grade of membership of x in (A, m).

Let x $\in$ A. Then x is not included in the fuzzy set (A, m) if m(x)=0, x is fully included if m(x)=1, and x is called fuzzy member if 0 < m(x) < 1.

The set x $\in$ A | m(x)>0 is called the support of (A, m) and the set x $\in$ A | m(x)=0 is called its kernel.

Time Series: A series of observations made sequentially in time constitute. In time domain analysis, a time series is represented by a mathematical model G(t) = O(t) + R(t), where O(t) represents a systematic or ordered part and R(t) represents a random part. The fact is that the two components cannot be observed separately and may involve several parameters.

Fuzzy Time Series: If fuzzy set F(t) is caused by more fuzzy sets; F(t-n), F(t-n+1)......... F(t-1), the fuzzy relationship is represented by A_{i1}, A_{i2} ....A_{in} → A_j, here F(t-n)= A_{i1}, F(t-n+1) = A_{i2} and so on F(t-1) = A_{in}. The relationship is called nth order fuzzy time series model.

Average Forecasting Error Rate: AFER can be defined as

$$AFER = (\sum_{t=1}^{n} (| A_t - F_t | / A_t)) / n * 100\%$$ (2)

where, A_t is actual value and F_t is forecasted value of time series data at time t.

Mean Square Error: MSE can be defined as

$$MSE = \sum_{t=1}^{n} (A_t - F_t)^2 / n$$ (3)

where, A_t is actual value and F_t is forecasted value of time series data at time t n is total number of time series data.

3.3 Algorithm Steps

Step 1: Event discretization function is calculated for given time series data t=1 to n:

Step 2: Define the universe of discourse on RoC as U and partition it into equal intervals say; u_1, u_2, u_3, u_4.................... u_n of equal lengths.

Step 3: Call frequency density distribution procedure.

Step 4: Define each fuzzy set F_i based on the re-divided intervals and fuzzify the time series data where fuzzy set F_i denotes a linguistic value of the RoC represented by a fuzzy set as in [19]. We use a triangular membership function to define the fuzzy sets F_i [20].

Step 5: After fuzzification of historical data, establish the fuzzy logic relationships using rule:

Rule: If Fj is the fuzzy production at time period n, Fi is the fuzzify production at time period n -1 and F_k is the fuzzify production at time period n +1 then the fuzzy logical relation is denoted as Fi→Fj and Fj→F_k. Here, Fj is called current state, Fi is the previous state and F_k is next state. RoC_j is percentage change at time frame n, RoC_k is percentage change at time frame n-1 and ROC_k is percentage change at time frame n+1.

Step 6: Let us assume the fuzzify value of RoC at particular time period is F_j, calculated in step5. Approximate value of RoC can be generated at same time j using defuzzification formula[20]. In this formula f_{j-1}, f_j, f_{j+1} are the mid points of the fuzzy intervals F_{j-1}, F_j, F_{j+1} respectively. F_{val} is defuzzify value of F_j. Above formula fulfills the Axioms of Fuzzy set like monotonicity, boundary condition, continuity and idempotency.

Step 7: Optimized forecasting of data F_j for the time period n and onwards is done as follows: For y=2 to ... Y (end of time series data). Call optimization fuzzy relationship procedure

Step 8: Calculate Forecasted value as:

$$Forecast_{val} = (x(t)_y * FRoC_{y+1}) + x(t)_y \qquad (4)$$

where, y=1 to n-1. Here, $Forecast_{val}$ is forecasted value at time y+1. $x(t)_y$ is the actual value at time y and $FRoC_{y+1}$ is corresponding value obtained at step7.

4   Experimental Results

Dynamic computations of above proposed method is being implemented on the time series data of enrollments of University at Alabama. Stepwise results obtained are as follows:

Step 1: Histological Data is given year wise. Calculate RoC_{t+1} of year 1971 and onwards.

Step 2: Define the universe of discourse U and partition it into intervals $u1$, $u2$... un of equal length. The ROC of enrollment from year to year is given in Table 1 and ranges from -5.83% to 7.66%. For example, take the universe of discourse to be U = [-6, 8] and partition U into seven equal intervals.

Step 3: The weighted frequency density based distribution and partitioning of ROC is given in Table 2.

Table 1. RoC of Every Year

Year	Enroll	RoC$_{t+1}$	Year	Enroll	RoC$_{t+1}$
1971	13055				
1972	13563	3.89%	1983	15497	0.41%
1973	13867	2.24%	1984	15145	-2.27%
1974	14696	5.98%	1985	15163	0.12%
1975	15460	5.20%	1986	15984	5.41%
1976	15311	-0.96%	1987	16859	5.47%
1977	15603	1.91%	1988	18150	7.66%
1978	15861	1.65%	1989	18970	4.52%
1979	16807	5.96%	1990	19328	1.89%
1980	16919	0.67%	1991	19337	0.05%
1981	16388	-3.14%	1992	18876	-2.38%
1982	15433	-5.83%			

Table 2. Frequency Distribution

Intervals	Number of Data
[−6.0,−4.0]	1
[−4.0,−2.]	3
[−2.0,0.00]	1
[0.00,2.00]	7
[2.00,4.00]	2
[4.00,6.00]	6
[6.00,8.00]	1

Table 3. Fuzzification of Interval

Linguistic	Intervals
F_1	[−6.0,−4.0]
F_2	[−4.0,−2.0]
F_3	[−2.0,0.00]
F_4	[0.00.0.5]
F_5	[0.50,1]
F_6	[1.00,1.5]
F_7	[1.50,2]
F_8	[2.00,]
F_9	[3.00,4]
F_{10}	[4.00,4.67]
F_{11}	[4.67,5.33]
F_{12}	[5.33,6]
F_{13}	[6.00,8]

Step 4: Define each fuzzy set F_i on the re-divided intervals and fuzzify the time series data where fuzzy set F_i denotes a linguistic value of the RoC represented by a fuzzy set is shown in Table 3.

Step 5: Based on the fuzzify RoC obtained in step 4, derived fuzzy logical relationship rules is shown in Table 4.

Table 4. Fuzzy Relationships

Year	RoC$_{t+1}$	Fuzzy	FLR	Year	RoC$_{t+1}$	Fuzzy	FLR
1971							
1972	3.89%	F_9	$F_9{\rightarrow}F_8$	1983	0.41%	F_4	$F_4{\rightarrow}F_1$, $F_4{\rightarrow}F_2$,
1973	2.24%	F_8	$F_8{\rightarrow}F_9$, $F_8{\rightarrow}F_{12}$,	1984	-2.27%	F_2	$F_2{\rightarrow}F_4$, $F_2{\rightarrow}F_4$,
1974	5.98%	F_{12}	$F_{12}{\rightarrow}F_{12}$, $F_{11}{\rightarrow}F_3$,	1985	0.12%	F_4	$F_4{\rightarrow}F_2$, $F_4{\rightarrow}F_{12}$,
1975	5.20%	F_{11}	$F_{11}{\rightarrow}F_{12}$, $F_{11}{\rightarrow}F_3$,	1986	5.41%	F_{12}	$F_{12}{\rightarrow}F_4$, $F_{12}{\rightarrow}F_{12}$,
1976	-0.96%	F_3	$F_3{\rightarrow}F_3$, $F_3{\rightarrow}F_7$,	1987	5.47%	F_{12}	$F_{12}{\rightarrow}F_{12}$, $F_{12}{\rightarrow}F_{13}$,
1977	1.91%	F_7	$F_7{\rightarrow}F_3$, $F_7{\rightarrow}F_7$,	1988	7.66%	F_{13}	$F_{13}{\rightarrow}F_{12}$, $F_{13}{\rightarrow}F_{10}$,
1978	1.65%	F_7	$F_7{\rightarrow}F_7$, $F_7{\rightarrow}F_{12}$,	1989	4.52%	F_{10}	$F_{10}{\rightarrow}F_{13}$, $F_{10}{\rightarrow}F_7$,
1979	5.96%	F_{12}	$F_{12}{\rightarrow}F_7$, $F_{12}{\rightarrow}F_5$,	1990	1.89%	F_7	$F_7{\rightarrow}F_{10}$, $F_7{\rightarrow}F_4$
1980	0.67%	F_5	$F_5{\rightarrow}F_{12}$, $F_5{\rightarrow}F_2$,	1991	0.05%	F_4	$F_4{\rightarrow}F_7$, $F_4{\rightarrow}F_2$,
1981	-3.14%	F_2	$F_2{\rightarrow}F_5$, $F_2{\rightarrow}F_1$,	1992	-2.38%	F_2	$F_2{\rightarrow}F_4$,,
1982	-5.83%	F_1	$F_1{\rightarrow}F_2$, $F_1{\rightarrow}F_4$,				

Table 5. Defuzzification of RoC

Year	Enroll	Fuzzy	F_{val}	Year	Enroll	Fuzzy	F_{val}
1971	13055						
1972	13563	F_9	3.33%	1983	15497	F_4	0.48%
1973	13867	F_8	2.67%	1984	15145	F_2	-2.14%
1974	14696	F_{12}	5.76%	1985	15163	F_4	0.48%
1975	15460	F_{11}	4.96%	1986	15984	F_{12}	5.76%
1976	15311	F_3	2.38%	1987	16859	F_{12}	5.76%
1977	15603	F_7	1.71%	1988	18150	F_{13}	6.49%
1978	15861	F_7	1.71%	1989	18970	F_{10}	4.22%
1979	16807	F_{12}	5.76%	1990	19328	F_7	1.71%
1980	16919	F_5	54.00%	1991	19337	F_4	0.48%
1981	16388	F_2	-2.14%	1992	18876	F_2	-2.41%
1982	15433	F_1	-4.10%				

Step 6: Defuzzified value in terms of RoC is presented in Table 5.

Step 7 and Step 8: Forecasted value of time series data is shown in Table 6, along with MSE and AFER.

Table 6. Calculations for forecasting, MSE and AFER

Year	Enroll.	F_{val}	FRoC	Fore. Enroll	$(A_i - F_i)^2$	$|A_i - F_i|/A_i$
1971	13055					
1972	13563	3.33%	3.33%	13489.732	5368.4929	0.0054021
1973	13867	2.67%	2.49%	13900.719	1136.95073	-0.0024316
1974	14696	5.76%	5.49%	14628.298	4583.52018	0.0046068
1975	15460	4.96%	4.96%	15424.922	1230.49415	0.002269
1976	15311	2.38%	0.65%	15560.49	62245.2601	-0.0162948
1977	15603	1.71%	1.38%	15522.292	6513.81355	0.0051726
1978	15861	1.71%	1.20%	15790.236	5007.5437	0.0044615
1979	16807	5.76%	5.67%	16760.319	2179.14377	0.0027775
1980	16919	0.54%	0.82%	16944.817	666.538143	-0.0015259
1981	16388	-2.14%	-2.66%	16468.955	6553.64726	-0.0049399
1982	15433	-4.10%	-4.68%	15621.042	35359.6433	-0.0121844
1983	15497	0.48%	0.30%	15479.299	313.325401	0.0011422
1984	15145	-2.14%	-2.36%	15131.271	188.490933	0.0009065
1985	15163	0.48%	0.12%	15163.174	0.030276	-1.148E-05
1986	15984	5.76%	5.76%	16036.389	2744.58637	-0.0032776

Table 6. (*continued*)

1987	16859	5.76%	5.49%	16861.522	6.35846656	-0.0001496
1988	18150	6.49%	6.16%	17897.514	63748.9782	0.0139111
1989	18970	4.22%	4.48%	18963.12	47.3344	0.0003627
1990	19328	1.71%	1.71%	19294.387	1129.83377	0.0017391
1991	19337	0.48%	-0.45%	19241.024	9211.39258	0.0049633
1992	18876	-2.41%	-2.41%	18870.978	25.2174709	0.000266
					9917.17122	0.34%

5   Performance Evaluation and Comparison

In this section, we evaluated the forecasting performance of our proposed model on enrollment data of University of Alabama and compared the results with previous selective models[6,9,11,17,18,19,20]. All previous models used same enrollment data set as benchmark The Forecasting accuracy is measured in terms of mean square error (MSE) and average forecasting error rate (AFER). Lower value of MSE and AFER are measure of higher forecasting accuracy. It can be observed from Table 7 that the obtained value of MSE and AFER is lowest in case of proposed model. The comparative study of MSE, AFER and the forecasted values obtained by our designed model clearly indicates its superiority over already existing soft computing time series models.

Table 7. Actual vs. Forecasted Enrollment of Models

Year	Song Chissom	Chen	Hwang Chen & Lee	Jilani, Burney & Ardi	Hur-ang	Singh	Jilani & Burney	Jilani, Burney & Ardi -adv	Mered-ith & John	Proposed Method
1971	–	–	–	14464	–		–	13579	–	
1972	14000	14000	–	14464	14000		–	13798	13410	13489.732
1973	14000	14000	–	14464	14000		–	13798	13932	13900.719
1974	14000	14000	–	14710	14000	14286	14730	14452	14664	14628.298
1975	15500	15500	–	15606	15500	15361	15615	15373	15423	15424.922
1976	16000	16000	16260	15606	15500	15468	15614	15373	15847	15560.49
1977	16000	16000	15511	15606	16000	15512	15611	15623	15580	15522.292
1978	16000	16000	16003	15606	16000	15582	15611	15883	15877	15790.236
1979	16000	16000	16261	16470	16000	16500	16484	17079	16773	16760.319
1980	16813	16833	17407	16470	17500	16361	16476	17079	16897	16944.817
1981	16813	16833	17119	16470	16000	16362	16469	16497	16341	16468.955
1982	16789	16833	16188	15606	16000	15735	15609	15737	15671	15621.042
1983	16000	16000	14833	15606	16000	15446	15614	15737	15507	15479.299
1984	16000	16000	15497	15606	15500	15498	15612	15024	15200	15131.271
1985	16000	16000	14745	15606	16000	15306	15609	15024	15218	15163.174
1986	16000	16000	15163	15606	16000	15442	15606	15883	16035	16036.389
1987	16000	16000	16384	16470	16000	16558	16477	17079	16903	16861.522

Table 7. (*continued*)

1988	16813	16833	17659	18473	17500	18500	18482	17991	17953	17897.514
1989	19000	19000	19150	18473	19000	18475	18481	18802	18879	18963.12
1990	19000	19000	19770	19155	19000	19382	19158	18994	19303	19294.387
1991	19000	19000	19928	19155	19500	19487	19155	18994	19432	19241.024
1992	–	19000	15837	18473	19000	18744	18475	18916	18966	18870.978
MSE	775687	321418	226611	227194	86694	90997	82269	41426	21575	9917.16
AFER	4.38%	3.12%	2.45%	2.39%	1.53%	1.53%	1.41%	1.02%	0.57%	0.34%

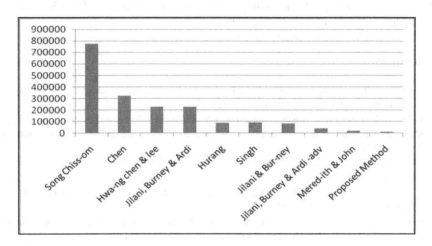

Fig. 1. Comparitive Study of MSE

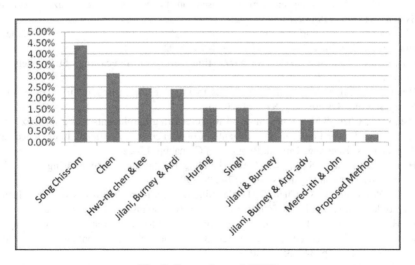

Fig. 2. Comparison of AFER

6 Conclusions

New technique introduced achieves the best accuracy having least mean square error among all related work in the field of forecasting till date. Pioneered dynamic computational method can be employed as an accurate and reliable means for estimating and predicting data. Presented model can also be considered as a strong standard methodology for better resource allocation, planning and management in Institutes. Proposed model pays special attention to dynamic computations. In this study, we have proved that joint consideration of suggested factors significantly improved the accuracy of forecasting model. In future, proposed model can be extended to optimize with genetic algorithm.

References

1. Garg, B., Beg, M.M.S., Ansari, A.Q.: Inferential historical survey of time series predication using artificial neural network (2010)
2. Zadeh, L.A.: Fuzzy sets. Information and Control 8(3), 333–353 (1965)
3. Song, Q., Chissom, B.S.: Fuzzy time series and its models. Fuzzy Sets and Systems 54, 269–277 (1993)
4. Song, Q., Chissom, B.S.: Forecasting enrollments with fuzzy time series Part I. Fuzzy Sets and Systems 54, 1–9
5. Song, Q., Chissom, B.S.: Forecasting enrollments with fuzzy time series: Part II. Fuzzy Sets and Systems 62, 1–8 (1994)
6. Song, Q.: A note on fuzzy time series model selection with sample autocorrelation functions. Cybernetics and Systems: An International Journal 34, 93–107 (2003)
7. Chen, S.M., Hsu, C.-C.: A new method to forecasting enrollments using fuzzy time series. International Journal of Applied Science and Engineering 2(3), 234–244 (2004)
8. Chen, S.M.: Forecasting enrollments based on fuzzy time series. Fuzzy Sets and Systems 81, 311–319 (1996)
9. Chen, S.M.: Forecasting enrollments based on high-order fuzzy time series. Cybernetics and Systems: An International Journal 33, 1–16 (2002)
10. Huarng, K.: Heuristic models of fuzzy time series for forecasting. Fuzzy Sets and Systems 123, 369–386 (2002)
11. Huarng, K.: Effective lengths of intervals to improve forecasting in fuzzy time series. Fuzzy Sets and Systems 12, 387–394 (2001)
12. Hsu, C.C., Chen, S.M.: A new method for forecasting enrollments based on fuzzy time series. In: Proceedings of the Seventh Conference on Artificial Intelligence and Applications, Taichung, Taiwan, Republic of China, pp. 17–22
13. Hwang, J.R., Chen, S.M., Lee, C.H.: Handling forecasting problems using fuzzy time series. Fuzzy Sets and Systems 100, 217–228 (1998)
14. Lee, L.W., Wang, L.W., Chen, S.M.: Handling forecasting problems based on two-factors high-order time series. IEEE Transactions on Fuzzy Systems 14(3), 468–477 (2006)
15. Li, H., Kozma, R.: A dynamic neural network method for time series prediction using the KIII model. In: Proceedings of the 2003 International Joint Conference on Neural Networks, vol. 1, pp. 347–352 (2003)
16. Melike, S., Degtiarev, K.Y.: Forecasting enrollment model based on first-order fuzzy time series. In: Proceedings of World Academy of Science, Engineering and Technology, vol. 1, pp. 1307–6884 (2005)

17. Jilani, T.A., Burney, S.M.A., Ardil, C.: Fuzzy metric approach for fuzzy time series forecasting based on frequency density based partitioning. In: Proceedings of World Academy of Science, Engineering and Technolog, vol. 23, pp. 333–338 (2007)
18. Jilani, T.A., Burney, S.M.A., Ardil, C.: Multivariate high order fuzzy time series forecasting for car road accidents. International Journal of Computational Intelligence 4(1), 15–20 (2007)
19. Singh, S.R.: A computational method of forecasting based on fuzzy time series. International Journal of Mathematics and Computers in Simulation 79, 539–554 (2008)
20. Stevenson, M., Porter, J.E.: Fuzzy time series forecasting using percentage change as the universe of discourse. In: Proceedings of World Academy of Science, Engineering and Technology, vol. 55, pp. 154–157 (2009)

A Novel Relation-Based Probability Algorithm for Page Ranking in Semantic Web Search Engine

C. Deisy, A.M. Rajeswari, R.M. Indra,
N. Jayalakshmi, and P.K. Mehalaa Devi

Department of Computer Science and Engineering,
Thiagarajar College of Engineering, Madurai, TamilNadu
{cdcse,amrcse}@tce.edu,
{indrarm31,jayalakshmi.jayanthi,magipk}@gmail.com

Abstract. Web search engines are essential nowadays as if we, the end users want to know or gather certain information on any field is done by web. It's a common tendency that people don't prefer waiting for an information or searching all down many pages to obtain certain information. But as the information growth is tremendous the result set obtained by the search engines provide a burden of useless pages. It is even more irritable for the end users when pages with useless contents are ranked above. To resolve this, in this paper we have proposed a relation-based probability algorithm for page ranking for the semantic web search engines. To provide the necessary information to the end user the relevance of the pages must be obtained. Relevancy is measured as the probability that a retrieved resources actually contains the information needed by the user.

Keywords: Semantic web, relevancy factor, query formulation, normalised relevancy factor.

1 Introduction

In the last few years with the massive growth of the world wide web, there exists a huge information explosion that are accessible to the internet users. Nevertheless, due to the same fact of increased information in the web it is very difficult to obtain the needed results through a search engine just by specifying the keywords. Although search engine serves to be the most important tool used to organize and extract information from the web, it is not uncommon that most renowned web search engines return result set that do not satisfy the user needs. It is because of the fact that the search engines rely mainly upon the presence of the keywords (that were mentioned in the query) in the page. Also, it is worth noting that some statistical algorithms have been applied to tune the result set that is returned from the basic keyword search algorithm. This type of algorithms is mainly employed to rank pages based on the relevancy that exist between the user specified facts and the one that exist in the pages. But in some cases these algorithms do not suffice.

S. Dua, S. Sahni, and D.P. Goyal (Eds.): ICISTM 2011, CCIS 141, pp. 138–148, 2011.

Better results can be obtained with the use of the semantic web where in each page would possess semantic meta data that would provide additional information about the page itself. The description that is provided in each page would be define a set of concepts and a set of relations that exist among them [1].

This paper provides an insight into how these relations can be exploited to define a ranking strategy for semantic web search engines which serves as the enhancement to [1].This ranking relies mainly upon the underlying ontology, user specified queries and also the pages that are to be ranked which makes it less complex to rank pages when compared with other similar algorithms.

In Section 2 , we provide a brief introduction about the existing algorithms and the problems faced on following it; Section 3 deals with the basic idea that is being proposed; Section 4 deals with the methodology that is used to arrive at the formula . Experimental results are dealt in the last section.

2 Related Works in Web Search

There have been a good number of algorithm that has been proposed for increasing the search efficiency both with relevancy as the factor that decides and also the time complexity. Some of the proposals has been listed below.

A ranking methodology based on the hierarchy that exist between pages has been proposed in [2]. It assigns a page keyword rank for each page based on keyword extraction. Both the page keyword rank and the level value is used for ranking the pages. The applicability of this method for all applications has not been dealt with the author.

A solution for ranking pages based on customer settings has also been proposed [3]. In this algorithm the author proposes a solution that makes use of certain ranking factors that specify document attributes. Based on these weighted attributes and the user query ranking, a relevant value is given for each document and thereby exploits user configurations. The generality of such method might be a bit difficult.

An enhancement to the existing page ranking algorithm that provides preference to aged web documents has also been proposed in [4]. The modification here is that the author of this paper proposes a solution that takes into account both the timestamp and link that exists among pages which ensure unbiased ranking. This algorithm is mainly based upon the modification time returned by the http response for ranking pages. To what extent this modification time factor would be reliable is a matter of concern here. In the field of semantic web search engines various researchers [5]-[8] proposed different   idea to exploit relations that exist between concepts for a particular field and ranks pages based on the relevancy that exist among the page to be ranked with a reference graph that defines all concepts and relations that exist between them in a particular field. As an enhancement to the above algorithm we propose a solution that takes into account not only the relevancy factor that is determined as the prime factor for ranking but also the hit count of the words given in the query.

3 Overview of the Ranking Strategy

The basic idea behind our ranking strategy is discussed here. The overall architecture of the prototypal search environment used is presented first.

3.1 Relation-Based Search Engine Components

A controlled Semantic Web Environment is constructed by the well known travel.owl ontology written in OWL language[1] after certain modifications to make it more suitable for demonstrating system functionality. Certain web pages in the field of tourism are downloaded and RDF is embedded into them based on the ontology above. This forms the knowledge data base. A query interface and the true search engine module embedding our ranking logic is created. A graphics User Interface allows for the definition of a query , which is passed onto the relation-based search logic. Finally the ordered result set is presented to the user. The ranking logic is explained in the following sections.

3.2 Introduction to Relation Based Ranking

Let us see an example of how the proposed algorithm works and in what way is this algorithm different from other such algorithm. The information presents below is with reference to [1]. Let us assume that the user specifies the keyword "hotel" and the keyword "Rome" in the query interface. The associated concepts with the keyword will be "destination" for Rome and "accommodation" for hotel. But the exact relation that is in the user's mind between the two concepts cannot be predicted easily but we can be sure that there would be at least one relation that exist between the two. As a first step all annotated pages containing the keywords are chosen followed by analyzing the relation that exist between the concepts that is given in the description is analyzed. The pages are given weight based on the relations that are present between the concepts related to the keywords.

3.3 Basic Idea

After identifying the number of relations that exist between the concepts , the larger is the number of relations linking each concept with other concept given the total number of relations that exist between the concepts in the ontology the higher is the probability that this page exactly contains those relations that are of interest to the user and as a consequence, that this page is actually the most relevant to the user query[1],[9-11]. This probability measure can be calculated easily with the help of the graph-based description of the ontology (ontology-graph), of the query (query subgraph) and of each annotated page containing queried concepts. In the following, the ontology graph, query subgraph, annotation graph and page sub graph notions is explained with examples in the next section.

3.4 Graph-Based Notation and Methodology

In all graphs, the concepts are represented as queries and the relations that exist among the concepts are treated as edges. The ontology graph is shown in Fig. 1, which is a part of the tourism.owl file with certain modifications to well suit the application. Examples of annotation graph with the keywords and concept that exist is shown in Fig. 2. In a query subgraph , nodes are represented by concepts that has been specified in the query. Nodes/concepts are linked by edge only if there exist at least one relation among those concepts described in the ontology[12-16]. Similarly, the page subgraph is built based on the annotation associated to the page itself.

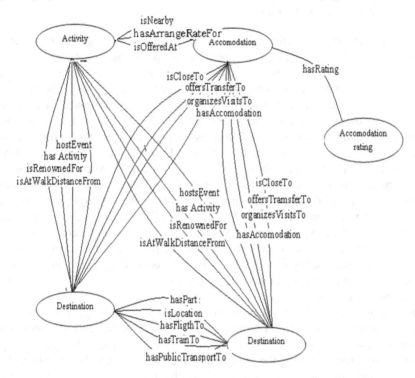

Fig. 1. Part of the Ontology graph of travel owl [1]

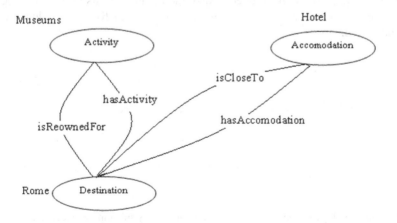

Fig. 2. Annotation Graph for Page 1 Given the Keywords Rome, Museum, and Hotel

4   Relation-Based Ranking Formal Model

A formal model for the proposed ranking strategy will be provided, taking into consideration all the critical situations that could be envisioned.

4.1   Graph-Based Formalization

Starting from the ontology defined for a domain, a graph based representation can be designed where OWL classes are mapped into graph vertices and OWL relation properties are the graph edges. Thus, the existing relations between couples of concepts in the domain are the connected vertices in the graph. This is the ontology graph G and can be defined $G(C,R)$ where C is the set of concepts and R is the set of Relations. An ontology graph is illustrated in Fig. 3a. Similarly, a keyword will contain a collection of keywords and their associated concepts. So a query can be represented as $Q = \{(k,c)\}$ where k is the keywords and c is the related concepts. Given particular query the query subgraph $Gq(Cq,Rq)$ where Cq is the set of concepts related to the query and Rq is the set of relations that exist between the concepts mentioned in the query. The query subgraph for the ontology graph given in Fig. 3a with only the concepts c1and c2 are shown in Fig. 3b. Similarly, based on the annotations in each page graph that depicts the relation between the concepts in that page can be created for each page which forms the page subgraph and is shown in Fig. 3c and Fig. 3d. The page subgraph can be denoted as $Gq,p(Cq,p , Rq,p)$ where Cq,p are the set of concepts that exist in page p and Rq,p are the relations that exist between the concepts of that page.

The aim of this paper is to demonstrate that given an ontology graph G and a query subgraph Gq, is to define a ranking strategy based on the relevance score that is calculated using the concepts and the relations within that page. A page subgraph Gq,p is constructed exploiting the information available in the web annotation. The relevance score for each page is defined as the ratio between the number of relations that exist among the concepts in the page subgraph to that of the ontology graph with respect to the given query. Also, the number of hits of each word given in the query is calculated for each page and is normalised and added to the relevance score.

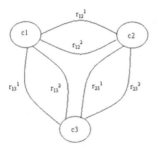

Fig. 3a. Ontology Graph

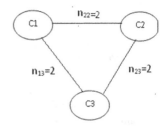

Fig. 3b. Query subgraph with c1, c2 and c3

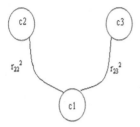

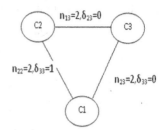

Fig. 3c. Page annotation Graph for Page1   **Fig. 3d.** Page1's Sub Graph

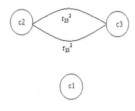

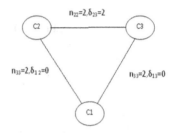

Fig. 3e. Page Annotation Graph for Page1   **Fig. 3f.** Page1's Sub Graph

4.2 Relevance and Semantic Relations

To start up with let us imagine that the user is interested in pages containing three generic keywords k1,k2,k3 related to the concepts c1,c2 and c3 respectively. Let us assume that there exists only two pages p1 and p2 containing all the keywords specified by the user . This represents the initial result set for the given query. The page subgraph and the annotation graph for page 1 is illustrated in Fig 3c and Fig 3d. The page subgraph and the annotation graph for page 2 is illustrated in Fig. 3e and Fig. 3f. For page p1, the concepts c2 and c3 are related to c1 through a single relation, while in the second page there exists two relations linking c3 to c1,with no link between c2 and c1. Since we cannot assume which could be the concepts or the relations more important with respect to user query, we can provide a significant measure of page relevance by computing the probability that a page is the one of interest to the user by calculating the probability that c2 is linked to c1 and c3 is related to c1 through the relations in the user's mind. The probability can be calculated as the $P(\ r_{ij},p)=\delta_{ij}/\eta_{ij} = \tau_{ij}$.We call this relational probability. Thus for the first page we have the $P(r_{12},p_1)=\delta_{12}/\eta_{12}=\tau_{12}=1/2$ and $P(r_{13},p_1)=\delta_{13}/\eta_{13}=\tau_{13}=1/2$. For the second page we have $P(r_{12},p_2)=\delta_{12}/\eta_{12}=\tau_{12}=0$ and $P(r_{13},p_2)=\delta_{13}/\eta_{13}=\tau_{13}=1$. Based on the considerations above we can compute the joint probability. $P(Q,p)=P((r_{12},p)\cap (r_{13},p))$. The dependency on Q is due to the fact that only concepts given in Q are taken into account. Since the events (r_{12},p) and (r_{13},p) are not correlated, $P(Q,p)$ can be rewritten as $P(Q,p)=P((r_{12},p) \cdot (r_{13},p))$. Thus, for the specific example being considered, it is $P(Q,p1)=1/4$ and $P(Q,p2)=0$,respectively, for the first and second page. This allows placing the first page before the second one in the ordered result set. Thus the final expression $P(Q,p)$ can be written as

$$P(Q,p)=P(r^P_{12}).P(r^P_{23})+P(r^P_{12}).P(r^P_{13})+P(r^P_{23}).P(r^P_{13}), \qquad (1)$$

144 C. Deisy et al.

According to the definition of relation probability, it is

$$P(Q,p)=[\tau_{12}.\tau_{23}+\tau_{12}.\tau_{13}+\tau_{23}.\tau_{13}] \tag{2}$$

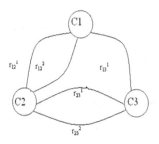

Fig. 4. Ontology Graph

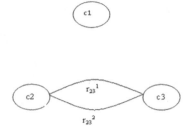

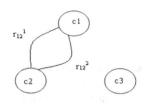

Fig. 5a. Annotation graph for page1

Fig. 5b. Annotation graph for page2 where c1 is not connected where c3 is not connected

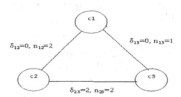

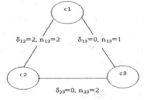

Fig. 5c. Page Sub Graph for Page1

Fig. 5d. Page Subgraph for Page 2

We consider again an example represented by two pages depicted in Fig.4 whose ontology graph is figure 5 , where concepts c1 (in the first page) and concepts c3 (in the second page) do not show any relations with the remaining concepts.. If we compute P(Q,p1) and P(Q,p2) using (2), we get a relevancy score equal to zero. Based on the relevancy score definition provided above, in order to find the score different than zero allowing each page to be ranked with respect to other pages, we have to relax the condition of having each concept related to each other concept. We provide a modification to rectify the zero relevancy score by using normalization of hit count which is then added to the relevancy score obtained from (1). The normalization of hit count is given by,

Normalization of hit count for a page = no of occurrences of each word / n (3)

Here we assume n to be 10,000 as of to provide a low normalization relevancy score for pages which do not contain the relation between specified concepts to that of the pages that contain the relation between specified concepts. Hence the normalized relevancy score $ps_{Q,p}$ is given by,

$$ps_{Q,p} = ([\tau_{12} \cdot \tau_{23} + \tau_{12} \cdot \tau_{13} + \tau_{23} \cdot \tau_{13}] + [\text{no of occurrences of each word of each page}/ n]) \tag{4}$$

which can also be represented as,

$$ps_{Q,p} = \text{Equation (2)} + \text{Equation (3)}.$$

The workflow for the above mentioned algorithm to rank web pages is given in Fig. 6.

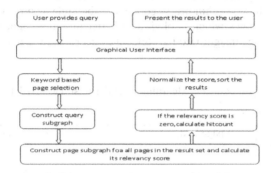

Fig. 6. Flow Chart from Query Definition to Presentation of Results

Table 1. Definition of Symbol

Symbol	Definition
$G(C,R)$	Ontology graph
$C=\{c_1,c_2,\ldots\ldots c_n\}$	Set of concepts constituting the vertices of the ontology
$R = \{R_{ij} \mid i=1,\ldots.,n, j=1,\ldots.,n,j>i\}$	Set of relations constituting the edges of the ontology graph
$R_{ij}=\{r_{ij}^{1},r_{ij}^{2},\ldots.,r_{ij}^{m},m<n\}$	Set of relations between concepts c_i and c_j in $G(C,R)$
$Q=\{(k_t,c_t)\}$	Query as a collection of pairs(keywords, concepts)
$G_Q(C_Q,R_Q)$	Query subgraph for Query Q
$C_Q=\{c_t \mid (k_t,c_t) \in Q\}$	Set of concepts constituting the query subgraph
$R_Q=\{R_{ij} \mid 1 \leq I,j \leq n,j>i\}$	Set of relations constituting the query subgraph
$R_{ij} = \{r_{ij} \mid c_i,c_j \in C_Q\}$	Set of relations between c_i and c_j in the query subgraph
$\eta_{ij} = \mid R_{ij}\mid$	Number of relations between c_i and c_j in $G_Q(C_Q,R_Q)$
$G_{Q,P}(C_{Q,P},R_{Q,P})$	Page subgraph for page p
Symbol	**Definition**
$C_{Q,P}= \{ct \mid ct \in C_Q\}$	Set of concept of page subgraph
$R_{Q,P}=\{r_{ij} \mid ci,c_j \in G_{Q,P}\}$	Set of relations of page subgraph
$\delta_{ij} = \mid R_{ij}\mid$	Number of relations in each page
$_{Ij} = \delta_{ij}/\eta_{ij}$	Relational probability for r_{ij} in page p for the given query Q

5 Experimental Results and Comparisons

In this section, the applicability of our technique into real scenarios will be analyzed by conducting two types of evaluations aimed at measuring the performance in terms of accuracy.

5.1 Accuracy

The accuracy of the proposed technique has been evaluated against the result set generated by running the query "hotel", "rome", "historicalcenter" over Google on 25 February 2010. Web pages returned by Google are reported in their original order in Table 2. Based on the algorithm proposed above , the table also provides the ranking for the same set of pages which is obviously better than the result obtained earlier. The table also provides insight into the Relevancy score and the normalised hit count of each page.

Table 2. Accuracy of Our Ranking Algorithm over the First Twenty Entries of a Result Set Generated by Google

Google Ranking	Our Ranking	Web page URL	Relevancy score	Normalised Relevancy Score
1.	3	http://www.starhotels.com/hotel/metro pole_roma/starhotels_metropole.php	0.2244	0.1230
2.	4	http://maps.google.co.in/maps	0.2244	0.1185
3.	19	http://www.expedia.co.uk/Capitoline-Museum-Hotels.0-l6074002-0.Travel-Guide-Filter-Hotels	0	0.0055
4.	7	http://www.hotelscombined.com/City/Rome.htm	0.0918	0.0537
5.	10	http://www.traveleurope.it/rome/rome_vatican_city.shtml	0.0204	0.0186
6.	14	http://www.hotelamalia.com/	0	0.0082
7.	5	http://www.expedia.com/Montemartini-Museum-Hotels.0-l6074048-0.Travel-Guide-Filter-Hotels	0.1632	0.0870
8.	6	http://www.nh-hotels.com/nh/en/hotels/italy/rome.html	0.1632	0.0861
9.	2	http://www.holidaycityeurope.com/rome-trevi/index.html	0.2244	0.1253
10.	15	http://www.starwoodhotels.com/sheraton/property/overview/index.html?propertyID=497	0	0.0078

Table 2. (*continued*)

11.	20	http://www.hotelforumrome.com/	0	0.0028
12.	12	http://www.starhotels.com/hotel/metro pole_roma/starhotels_metropole.php?i dpag=186	0.0204	0.0172
13.	1	http://www.hotelscombined.com/City/ Rome.htm	0.3673	0.1967
14.	16	http://www.easytobook.com/en/italy/ rome/rome-hotels/	0	0.0064
15.	11	http://www.romaclick.com/	0.0204	0.0183
16.	8	http://www.romanholidays.com/ rome-map.htm	0.0918	0.0493
17.	18	http://www.albergodelsenato.it/	0	0.0061
18.	9	http://www.travel-to-italy.com/ rome/hotels/rome_hotels_3star.html	0.0918	0.0493
19.	17	http://www.scotthouse.com/	0	0.0061
20.	13	http://www.leonardihotels.com/index.jsp	0.0204	0.0134

6 Conclusions

The forthcoming web strategies in Semantic web would help in more effective searching without much of manual interference. In this paper, we propose a that can provide a normalized score for ranking the pages, by considering certain ontologies and annotations of web pages. Also the hits of the words given in the query that is in the page itself is also taken into consideration. The proposed algorithm is different from other semantic based algorithms as it does not take the entire knowledge base into account. Further enhancement includes support for scalability, support for multiple ontologies.

References

1. Lamberti, F., Sanna, A., Demartini, C.: A relation –based page rank algorithm for semantic web search engines. IEEE Transcations on Knowledge and data Engineering, 123–136 (2009)
2. Aleman-Meza, B., Halaschek, C., Arpinar, I., Sheth, A.: A context-aware semantic association ranking. In: Processions of the First International Workshop Semantic Web and Databases (SWDB 2003), pp. 33–50 (2003)
3. Anyanwu, K., Maduko, A., Sheth, A.: SemRank: ranking. In: The International Conference of World Wide Web (WWW 2005), pp. 117–127 (2005)

4. Baeza-Yates, V., Calderon-Benavides, L., Gonzalez-Caro, C.: The intention behind web queries. In: Crestani, F., Ferragina, P., Sanderson, M. (eds.) SPIRE 2006. LNCS, vol. 4209, pp. 98–109. Springer, Heidelberg (2006)
5. Berners-Lee, T., Hendler, J., Lassila, O.: The semantic web. Scientific American Journal (2001)
6. Brin, S., Page, L.: The anatomy of large-scale hypertextual web search engine. In: Proceedings of the Seventh International Conference of the World Wide Web (WWW 1998), pp. 107–117 (1998)
7. Cohen, S., Mamou, J., Kanza, Y., Sagiv, Y.: XSEarch: a semantic search engine for XML. In: The Proceedings of the Twenty-Ninth International Conference on Very Large Data Base, pp. 45–56 (2003)
8. Ding, L., Finin, T., Joshi, A., Pan, R., Cost, R.S., Peng, Y., Reddivari, P., Doshi, V., Sachs, J.: Swoogle: a search and metadata engine for the semantic web. In: The Proceedings of the Thirteenth ACM International Conference on Information and Knowledge Management (CIKM 2004), pp. 652–659 (2004)
9. Ding, L., Finin, T., Josh, A., Peng, Y., Pan, R., Reddivari, P.: Search on the semantic web. Computer 38(10), 62–69 (2005)
10. Ding, L., Kolari, P., Ding, Z., Avancha, S.: Using ontologies in the semantic web: a survey. Ontologies, 79–113 (2007)
11. Guha, R., McCool, R., Miller, E.: Semantic search. In: The Proceedings of the Twelfth International Conference on the World Wide Web (WWW 2003), pp. 700–709 (2003)
12. Gyongyi, S., Garcia-Molina, H.: Spam: its not just for inboxes anymore. Computer 38(10), 28–34 (2005)
13. Junghoo, C., Gracia-Molina, H., Page, L.: Efficient crawling through URL ordering. Computer Networks and ISDN Systems 30(1), 161–172 (1998)
14. Kapoor, S., Ramesh, H.: Algoritms for enumerating all spanning trees of undirected and weighted graphs. SIAM Journal of Computing 24, 247–265 (1995)
15. Lei, Y., Uren, V., Motta, E.: SemSearch: a search engine for the semantic web. In: Staab, S., Svátek, V. (eds.) EKAW 2006. LNCS (LNAI), vol. 4248, pp. 238–245. Springer, Heidelberg (2006)
16. Li, Y., Wang, Y., Huang, X.: A relation-based search engine in semantic web. IEEE Transactions on Knowledge and Data Engineering 19(2), 273–282 (2007)

SASM- An Approach towards Self-protection in Grid Computing

Inderpreet Chopra[1] and Maninder Singh[2]

[1] Research Scholar, Computer Science Department,
Thapar University, Patiala, India
[2] Associate Professor, Computer Science Department,
Thapar university, Patiala, India

Abstract. In recent years grid which facilitates the sharing and integration of large scale heterogenous resources has been recognized as the future framework of distributed computing. Modern software is plagued by security flaws at many levels. Implementation of an autonomic system provides an inherent self-managing capability to the system overcoming the shortcomings of the manual system. This paper presents a new self-protection model- SASM based upon few principles of genetics.

1 Introduction

The Grid is constantly growing and it is being used by more and more applications. However the grid services are increasing in complexity [4,5] in modern era, as these can vary over a huge number of parameters[3]. The complexity of today's distributed computing environment is such that the presence of bugs and security holes is unavoidable [13]. When a user's job executes, the job may require confidential message passing services [18].Hence, security is now a major concern for any IT infrastructure. Overall,most problems stem from the fact that human administrators are unable to cope up with the amount of work required to properly secure the computing infrastructure at the age of Internet [14]. Autonomic systems and self-managed environments provide security policies establishing promising trust between the server and the client. Hence, autonomic computing [7,8] presented and advocated by IBM, suggests a desirable solution to this problem [6].

Self-Protection enables the system with an ability to secure itself against attacks i.e. detect illegal activities and trigger counter measures in order to stop them. It also helps to overpower the limitations of manual management of the system i.e. reduced speed, increased chances of errors and unmanageability by human administrators.

This paper proposes a Self-Automated Security Model (SASM)- an agent enabling autonomic computing providing a promising solution to the system management troubles caused by increased complexity of large scale distributed systems. SASM is based upon few implications of the genetic algorithm. This new model adopts intelligent agents to dynamically organize system management with centralized control. Cooperative works are organized by dynamically

S. Dua, S. Sahni, and D.P. Goyal (Eds.): ICISTM 2011, CCIS 141, pp. 149–159, 2011.

associated relationships among autonomic elements (agents), including acquaintance, collaboration and notification.This self-organized model is more suitable for the grid environment as it is characterize by distributed, open and dynamic properties.

2   Self-protection

The main goal of self-protection system is to defend grid environment against malicious intentional actions, self- protecting systems would scan for suspicious activities and react accordingly without users being aware that such protection is in process [13]. Kind of vulnerabilities which need self-protection are: QoS Violation [16],DoS attacks [20],Insider Attacks[12], Remote to local(R2L)[18] and User to root (U2R). Some of the current self-protection techniques in use are:

- **Firewall**: Firewalls basically separate the protected network from the unprotected network [28] [29]. It screens and filters all connections coming from internet to the protected (corporate) network and vice versa, through a single concentrated security checkpoint.
- **Intrusion Detection System**: Intrusion detection systems (IDSs) often work as misuse detectors, where the packets in the monitored network are compared against a repository of signatures that define characteristics of an intrusion. Many grid based IDS systems have also been conceived, designed and implemented. [26] The basic components for grid based IDS systems include sensors which collects information to be further analysed by systems like SNORT [27].

Kenny and Coghlan [30] described a system SANTA-G (Grid enabled System Area Networks Trace Analysis) that allows querying of log files by using Relational Grid Monitoring Architecture (R-GMA). Their implementation queries Snort log files by using SQL. SANTA-G is composed of three elements: A Sensor, a QueryEngine and a Viewer GUI. Schulter et al. [31] describe different types of IDS systems. They proposed a high-level GIDS that utilizes functionality of lower-level HIDS and NIDS provided through standard inter-IDS communication. Michal Witold [32] in his thesis investigated the possibility of Grid-focused IDS, stressing on feature selection and performance of the system. Leu and Li [33] proposed Fault-tolerant Grid Intrusion Detection System (FGIDS) which exploits grids dynamic and abundant computing resources to detect malicious behaviors from a massive amount of network packets. In FGIDS, a detector can dynamically leave or join FGIDS anytime.

3   Self-Automated Security Model (SASM) for Grid

Considering the various problems and the related work that has already been done to avoid a compromise in the grid system services, no grid middleware intelligent enough to handle the failures and faults automatically has an yet been established. In an attempt to unify related researches on the issue a proposal for Self Protection- SASM has been given in this paper providing a way for automatically creating a security wall to protect against external attacks like DDoS.

3.1   SASM Architecture

SASM has the layered architecture based upon MVC- Model (Database layer), View (User Interface) and Controller (Business Logic).In SASM, there is an application layer (View), the agents and autonomic unit layer (Controller) and the database layer defines the model. In this section the details of SASM architecture are described.

Application Layer: It provides an interface for the user to interact with the grid environment and authenticates the originality of the user before letting the him access the grid. After authentication, it shows the different options to the user to submit, monitor and cancel the job.

Agents Layer: It consists of three types of agents and each perform their respective tasks. These are:

- *Filter Agent*: It helps to safeguard the system from unwanted requests and various security threats. Only the authenticated users can submit the jobs. These agents are configured to allow only a certain type of traffic to pass through them. Whenever they detect the illegal packets, they simply discard them. After all inspection is done, they publish the job request to the queues. For this agent, we use Snort to setup the distributed IDS [36][37]. Snort with its default settings for distributed network is configured. To make filter agent intelligent enough to handle new threats, we have used the Genetic Algorithm (GA) approach to define new rules. This is the new addition that we are doing in snort. Based upon the grid environment, through GA, new rules are generater in the snort rules database.

Genetic Algorithm: GA is based upon the principles of evolution and natural selection (Figure: 2). GA converts the problem in specific domain into model by using chromosome like data structure and evolve the chromosome using selection, crossing over (Figure: 1) and mutation [35]. The process of genetic algorithm usually begins with the population of chromosomes representing the gene pool. Different positions of each chromosomes are encoded as bits, characters or numbers and referred to as *Genes*. Genes can be randomly changed within a range during evolution of the species. An evaluation function is used to calculate the "goodness" of each chromosome. During evaluation, two basic operators- Crossing Over and Mutation are used, simulating the natural reproduction and mutation of species.

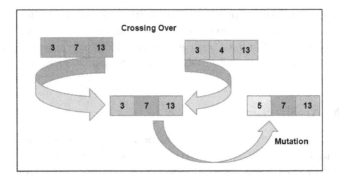

Fig. 1. Crossing Over and Mutation

In our model we use this approach for intelligently defining the new rules which are later transferred to various child resources attached to the central node (Figure: 2-b). Our filter agent GA for snort work like this:

```
SASM Filter Agent GA()
{
        Initialization;
        Evaluation;
        while termination criterion has not been reached
        {
                Selection and Reproduction;
                Crossover;
                Mutation;
                Evaluation;
        }
}
```

Initialization: It involves setting the parameters for the algorithm and creating the first generation of chromosomes. In this benchmark, each chromosome is represented by the set of genes. And the genes contains the description of the rules.

Let's take an example. Consider one sample rule *alert tcp any any->192.168.1.0/24 111* and it is represented as hexadecimal chromosome set (C0 A8 10 -1 18 -1 -1 6F)

Evaluation: This is accomplished by looking up the success rate of the rule representing the chromosome. If the rule is able to find the anomalous behaviour, its fitness is increased else decreased.

Selection and Reproduction: We use the simplest of all i.e. roulette wheel selection[34]. In roulette wheel selection, individuals are given a probability of being selected that is directly proportionate to their fitness. Two individuals are then chosen randomly based on these probabilities and their capability of producing offspring.

Crossover: In the crossover phase, all of the chromosomes with similar lengths are paired up, and with a probability they are crossed over. The crossover is

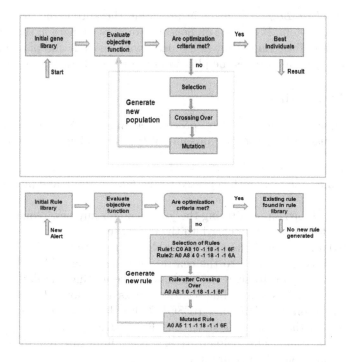

Fig. 2. Structure of GA [11] and GA Based Rule Generation

accomplished by randomly choosing a site along the length of the chromosome, and exchanging the genes of the two chromosomes for each gene past this crossover site. Say if we have the two parent chromosomes represented by (C0 A8 1 0 -1 18 -1 -1 6F) and (A0 A8 4 0 -1 18 -1 -1 6A), then the child chromosome can be (A0 A8 1 0 -1 18 -1 -1 6F)

Mutation: After the crossover, for each of the genes of the chromosomes, the gene can be mutated. When the crossover and mutations are completed, the chromosomes are once again evaluated for another round of selection and reproduction.

– *Data Agent*: Picks the job requests and related information, storing data into the database. Later this information is used for execution based upon the priority. Also these agents are responsible for data transfer from the user machine to the centralized server, so that the execution unit can pick up the data from the one single location. For data transfer, ftp protocol is configured that requests the users to upload the necessary data that their job will require for execution. The packets while transfer have to pass through the filter agent that keeps on scanning each packet for any harmful content.

– *Monitor Agent*: It keeps on monitoring the participating nodes in the grid, updating the resources in the database. It is also featured to pick information from the scheduler rather than the individual head nodes.

Autonomic Unit Layer: It helps in providing the autonomic capabilities to self-protect the system. It consists of: Sensor- monitors changes in the jobs or resources database, Processing Unit (PU)- deals with analysis and planning of job execution and Actuator- takes the plan from the PU and starts work execution.

3.2   SASM Working

Figure: 3 describes the complete process behind the SASM working and Figure: 4 discusses the basic flow in some of the cases. The basic process is as follows:

1. User submits the job which is passed to the *filtering unit*, and published into queues.
2. *Data agent* keeps on monitoring the job queue for any new requests, fetches the job details through fetcher into the database.
3. The job data is pushed into the JOB table. Along with this, the data needed by this job is transferred to the central location using FTP. Filtering unit plays its role again to scan the packets before getting the data to shared location. This helps in avoiding security checks again for different user data.
4. Sensor monitors the JOB and RESOURCES database and passes the information to the processing unit. The RESOURCES table is updated by the *monitor agent*. This helps to reduce the failure rate due to invalid old information about the resources.
5. *Processing unit* analyzes the information provided by sensor, and publishes the plan describing which jobs execution. This plan is shared to Actuator by the processing unit.
6. The *Actuator*, based upon the plan, starts passing the jobs to the meta scheduler (not shown in figure).

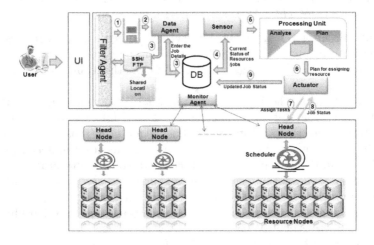

Fig. 3. SASM Detail Working

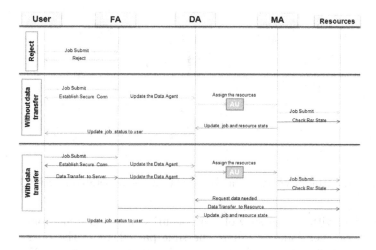

Fig. 4. SASM Work Flow

7. *Meta scheduler* then schedules the jobs to different head nodes (GRAM nodes in case of globus).
8. Actuator keeps monitoring the job status from the scheduler and updates the status into the JOB table.

4   Experimental Results

The SASM was implemented using the Snort. Snort is used to leverage existing distributed IDS capability to one step ahead by automating its rules generation process by the use of Genetic Algorithm. For testing the SASM in grid environment, we installed snort sensors on different resource nodes participating in the Grid Environmnet, then installed the database(MySql) on one machine, and deployed a Web server and ACID onto another machine. Figure 5-a is a graphical representation of the number of alerts raised with passage of time with default configurations.

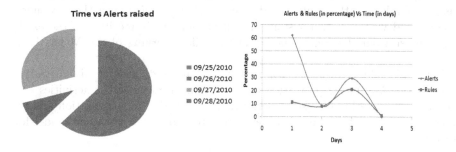

Fig. 5. Time vs Alerts Raised and Alerts & Rules (in percentage) Vs Time (in days)

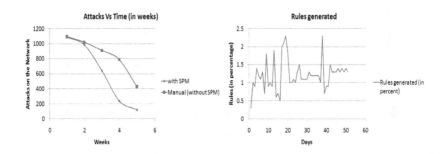

Fig. 6. Attacks Vs Time (in Weeks) and Rules Generated by SASM

We see that with the passage of time, security of the grid environment with SASM keep on increasing. As more the time passed, more the rules are auto generated with help of genetic features- crossing over and mutation. We observed the SASM for the next few days after its setup in grid environment and found that based upon the alerts raised, rules keep on adding into the rule database (Figure 5-b).

As the rule database grow, more attacks are handled by the SASM. We have found that with respect to the manual system for configuring rules, our SASM reduces the attacks rate (Figure 6-a).

5   Conclusions

In this paper a description of a new approach- SASM has been proposed to deal with failures and security threats with least manual administrative interference. This new approach has layered architecture comprising- Application Layer, Agent Layer and Autonomic Layer. SASM uses the Genetic Algorithm approach to provide the self-protection. The advantage of GA approach over existing approaches is that, SASM makes the system more robust.

As central node transfers rules to all the child nodes once they are created through crossover or mutation, in case of central node failure, the system will still keep on protecting the local system. Another advantage is that system is easily scalable. As the central node carries all the rules, once the new node is registered with the grid system, it will automatically receives all the existing rule set from the central node. The detailed working of SASM has been described.

The goals achieved by the proposed system are:Reduction of administrative complexities,Automatic protection of the system from malicious activities and automatic monitoring of system performance.With the implementation and results presented in the paper, we have tried to prove the prototype model for SASM.

SASM Algorithm

```
/*
Pass through the application layer
/
#VerifyUserData(){
        // validate the user
        If(user exist){
          If(PassThroughFilterAgent){
            // publish the job to queue
            JobId, UID, Priority, etc.
          }
        }
}
/*
filter agent working
/
#filterAgent(){
        GeneticRuleFormation();
        //transfer the rules to child nodes
}
#GeneticRuleFormation(){
        //Initialization
          Based upon the existing rules, form the chromosome structure(hexadecimal format)
        //Evaluation
          More the rule give the correct result, better is its fesibility factor
        while termination criterion has not been reached
          {
          //Selection and Reproduction
            Based upon the feasibility factor select the set of chromosome
          //Crossover
            two chromosomes are combined to reproduce the new offspring
          //Mutation and Evaluation
            some major change is done to the existing chromosome to form the new chromosome
          }
          //Update the rules database with new chromosome thus formed.
}
#monitorAgent(){
        // check the status of resources
        // we will not be checking the resources actually, based
        //upon the job statuses and time they are taking to execute
        // we are predicting the grid load.
        If(NumberofJobspending/hr¿AvgNumberof Executing job/hr){
          Performance is degrading
          }
}
#dataAgent(){
        //collects the job info from queue
        If(job needs data from user){
          //form the ssh session
          //transfer the data to central location
          }
          //update the job information in the job's database
}
#raiseAlert(){
        //at some time interval, agent will be called that will raise
        // alert for all the still cancelled jobs in the database
        Select job_id from job where state=cancelled
        If(count>0){
          //raise the alert
          }
}
```

References

1. Mohamed, Y.A., Abdulla, A.B.: Immune Inspired Framework for Securing Hybrid MANET. In: ISIEA. IEEE, Los Alamitos (2009)
2. Claudel, B., De Palma, N., Lachaize, R., Hagimont, D.: Self-protection for Distributed Component-Based Applications. Springer, Heidelberg (2006)
3. Nou, R., Julia, F., Torres, J.: The need for self-managed access nodes in grid environments. IEEE, EASe (2007)
4. Romberg, M.: The unicore grid infrastructure (2002), http://www.unicore.org
5. Sotomayor, B., Childers, L.: Globus Toolkit 4: Programming Java Services (2005)
6. Guo, H., Gao, J., Zhu, P., Zhang, F.: A Self-Organized Model of Agent-Enabling Autonomic Computing for Grid Environment. In: 6th Word Congress on Inteligent Control (2006)
7. Ganek, A.: Overview of Autonomic Computing: Origins. In: Evolution, Direction. CRC Press, Boca Raton (2004)
8. Kephart, J.O., Chess, D.M.: The Vision of Autonomic Computing. IEEE Computer (2003)
9. Humphrey, M., Thompson, M.: Security Implications of Typical Grid Computing Usage Scenarios. Cluster Computing 5(3) (2002)
10. Li, W.: Using Genetic Algorithm for Network Intrusion Detection. In: Proceedings of the United States Department of Energy Cyber Security Group (2004)
11. Hartmut, P.: Genetic and Evolutionary Algorithms: Principles, Methods, and Algorithms. In: Genetic and Evolutionary Algorithm Toolbox (2003)
12. Stopping insider attacks: how organizations can protect their sensitive information (September 2006), ibm.com/services
13. Claudel, B., De Palma, N., Lachaize, R., Hagimont, D.: Self-protection for Distributed Component-Based Applications. Springer, Heidelberg (2006)
14. Chopra, I., Singh, M.: Agent based Self-Healing System for Grid Computing. In: ICWET, pp. 31–35. ACM, New York (2010)
15. Foster, I., Kesselman, C., Nick, J.M., Tuecke, S.: The physiology of the Grid. Global Grid Forum (2002)
16. Hariri, S., Qu, G., et al.: Quality-of-Protection (QoP)-An Online Monitoring and Self-Protection Mechanism. IEEE Journal on Selected Areas in Communications 23(10) (2005)
17. Horn, P.: Autonomic computing: IBM's perspective on the state of information technology (2001), http://www.research.ibm.com/autonomic/
18. Humphrey, M., Thompson, M.: Security Implications of Typical Grid Computing Usage Scenarios. Cluster Computing 5(3) (July 2002)
19. Foster, I., Kesselman, C., Tuecke, S.: The anatomy of the Grid. John Wiley and Sons, Chichester (2003)
20. Wang, J., Liu, X., Chien, A.: Empirical Study of Tolerating Denial-of-Service Attacks with a Proxy Network. In: Proceedings of the 14th Conference on USENIX Security Symposium (2005)
21. Ferreira, L., Berstis, V., et al.: Introduction to Grid Computing with Globus, IBM, http://www.liv.ac.uk/escience/beowulf/IBM_globus.pdf
22. Montero, R.S.: The Grid Way Meta-Scheduler. In: Open Source Grid and Cluster, Oakland, CA (May 2008)
23. Parashar, M., Hariri, S.: Autonomic Computing: An Overview. Springer, Heidelberg (2005)

24. Tesauro, G., Chess, D.M., Walsh, W.E., Das, R., et al.: A Multi-Agent Systems Approach to Autonomic Computing. In: AAMAS 2004. ACM, New York (2004)
25. Kreibich, C., Crowcroft, J.: Honeycomb - Creating Intrusion Detection Signatures Using Honeypots. ACM SIGCOMM (January 2004)
26. Chakrabarti, A.: Grid Computing Security, ch. 6, p. 105. Springer, Heidelberg (2007)
27. Snort, https://edge.arubanetworks.com/article/leveraging-centralized-encryption-snort-part-1
28. Stephen, M., Sukumaran Nair, V.S., Abraham, J.: Distributed Computing Grids-Safety and Security. In: Security in Distributed, Grid, Mobile, and Pervasive Computing, ch. 14
29. Chakrabarti, A.: Grid Computing Security, ch. 8, p. 159. Springer, Heidelberg (2007)
30. Kenny, S., Coghlan, B.: Towards a Grid-Wide Intrusion Detection System. In: Advances in Grid Computing, pp. 275–284. Springer, Heidelberg (2005)
31. Schulter, A., Navarro, F., Koch, F., Westphall, C.: Towards Gridbased Intrusion Detection. In: 10th IEEE/IFIP, Network Operations and Management Symposium, NOMS 2006, pp. 1–4 (2006)
32. Witold Jarmo lkowicz, M.: A Grid-aware Intrusion Detection System, Technical University of Denmark, IMM-THESIS-2007-109
33. Leu, F.-Y., Li, M.-C., Lin, J.-C.: Intrusion Detection based on Grid. In: ICCGI 2006 (2006)
34. Srinivas, M., Patnaik, L.M.: Genetic algorithms: A survey. IEEE Computer 27(6), 17–26 (1994)
35. Li, W.: Using Genetic Algorithm for Network Intrusion Detection. In: Proceedings of the United States Department of Energy Cyber Security Group (2004)
36. Brennan, M.P.: Using Snort For a Distributed Intrusion Detection System. SANS Institute (2002)
37. Sallay, H., AlShalfan, K.A., Fredj, O.B.: A scalable distributed IDS Architecture for High speed Networks. IJCSNS 9(8) (August 2009)

Change Impact Analysis: A Tool for Effective Regression Testing

Prem Parashar[1], Rajesh Bhatia[2], and Arvind Kalia[3]

[1] Computer Science Department, Chitkara University, Barotiwala, Solan, India
[2] Computer Science Department, Thapar University, Patiala, India
[3] Computer Science Department, Himachal Pradesh University, Shimla, India
{prem.parashar,rbhatiapatiala,arvkalia}@gmail.com

Abstract. Change impact analysis is an imperative activity for the maintenance of software. It determines the set of modules that are changed and modules that are affected by the change(s). It helps in regression testing because only those modules that are either changed or affected by the suggested change(s) are re-tested. Change impact analysis is a complex activity as it is very difficult to predict the impact of a change in software. Different researchers have proposed different change impact analysis approaches that help in prioritization of test cases for regression testing. In this paper, an approach based on Total Importance of Module (TIM) has been proposed that determines the importance of a module on the basis of (i) user requirements, and (ii) system requirements. The results of the proposed algorithm showed that the importance of a module is an essential attribute in forming a prioritized test suite for regression testing.

Keywords: granularity, prioritization, maintenance, change impact, impact set.

1 Introduction

Software maintenance is a regular activity as the requirements of user changes frequently. The maintenance of software may be preventive maintenance, adaptive maintenance or extensive maintenance. Regression testing is one of the most important and complex software testing methods [2] because even a subtle change in software may require rerun of all its old test cases and test cases designed for the change [16]. The major challenge for regression testing is the permissible time budget [8, 10]. After maintenance, modified software is required to be re-installed at the earliest. Due to a limited time budget allowed for regression testing, it may not be possible to rerun the entire test suite designed for the original software and test suite designed for its modified components. For gaining confidence in maintenance, all those test cases that cover the changed components and the components that get affected by the changes should be rerun[21]. The impact set of modules generated after the suggested change(s) may be formulated with the help of change impact analysis. Change impact analysis helps in finding the changed modules and modules affected by the change. A fine grained change impact analysis technique is always preferred to generate the precise impact set [13]. The impact set generated thus facilitates in effective regression testing because only the modules that constitute the impact set are tested.

S. Dua, S. Sahni, and D.P. Goyal (Eds.): ICISTM 2011, CCIS 141, pp. 160–169, 2011.
© Springer-Verlag Berlin Heidelberg 2011

Several change impact analysis approaches are reported in the review of literature [20, 21] . Most of the approaches are based on two assumptions, i.e. (i) all modules have equal importance in software [12], and (ii) each fault has equal severity [3]. In real practice, the importance of a module may play a vital role in regression testing i.e. module with the highest priority should be tested first. This is the main motivation factor behind performing this study.

In this study, it has been assumed that (i) user requirements, (ii) the call graph of system, and (iii) the historical records about the execution of the module are three main factors that define the overall importance of module. To determine the importance of a module, three metrics i.e. User-defined Importance of Module (UIM), Dynamic Importance of Module (DIM), and History Value of Module (HVM) have been proposed. UIM is supposed to be defined by user in requirements, DIM of a module is supposed to be dependent on the structure of the call graph of system, and HVM is defined by the historical records of the execution of the module with respect to the execution of the system.

2   Review of Literature

Rajlich [13] et al. proposed an iterative impact analysis method that uses the different level of granularity in order to achieve high level of precision. The granularity of components was performed at three different levels i.e. class, class member and code fragment level. From the empirical study conducted, it was evident that proposed algorithm is effective when the level of granularity is grained to code fragment level. For dynamic change impact analysis of software, atomic changes in source code responsible for the behavioral changes of test cases were considered .The main concern of this technique is to identify a subset of the changes that actually affects the behavior of a test case [15]. The results of the case studies based upon above technique evident that most of changes were captured by the execution of half of test cases.

Tie [19] et al. proposed a component interaction trace based approach to trace out the dynamic change impact analysis. The static structure of the components and their interfaces through which they are integrated is represented with the help of collaboration diagram and the dynamic model is represented with the help of the sequence diagram describing the objects interactions. If the two interfaces are serving approximately same purposes then it was proposed to replace one interface by the other and the unique services provided by the former were introduced in the later.

Sherriff [17,18] et al. proposed a change impact analysis approach based upon singular value decomposition (SVD) to find the structure of the file association clusters and the amount of variation done by this cluster in the original system after a change. They explored the concept of SVD in finding the impact of change in a system that contains a number of executable and non-executable files. This technique has been found quite satisfactory if the level of granularity is file and provides encouraging results. This technique is suitable for finding file association clusters of source files as well as non-source files. For object-oriented software, a model-based approach [10] that takes two different models i.e. activity diagram and sequence diagram to establish mapping is used to find the impact of change. These models of object-oriented software are mapped on the basis of some well-founded rules of relationship. The main advantage of this approach is that it does not require source code of software.

Rao [14] et al. proposed a quantitative method to detect the impact set if a change is made in an artifact and also to detect an artifact that is affected by the change. The method uses design change propagation probability (DCPP) matrix. DCPP is based upon the degree of coupling between two artifacts. The DCPP method is a quantitative method. It helps in finding the propagation of change. URA used in this case along with the corresponding DCPP clearly depicts the static behavior of the system. The method can also be used to automate the change impact propagation.

Park [12] et al. proposed a method to identify multi dimensional dependencies on the basis of process slicing. Process slicing can handle multi-dimensional dependencies effectively. The proposed algorithm handles the cases where any change in one activity produces change in two or more artifacts. The main objective of the proposed algorithm is to find out and to test that part of the software which is affect by the specified change. The time budget for regression testing is usually less therefore it is not practical to rerun all the tests of the old software along with the additional tests for the changed part. German [5] et al. proposed a method of determining the impact of historical changes on a particular code segment that in terms makes some failures. They considered C language program as a code segment for their study and took all the functions into consideration that affect the performance of the function. They prepare a change impact graph (CIG) that clearly determines all the functions that are falling into the way to make this function affected.

Orso [10] et al. proposed an approach that uses the field data instead of synthesized data to find the change impact analysis and regression testing. The software is tested against the field data collected. In this study, different ways of measuring the impact of change on the user population has been studied. Since live data from the actual users are used to analyze the impact of change, therefore the approach immediately reflects the results whether the changed software is useful to specified class of users or not. The approach is adequate if the user population is limited. Orso [11] et al. performed some empirical studies to make the comparisons of dynamic impact analysis algorithms. In their studies two dependency based algorithms i.e. coverage dynamic impact algorithms and path dynamic impact algorithms are explored to make a comparison about the precision of these algorithms. The empirical studies are performed on small java programs. The level of granularity considered in the studies is of method level. This is one of the empirical studies that deals directly with the cost –benefit analysis. Nadi [9] et al. developed a framework that helps in determining the root cause analysis and change impact analysis for configuration management data bases (CMDB). CMDB is used to store the different types of information like hardware, software and the services provided by an organization. For each component, it records all relevant information. The framework proposed serves two main purposes, (i) it helps to find the root cause of a failure of a component, and (ii) it determines the set of components that are affected by a change in a particular component.

Breech [1] et al. proposed an approach to whole program path-based dynamic change impact analysis. This approach is completely online. The method used in this case assume that any function that is either called directly or indirectly by the changed function or executed after a call to the changed function will constitute impact set. The approach used in this case is safe as it deals with the set of those functions which are potentially affected by a changed function but sometimes it generates imprecise impact set.

Yu and Rajlich [22] studied hidden dependencies among various software components that may cause malfunctioning of the software. They proposed an algorithm that warns about the hidden dependencies in the components that may be caused due to the improper structure of the components. Ma [7] et al. developed an algorithm that helps to find the relative importance of modules present in the system. One of the main objectives of change impact analysis is to trace out those modules that are observing the change or are affected by the proposed change. In real practice, when subtle change is made in one module that affect different parts of the software. The parts may be technical or non technical. After change impact analysis the subset of the modules that have been surfaced out as impact set need to be tested. Due to limited time and other resource constraints it is not easy to test all the modules of the impact set. There is always a tendency from the developer side to test those modules first which detect maximum faults. Ma [7] et al. proposed an effective algorithm that determines the importance of a module to the system. The algorithm successively deals with the hierarchical structure of the module. It deals with the acyclic graph of module effectively.

English [3] et al. studied the fault detection and prediction in open-source software. Their study is mainly content based. They leveraged some popular metrics like Number of methods in a class (NMC), depth of inheritance tree (DIT) number of children (NOC), coupling between object classes (CBO) and response for a class (RFC). The results of the study are encouraging as they strengthen Pareto's law i.e. about 80%of the faults/issues are due to 20% of the code. They have tried to trace out dependent and non-dependent variable to some extent.

Engstrom [4] et al. conducted a longitudinal study on the various techniques used for regression test selection. Since this review contains research papers from the start of the era of software testing, it gives the future researchers as well as developers a systematic review of how the regression testing test selection methods varied from software to software and from time to time. The techniques summarized certainly helps the industry and academia to pursue further research in the area.

3   Objectives

The broad objective of this study is to prioritize a test suite on the basis of the importance of changed /affected modules. The specific objectives of the study are:

1. To analyze the significance of the importance of a module when we prioritize a test suite for regression testing.
2. To analyze whether proposed algorithm is effective for prioritization of test cases for regression testing.

4 Research Methodology

Change impact analysis is before maintenance activity of software. It gives a clear clue to maintenance manager about (i) whether the required changes should be incorporated in the existing software or (ii) should it be re-engineered. Each module that constitutes software has its own importance. The importance of a module is defined in many ways. In the proposed approach, the importance of a module has been assumed to be defined by the user, by the developer or by the history of its use. For example, ATM system of a bank uses cash withdrawal module most of the time as compared to mini statement module. Therefore according to proposed approach cash withdrawal module is more important as compared to mini statement module. Importance of a module plays a significant role while it is maintained. The most important modules of system should be maintained with utmost care so that at least they should not malfunction at any point of time during execution. At the time of regression testing, those modules that have high priority values to the software are tested first. Total Importance of Module m_i (TIM (m_i)), is given by:

$$TIM\ (m_i) = UIM\ (m_i)*DIM\ (m_i)*HVM\ (m_i) \tag{1}$$

In Equation (1), $UIM\ (m_i)$, $DIM\ (m_i)$, and $HVM\ (m_i)$ represent the static importance, dynamic importance, and historical value of the module m_i respectively. UIM of a module is a numeric value assigned to it by user. The value ranges from 1 to 5. If a module has a UIM value of 5, it means it is the most important, and one means it is the least important. DIM of a module is calculated from the Call Graph (CG) of software. It is calculated as:

$$DIM\ (m_i) = TD\ (m_i)/TD\ (CG) \tag{2}$$

In Equation (2), $TD\ (m_i)$ and $TD\ (CG)$ represent total degree of module m_i and total degree of call graph respectively. $HVM\ (m_i)$ is calculated from the historical records of execution of software. It is calculated as:

$$HVM\ (m_i) = THC\ (m_i)/THC\ (CG)*HTI \tag{3}$$

In Equation (3), $THC\ (m_i)$, $THC\ (CG)$, and HTI represent the total number of calls made to module m_i in a history time interval, total calls made to all modules in a history time interval, and a history time interval respectively. For example, if a module is called 10 times in previous five executions and the total modules called during these five executions are 50 then HVM (m_i) will be 1.0. HTI represents the total number of previous executions taken into account while finding $HVM\ (m_i)$. It is practically difficult to maintain the records of all the previous executions of large software which is running from years. HTI limits time interval to some well defined number of executions, so that the better idea of its recent importance could be drawn.

After modification of software, two module sets are generated by change impact analysis i.e. a changed set and an affected set. If a module is changed then its entire neighboring module in the call graph are affected by the change. The neighboring modules constitute the estimated impact set. Depending upon the importance, changed modules and impacted modules are tested. A simple algorithm for this has been described in Fig. 1. In this algorithm, a term 'cover a module' has been used which signifies that the module has been tested with all test cases required for it, and the bugs have been fixed. Simple mathematical set operations like A U B, A-B have been used to make the proposed algorithm understandable.

---

1. Let TCM, is impact set and TS, is a test set array where TS[i] contains the number of test cases required to cover a module M[i].
2. Time, is an array where Time[i] stores maximum time required to cover a module M[i].
3. Calculate TIM value for each module of TCM with the help of Equations (1), (2), and (3).
4. Sort the modules of TCM according to TIM values (in descending order)
5. Rearrange TS, and Time elements according to TCM. e.g. if nth. element of original TCM has become first element in the sorted TCM, them Time [1] =Time[n] and   TS [1] =TS[n].
6. Take t=0, i=0   //t represents the time required to execute the selected module(s).
7. While ((TCM $\neq \Phi$)and(t $\leq$ TB)) //TB permissible time budget.
8. {
9.   if(t+ Time[i]$\leq$ TB)
10.   {
11.   (i)   P = P U TS[i]   // P: prioritized test suite
12.   (ii) t = t+ Time[i]
13.   (iii) TCM=TCM-M[i] // module present at i^{th} location of TCM will be removed.
14.   }
15. Else
16.   TCM=TCM-M[i]
17.   }

Fig. 1. TIM Test Case Prioritization Algorithm

5 Analysis

For the experimental setup, a small menu driven C language program, for basic mathematical operations, has been considered. The program is presented in Fig. 2. Suppose that the line no. 29 (highlighted) of program is modified (modified statement is written in bold and the old statement is written in comments). This change in the program affects the modules main(), sub() and div(). Thus, TCM consists of {main (), sub (), ()}. The CG of this program is represented in Fig. 3.

Module div() (represented with '*') is the changed module and the modules sub(), and main() (represented with '**') are affected modules. Modules add (), and mul() are unaffected by the change. Let us assume that there are six test cases in the test suite: T1, T_2, T_3, T4, T_5, and T_6. Let time taken(in seconds) to execute test cases T_1, T_2, T_3, T_4, T5, and T_6 be 3, 2, 4,1, 5,and 2 seconds, respectively.

```
1.  #include <stdio.h>
2.  void main(void)
3.  {
4.    int a, b, sum,subt , mult,divd ;
5.    char choice;
6.    int add(int,int);   int sub(int,int); int mul(int, int);  int div(int,int);
7.    scanf("%d%d%c", &a,&b, &choice);
8.    switch(choice)
9.    {
10.   case 'a':       printf("Addition of a and b=%d\n", add(a,b)); break;
11.   case 's' :      printf("subtraction of a and b=%d\n",sub(a,b)); break;
12.   case 'm':       printf("Multiplication of a and b=%d\n",mul(a,b)); break;
13.   case 'd' :      printf("division of a by b=%d\n", div(a,b)); break;
14.   default:        printf("Invalid Choice\n");
15.   }
16. }   /* end main() */
17. int add(int x,int y) {return(x+y);}
18. int sub(int x,int y) { return (x+y);}
19. int mul ( int x, int y)
20. {
21.     int i=1,p=0;
22.     for(i=1;i<=y;i++)
23.     p=p+add(x,0);
24.     return p;
25. }              /* end mul()  */
26. int div ( int x, int y)
27. {
28.   int p=0;
29.   while(x>=y)    /* while(x>y) is the original statement */
30. {
            a.   x=sub(x,y);    b.   p=p+1;
31. }
32. return p;
33.          }          /* end div() */
```

Fig. 2. C language Program for Basic Mathematical Operations

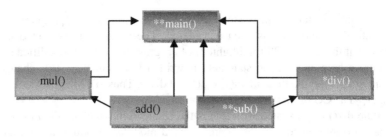

Fig. 3. Call Graph of Program (Fig. 2)

The assumed test case coverage of modules of Fig. 3 is shown in Table 1. From Table 1, it can be derived from Table 1 that test suite required for the execution of changed module and affected modules will be consist of T_2, T_3, T4, T_5, and T_6. In this example, TCM array will store values {div(), sub(), main()}, TS will store three test suites corresponding to each module present in TCM i.e. {{ T_4,T_6},{ T_2, T_6},{ T_3,T_4, T_5 }}. Time array will store corresponding values of execution time of respective test suites i.e. Time {3, 4, 10}.

Table 1. Test Coverage Representation of Modules (Fig. 3)

Module	Test cases required for coverage	Time taken(in seconds)
main()	T_3,T_4, T_5	10
add()	T_1,T_5	8
sub()	T_2, T_6	4
mul()	T_3,T_5	9
div()	T_4,T_6	3

TIM for these modules can be calculated by using Equations (1), (2), and (3). HTI value has been considered as 20, and both modules sub () and div () are called 5 times in 20 executions of program. TIM values of the required modules have been tabulated in Table 2. From TIM values, it can be concluded that module main () is the most important and module sub () has the least importance. Thus, TCM will have modules arranged in order {main (), div (), sub ()}. Accordingly, TS will also get rearranged and the element will be {{T3, T4, T_5}, {T_4, T_6}, {T2, T_6}}. Time array will have vales {10, 3, 4}. The prioritized test set P for this example will contain test cases in order T_3, T_4, T_5, T_6, and T_2 according to the importance of modules of estimated impact set.

Table 2. TIM Values of Changed/Affected Modules

Module	UIM	DIM	THC(Module)	HVM	TIM
main()	4	0.333333	20	20	26.66667
sub()	2	0.166667	5	5	1.666667
div()	3	0.166667	5	5	2.5

According to the proposed algorithm represented in Fig. 1, depending upon the value of time budget available for regression testing, all test cases for selected module/modules will be executed. For example, if time budget is 10, only module main () will be tested.

5.1 Major Findings

For the evaluation of the proposed algorithm (Fig. 1) a small C language program (Fig. 2) , has been considered. From the results of this experiment, it has been observed that the importance of a module plays a significant role in prioritizing a test suite for regression testing. UIM, DIM and HVM metrics proposed in this paper facilitates in the calculation of TIM of a module. Even, if the time budget allowed is greater than the time required to execute the prioritized test suite, prioritization of test cases is important for additional coverage of modules. Further, it has been observed that there is the least possibility for any two modules to have the same value of TIM. This will help to reduce the random selection of test cases at the time of prioritization.

5.2 Threats to Validity

Main threat to validity is that two main contributors towards TIM, i.e. UIM and HVM are decided by user and developer respectively. Though, the metrics, TIM, UIM, DIM, and HVM are effectual to find the importance of a module, still, Generalization of the results of proposed algorithm requires more empirical studies. The correlation of metrics UIM, DIM, and HVM used in the algorithm to generate TIM, needs extensive investigational studies for its universal acceptance.

6 Conclusions

In prioritization of test cases for regression testing, the role of a module is very significant. The modules of *TCM* set should be tested according to the order of their importance. *TIM* algorithm, present in the current study, prioritizes test cases on the basis of the importance of their respective modules. The metrics proposed in this study played an imperative role in finding the importance of modules. The current study can be extended further by applying *TIM* algorithm to programs of different sizes, types, and complexities.

References

1. Breech, B., Danalis, A., Shindo, S., Pollock, L.: Online impact analysis via dynamic compilation technology. In: ICSM 2004, September 11-14 (2004)
2. Chechik, M., Winnie, L., Nejati, S., Cabot, J., Diskin, Z., Eaterbrook, S., Sabetzadeh, M., Salay, R.: Relationship-based change propagation: a case study. In: MiSE 2009, May 17-18 (2009)
3. English, M., Exton, C., Rigon, I.: Fault detection and prediction in an open-source Software project. ACM, New York (2009)
4. Engstrom, E., Skoglund, M., Runeson, P.: Empirical evaluation of regression test selection techniques: a systematic review. In: ESEM 2008, October 9-10 (2008)
5. German, D.M., Robles, G., Hassan, R.E.: Change impact graph: determining the impact of prior code changes. In: IEEE Working Conference on Source Code Analysis and Manipulation, SCAM (2008)
6. Kagdi, H.Z., Jonathan, I.M.: Software-change prediction: estimated+ actual. In: Proc. IEEE Workshop on Software Evolvability, pp. 38–43 (2006)
7. Ma, Z., Zhao, J.: Test case prioritization based on analysis of program structure. In: APSEC (2008)
8. Maia, C.L.B., Refael, A.F.D.C., Fabricio, G.D.F.: Automated test case prioritization with reactive GRASP. In: Advances in Software Engineering. Hindawi Publishing Corporation (2010)
9. Nadi, S., Holt, R., Davis, I., Mankovskii., S.: DRACA: decision support for root cause analysis and change impact analysis for CMDBs (2009)
10. Orso, A., Apiwattanapong, T., Law, J., Rothermel, G., Harrold, M.J.: Leveraging field data for impact analysis and regression testing. In: ESEC/FSE 2003, September 1-5 (2003)
11. Orso, A., Apiwattanapong, T., Law, J., Rothermel, G., Harrold, M.J.: An empirical comparison of dynamic impact analysis algorithms. In: ICSE 2004 (2004)

12. Park, S., Kim, H., Bae, D.H.: Change Impact analysis of a software process using process slicing. In: QSIC (2009)
13. Rajlich, V., Patrenko, M.: Variable granularity for improving precision of impact analysis. In: ICPC (2009)
14. Rao, A.A., Reddy, K.N.: Detecting bad smells in object oriented design using design Change propagation probability matrix. In: IMECS 2008, Hong Kong (March 2008)
15. Ryder, B.G., Ren, X., Shah, F., Tip, F., Chesley, O., Chianti, J.: A tool for change Impact analysis of java program. In: OOPSLA 2004, October 24-28 (2004)
16. Sanjeev, A.S.M., Wibowo, B.: Regression test selection based on version changes of components. In: Proceedings of APSEC 2003 (2003)
17. Sherriff, M., Lake, M., Williams, L.: Prioritization of regression tests using singular value decomposition with empirical change records. In: International Symposium on Software Reliability Engineering (November 2007)
18. Sherriff, M., Williams, L.: Empirical software change impact analysis using singular value decomposition. In: ICST (2008)
19. Tie, F., Maletic, J.I.: Applying dynamic change impact analysis in component-based Architecture design. In: ACIS International Conference on Software Engineering (2006)
20. Walcott, K.R., Kapfhammer, G.M., Soffa, M.L., Roos, R.S.: Time-aware test suite prioritization. In: Proceedings of ISSTA 2006, July 17-20 (2006)
21. Yoo, S., Harman, M.: Regression testing minimization, selection, and prioritization: a survey. Software Test. Verif. Reliab. (2007)
22. Yu, Z., Rajlich, V.: Hidden dependencies in program comprehension and change Propagation. In: IWPC (2001)

Multi-feature Fusion for Closed Set Text Independent Speaker Identification

Gyanendra K. Verma

Indian Institute of Information Technology, Allahabad
Jhalwa, Allahabad, India
gyanendra@iiita.ac.in

Abstract. An intra-modal fusion, a fusion of different features of the same modal is proposed for speaker identification system. Two fusion methods at feature level and at decision level for multiple features are proposed in this study. We used multiple features from MFCC and wavelet transform of speech signal. Wavelet transform based features capture frequency variation across time while MFCC features mainly approximate the base frequency information, and both are important. A final score is calculated using weighted sum rule by taking matching results of different features. We evaluate the proposed fusion strategies on VoxForge speech dataset using K-Nearest Neighbor classifier. We got the promising result with multiple features in compare to separate one. Further, multi-features also performed well at different SNRs on NOIZEUS, a noisy speech corpus.

Keywords: Multi-feature fusion, intra-modal fusion, speaker identification, MFCC, wavelet transform, K-Nearest Neighbor (KNN).

1 Introduction

Multiple information fusion is a new challenge nowadays. Information fusion is defined as combine information from multiple sources in order to achieve higher performance than the performance achieved by means of a single source [1]. There are basically two fusion categories. Intra-modal fusion [2, 3]: this is the fusion of different features of the same modal and Multimodal Fusion [4, 5]: this is the fusion of different modalities e.g. combined face, speech; fingerprint etc. this paper is based on the intra-modal fusion. Further the information can be fused at signal level, feature level and decision level. In signal level, data acquired from different sources to be fused directly after preprocessing. Feature level: fusion of multiple features is performed in this case. Decision level: the output of multiple classifiers based on a set of match score is fused. Complementary information is useful in fusion process as it enhance the confidence in decision [6].

Various methods have been developed for speaker identification such as Mel Frequency Cepstral Coefficients (MFCC), Linear Predictive Coding (LPC), Linear Prediction Cepstral Coefficients (LPCC) and Gabor etc. however there are still open

S. Dua, S. Sahni, and D.P. Goyal (Eds.): ICISTM 2011, CCIS 141, pp. 170–179, 2011.

problems which arise into real application. Longbiao Wang et al. [7] proposed a combined approach using mfcc and phase information for feature extraction from speech signal. Most of the algorithms considered only single features or directly combined features. A multi-features based fusion method is proposed in this study for closed set text independent speaker identification in order to improve the performance of the system. Closed-set identification only considers the best match from the enrolled speakers. We have used two feature extraction approaches namely MFCC and Wavelet Transform. The cepstral representation is a better way to represent the local spectral properties of a speech signal [8] whereas wavelet transform capture frequency variation across time. Features obtained from the above two approaches was fused at feature and decision level. A combined feature is generated by fusion of different features of the same speech at feature level fusion. At decision level, a final score is calculated using weighted sum rule by taking matching results of different features. VoxForge and Noizeus speech corpus has been used to evaluate the fusion schemes. This study contributes to development of new data fusion methods in signal processing and information extraction. The multi-feature fusion approach can be beneficial to many application of pattern recognition in order to enhance the performance of the system under consideration of pros and cons of the system. A general architecture of feature and decision level information fusion is illustrated in Figs.1a and b.

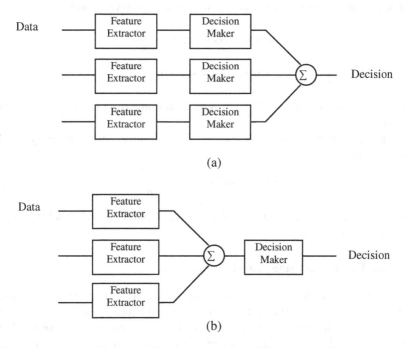

(a)

(b)

Fig. 1. A general architecture of Information Fusion (a) Feature level (b) Decision level

The rest of the paper is organized as follows: review of feature extraction techniques are given in Section 2. Proposed fusion approach is described in Section 3. Experiment results and discussion about proposed work is described in Section 4. Concluding remark is given in Section 5.

2 Feature Extraction

Feature extraction is an important phase in any pattern recognition problem. In our study the features are obtained by applying two approaches named MFCC and wavelet transform. Wavelet transform is able to perform local analysis to capture the local information of a signal at multi-resolution. Feature vectors extraction process using MFCC and wavelet transform are described below.

2.1 Feature Extraction Using MFCC

The process of calculating the MFCC consists of the following steps.

• Framing: In this step the speech signal segmented into N samples with overlapping frames.
• Windowing: To spectral analysis of the speech signal in order to minimize the spectral distortion. Generally Hamming window is used as given below

$$W(n) = 0.54 - 0.46 \times \cos(2 \times \frac{n}{N-1}) \tag{1}$$

where 0 <= n <= N-1

• Fast Fourier Transform (FFT): FFT converts each frame from the time domain into the frequency domain.
• Mel-frequency wrapping: the Mel can be obtained for a given frequency by the formula given below.

$$Mel(f) = 2595 \times \log_{10}(1 + \frac{f}{700}) \tag{2}$$

• Cepstrum: Finally take the discrete cosine transform (DCT) to the signal in order to obtain MFCC coefficients.

2.2 Feature Extraction Using Wavelet Transform

Wavelet transform provides a compact representation that shows the energy distribution of the signal in time and frequency [9]. We have used Discrete wavelet transform to decomposes the signal into multilevel successive frequency band utilizing two sets of function called scaling function ϕ and wavelet functions (ψ) associated with low pass and high pass filters respectively [10]. Filters are obtained as

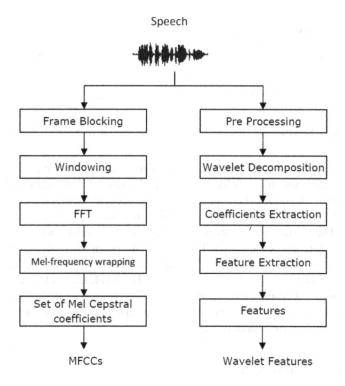

Fig. 2. Feature extraction process

weighted sum of scaled and shifted versions of scaling function itself. Information captured by wavelet transform depends on properties of wavelet function family like Daubechis, Symlet, Biorthogonal, Coiflet etc and properties (waveform) of target signal. Information in signal extracted by wavelet transforms using different family of wavelet function need not to be same. It is required to choose or evaluate wavelet function that provides more useful information for particular application. Signal at various scales and translations providing multi resolution time-frequency representation, as show in Fig. 3.

In Discrete wavelet decomposition of signal, the output of high band pass filter and low band pass filter can be represented mathematically by the Equations 3 and 4.

$$Y_{high}[k] = \sum nX[n]g[2k-1] \tag{3}$$

$$Y_{low}[k] = \sum nX[n]h[2k-1] \tag{4}$$

where Y_{high} and Y_{low} are the outputs of the high band pass and low band pass filters, respectively.

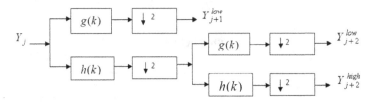

Fig. 3. Schematic of Discrete Wavelet decomposition of a speech signal

In order to extract wavelet coefficients, the speech signal is passed into successive high pass and low pass filter. Selection of suitable wavelet and number of levels of decomposition is important. For one dimensional speech signal Daubechis wavelet family provides good results for non-stationary signal analysis [11] so we have used it in our study. The feature vectors obtained from six level wavelet coefficients provides compact representation of the signal. The coefficients occur in whole bandwidth from low frequency to high. The original signal can be represented by the sum of coefficients in every sub band, which is cD6, cD5, cD4, cD3, cD2, cD1. Feature vectors are obtained from the detailed coefficients applying common statistics and entropy. The discriminatory property of entropy features makes it suitable to extract frequency distribution information [12].

3 Proposed Fusion Approach

We proposed two fusion approaches for multiple features. The first approach fuse information at feature level (Fig. 4) and other one fuse information at decision level (Fig. 5). Low level features of the speech signal are extracted independently using mfcc and wavelet transform analysis described in Sections 2.1 and 2.2, respectively. The fusion strategies are discussed below.

3.1 Feature-Level Fusion

In feature level information fusion the features obtained from both approaches are organized in such a way that the mfcc features remain in the first half of feature vector and the wavelet features are into the second half of the feature vector. Let the features obtained from mfcc coefficients are $F_{mfcc} = (f_{m1}, f_{m2}, \ldots f_{mn})$ and from wavelet coefficients are $F_{wav} = (f_{w1}, f_{w2}, \ldots f_{wn})$ then the fused feature vector can be given as $F_{fusion} = [F_{mfcc}, F_{wav}]$

$$F_{fusion} = \{f_{m1}, f_{m2}, \ldots f_{mn}, f_{w1}, f_{w2}, \ldots f_{wn}\}$$

3.2 Decision-Level Fusion

In decision-level, we start the procedure with normalization of the scores obtained from different feature extraction approaches. Normalization is performed to map the

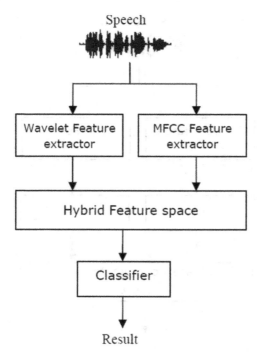

Fig. 4. Feature level fusion architecture

values of different classifiers in common range. Min-max normalization is used here. The threshold value for different classifier is different so we further rescale the matching score in order to obtain the same threshold value for different classifier. A speaker is accepted only within threshold range otherwise rejected. Finally the scores are combined using sum rule. The sum rule method of integration takes the weighted average of the individual score values. Let x_1, x_2, x_n are the weighted scores corresponding to classifier 1, 2 and n. Then the fusion equation can be given by Equation 5.

$$S_{comb} = \frac{\sum_{i=1}^{n} X_i}{n} \tag{5}$$

4 Experimental Results and Discussion

VoxForge corpus [13]: It containing more than 200 speaker profiles of males and females and each profile contains 10 speech samples. The sampling frequency was kept 8 KHz and bit depth 16. The duration of speech samples range between 2 - 10 second. All the speech files are in wav format.

Speech

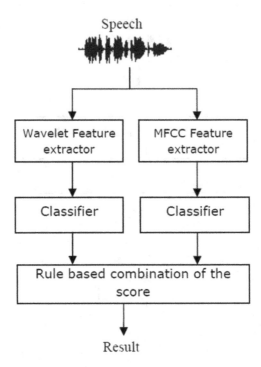

Result

Fig. 5. Decision-Level Fusion Architecture

NOIZEUS [14]: The noisy database contains 30 IEEE sentences (produced by three male and three female speakers) the speech was recorded from different places of crowd of people, car, exhibition hall, restaurant, street, air port, train station and train. The noises were added to the speech signals at SNRs of 15dB, 10dB and 5dB. All files are in wav format (16 bit PCM, mono).

The experiments comprised of two modules: training and testing were performed on standard VoxForge speech corpus and NOIZEUS. Five speech samples were used for training and another five for testing purpose. Total 33 and 30 dimensional feature vectors was obtained from MFCC and Wavelet decomposition respectively as described in section II a and b. Min-max algorithm has been used for feature set normalization in order to improve the identification accuracy before the classification for large dataset. All the experiments are performed on Mat Lab 7.6 (R2008b).

For classification purpose speech samples of same speaker is assigned same class. In this way five speech samples assigned the same class and so on such that speaker A = A1, A2, A3, A4, A5 assign class "1" and for speaker B = B1, B2, B3, B4, B5 assign class "2". In this way the whole training data is grouped in class. Euclidean distance is used to calculate the distances among vectors using KNN algorithm. The performance result of discrete wavelet and mfcc features is shown in Table 1 for standard VoxForge speech corpus. The proposed design of the speaker identification system uses 33 dimensional wavelet features and 30 dimensional mfcc features with 10 samples of 200 speakers. The parameters of the proposed design are the result of

Table 1. Classification results with multi-feature

No. of Speakers	Classification Rate (%)		
	Wavelet	MFCC	Fusion
10	100	96	100
20	99.0	91	100
30	94.0	84	95
40	89.5	81	92
60	87.0	74.6	90.6
80	88.0	73	92.5
100	89.4	73.6	93.2
120	88.1	72.8	91.8
140	85.8	68.4	90.6
160	85.5	68	90.4
180	83.7	67.33	90.4
200	83.9	66.8	90.2

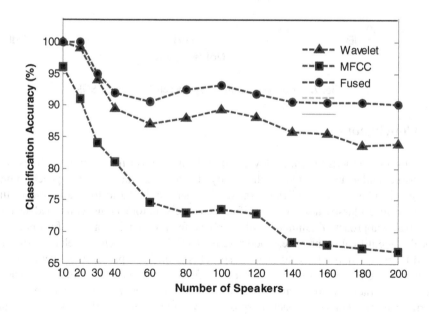

Fig. 6. Performance Graph

evaluation of different speaker identifications designs, evaluated using the VoxForge and Alternative corpora. The classification accuracy of fused features is 90.2% with 200 speakers.

A threshold θ is used to assign the class of speakers, here θ = 0.60. If the query samples matched and all matched samples belongs to same class and if x (score) > θ then assign the query sample to that class. The classification results are shown in Table 1 and the corresponding performance graph is illustrated in Fig. 6. The performance of the system with noisy dataset at different SNRs is illustrated in Fig. 7.

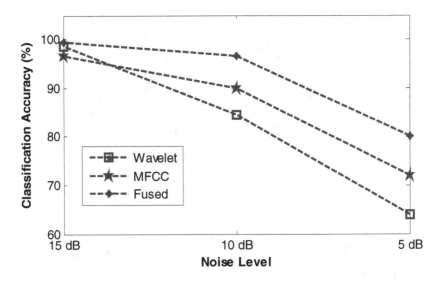

Fig. 7. Performance with Noisy Speech at Different SNRs

5   Conclusions

New fusion methods at feature level and at decision level for multiple features were discussed and experimented in this study. Features extracted from speech signals using MFCC and Wavelet Transform were either combined to form a hybrid feature space before classification or classified separately before being combined using a rule-based approach. Examination of the above feature fusion strategies for improving speaker identification provides good results. MFCC and Discrete wavelet transform is used to extract multiple features from speech signal. Multiple features were fused at feature as well as decision level. KNN classifier has been used for similarity measure between extracted features and a set of reference features. All the experiments were performed on standard speech corpus i.e. VoxForge and NOIZEUS. The results obtained from fusion scheme shows significant increment in performance of the system.

References

1. Multimodel Data Fusion, http://www.multitel.be/?page=data
2. Marcel, S., Bengio, S.: Improving face verification using skin color information. In: 16th International Conference on Pattern Recognition, pp. 378–381 (2002)
3. Czyz, J., Kittler, J., Vandendorpe, L.: Multiple classifier combination for face-based identity verification. Pattern Recognition 37(7), 1459–1469 (2004)
4. Wang, Y., Tan, T., Jain, A.K.: Combining face and iris biometrics for identity verification. In: Kittler, J., Nixon, M.S. (eds.) AVBPA 2003. LNCS, vol. 2688. Springer, Heidelberg (2003)
5. Hong, L., Jain, A.K., Pankanti, S.: Can multi-biometrics improve performance? In: Technical Report MSU-CSE-99-39, Department of Computer Science, Michigan State University, East Lansing, Michigan (1999)
6. An Introduction to Data Fusion, Royal Military Academy, http://www.sic.rma.ac.be/Research/Fusion/Intro/content.html
7. Wang, L., Minami, K., Yamamoto, K., Nakagawa, S.: Speaker identification by combining MFCC and phase information in noisy environments. In: 35th International Conference on Acoustics, Speech, and Signal Processing, Dallas, Texas, U.S.A. (2010)
8. Patel, I., Srinivas Rao, Y.: A Frequency Spectral Feature Modeling for Hidden Markov Model Based Automated Speech Recognition. In: Meghanathan, N., Boumerdassi, S., Chaki, N., Nagamalai, D. (eds.) NeCoM 2010. CCIS, vol. 90, pp. 134–143. Springer, Heidelberg (2010)
9. Dutta, T.: Dynamic time warping based approach to text dependent speaker identification using spectrograms. Congress on Image and Signal Processing 2, 354–360 (2008)
10. Tzanetakis, G., Essl, G., Cook, P.: Audio analysis using the discrete wavelet transform. In: The Proceedings of Conference in Acoustics and Music Theory Applications, Skiathos, Greece (2001)
11. Liu, Y., Shengjun, L., Dongsheng, Z.: Feature Extraction and classification of lung sounds based on wavelet coefficients. In: Proceeding of the 6th International Progress, Wavelet Analysis and Active Media Technology, Chongqing, China, pp. 773–778. World Scientific, Singapore (2005)
12. Toh, A.M., Togneri, R., Northolt, S.: Spectral entropy as speech features for speech recognition. In: The Proceedings of PEECS, Perth, pp. 22–25 (2005)
13. VoxForge Speech Corpus, http://www.voxforge.org
14. NOIZEUS: A Noisy Speech Corpus for Evaluation of Speech Enhancement Algorithms, http://www.utdallas.edu/~loizou/speech/noizeus/

An Efficient Metamorphic Testing Technique Using Genetic Algorithm

Gagandeep Batra and Jyotsna Sengupta

Department of Computer Science, Punjabi University, Patiala, India
{gdeep.pbi,jyotsna.sengupta}@gmail.com

Abstract. Testing helps in preserving the quality and reliability of the software component thus ensuring its successful functioning. The task of testing components for which the final output for arbitrary input cannot be known in advance is a challenging task; as sometimes conditions or predicates in the software restrict the input domain Metamorphic testing is an effective technique for testing systems that do not have test oracles. In it existing test case input is modified to produce new test cases in such a manner that they satisfy the metamorphic relations. In this paper, we propose a genetically augmented metamorphic testing approach, which integrates genetic algorithms into metamorphic testing, to detect subtle defects and to optimize test cases for the component. We have further verified metamorphic testing results by all path coverage criteria information, which is generated during the metamorphic testing of the program and its mutants. The effectiveness of the approach has been investigated through testing a triangle type determination program.

Keywords: Metamorphic testing, test oracle problem, genetic algorithm.

1 Introduction

Software testing generally refers to test case selection and result evaluation based on functional and structural model of the software component. However, it is difficult to evaluate result and detect errors, faults, defects or anomalies in many domains like scientific systems, embedded systems, fuzzy systems etc; because there is no reliable "test oracle" to indicate what the correct output should be for arbitrary input. This category of software systems with no reliable test oracle available are known as "non-testable programs." There are a wide range of applications of metamorphic testing such as numeric, graph theory, computer graphics, compilers and interactive software [3].

Testers usually apply static testing (through static analyzers) or dynamic testing techniques to assure the correctness of a software application. The evaluation criteria of the functional or structural correctness must be predefined. However, the correctness of the program on the test suit does not imply its correctness on the whole input domain of the program. In this paper, we restrict our attention to class of programs for which no test oracle is available. An oracle is any program, process, or body of data that specifies the expected outcome of a set of tests as applied to a tested object. Even when the oracle is available, testing still remains an expensive activity.

S. Dua, S. Sahni, and D.P. Goyal (Eds.): ICISTM 2011, CCIS 141, pp. 180–188, 2011.

The contributions of this paper include: (1). We present a new metamorphic testing approach called genetically augmented metamorphic testing, which extends the metamorphic testing with genetic algorithms to generate optimized test data and test the systems that do not have test oracles. (2) To evaluate the quality of metamorphic relations by satisfying, test coverage criteria. (3) an implementation framework; (4) case study in which we apply our approach to non-testable program and its mutants to demonstrate the effectiveness of the technique. The rest of this paper is organized as follows: Section 2 gives a brief description of metamorphic testing, metamorphic relations, and the concept of genetic algorithms, Section 3 provides the framework for implementation, Section 4 incorporates case study and finally, Section 5 describes the summary and future work.

2 Background

2.1 Metamorphic Testing

Metamorphic testing is a property-based testing strategy. It states that, even if a test case (known as the *original* test case) does not reveal any failure, follow-up test cases should be constructed to check whether the software satisfies some necessary conditions of the target solution of the problem. These necessary conditions are known as Metamorphic Relations. A metamorphic relation is the relationship among the test input data and could be used to create additional test inputs based on existing one and to predict the relations among the test outputs [5].

Metamorphic Testing is different from other testing techniques in the following ways:

1. Testers provide a probabilistic oracle, in order to estimate whether the corresponding output is likely to be correct in traditional testing, whereas Metamorphic testing is designed as a property-based test case selection strategy rather than providing some form of alternative oracle.
2. Metamorphic properties can identity relations and can be expressed in other forms such as inequalities which are not covered by other techniques.
3. Conventional testing techniques make use of Data diversity as a fault-tolerance technique designed to process the original test cases in which only the re-expressed forms of the original test cases are to be considered; while metamorphic testing is a fault based detection technique.

Many researchers are using the concept of metamorphic testing in different domains. Chen et. al [1] demonstrate the use of metamorphic testing to solve partial differential equation programs. It helps to alleviate the oracle problem in testing of numerical software. An automated framework to check against a class of metamorphic relations is based on the assumption that a complete oracle should be available during the testing process [2]. Metamorphic testing is also applied to implement shortest and critical path algorithm [4]. W.K. Chan et. al [9] proposes a metamorphic approach to online testing of service-oriented software applications. It involves executing test cases by metamorphic services that encapsulate the services under test.

2.2 Genetic Algorithms

Genetic Algorithm (GA) is a search technique used in computing to find exact or approximate solutions to optimization and search problems. It is generally used in situations where the search space is relatively large and cannot be traversed efficiently by classical search methods. Genetic algorithms searching mechanism starts with a set of solutions called a population. One solution in the population is called a chromosome. The search is guided by a survival of the fittest principle. The search proceeds for a number of generations, for each generation the fitter solutions (based on the fitness function) will be selected to form a new population. A large gap between a technique based on genetic algorithms and those based on random test generation has been found [5]. The problem of generating test data has been reduced to the problem of function minimization. An automatic test data generation technique that uses a genetic algorithm (GA), guided by the data flow dependencies in the program, has also been proposed [6]. A method for optimizing software testing efficiency is proposed by Srivatsva et. al [8]. They have identified the most critical path clusters in a program by developing variable length Genetic Algorithms that optimize and select the software path clusters which are weighted in accordance with the criticality of the path. We have also adopted the same approach. The GA conducts its search by constructing new test data from previously generated test data that are evaluated as effective test data.

3   Approach

In this paper, we present an extended metamorphic testing technique called *genetically augmented metamorphic testing*, which is based on generating optimized set of test cases for checking the metamorphic relations to prove correctness of the program. Our approach is used for testing of systems that do not have test oracles, and its effectiveness is investigated through testing a triangle type determination program.

The steps of the approach are as follows:

1. *Identify Metamorphic Relations:* We initially identify the metamorphic relations based on behaviors of a given problem, and then conduct metamorphic testing on the system to check these relations. Therefore good knowledge of the problem domain is necessary for an effective application of metamorphic testing Metamorphic relations are not limited to identity relation. It includes inequalities, sub-assumption relations and convergence properties.
2. *Creation of program mutants:* Mutation analysis is a fault-based technique to assess the quality of a test suit. Faults are seeded systematically into the implementation which is to be investigated. By inserting small changes in the program, mutants are created. Currently only first order mutants are being created by inserting single fault.
3. *Application of genetic algorithm on test data:* Genetic Algorithms provide an effective method for optimizing test data by identifying the most critical path clusters in a program. For this, a control flow graph of program has been constructed. To initiate the process, parents (or initial test cases) are selected

according to a probability distribution based on the individual's fitness values. Crossover is performed according to a crossover probability pc, which is an adjustable parameter. For each entry in the new data set, bit-wise random numbers are generated. The process is iterated until the best solution in current population is obtained.

4. *Conduct Metamorphic Testing:* Once above steps are completed, the original program and its mutants are tested with the optimized set of test data generated using genetic algorithms. The metamorphic relations and coverage criteria are checked. To analyze result of mutation testing i.e to measure the quality of test data mutation score is computed. Any variation in result from what is expected indicates defects in the system implementation or input data.

$$\text{Mutation Score MS(t)} = M_k / (M_t - M_q)$$

where T is the test data; M_k is the number of mutants detected by T and M_t is the total number of mutants and M_q the number of equivalent mutants that remain undetected.

5. *Computation Of Fault Detection Ratio:* To find out the extent to which a testing method can detect a fault is by calculating Fault Detection ratio FD, it is the percentage of test cases that could detect certain mutant M

$$FD(M, T) = N_f / (N_t - N_e)$$

where N_f is the number of times the software or the component under test fails, N_t is the number of tests and N_e is the number of infeasible tests. For metamorphic testing N_t is product of number of source test cases and the number of metamorphic relations N_m, as each test case should be executed once for each metamorphic relation. N_e is the total number of times that test cases could not satisfy the requirement of a metamorphic relation.

4 Experimental Study

We conducted an experimental study on triangle type determination program 'Tritype'(Program 1) in order to explain the genetically augmented metamorphic testing and investigate the effectiveness of the approach. We assumed that no test oracle exists for the program hence, we started with identifying Metamorphic relations. Five different MRs are identified based on the properties of the triangle (Table 1)[8]. Same set of mutants are considered. Major change is achieved after creation of test cases by the use of genetic algorithm.

```
0. int a,b,c;
1. int match = 0;
2. if (a = b)
3. match = match + 1;
4. if (a == c)
```

5. match = match + 2;
6. if (b == c)
7. match = match + 3;
8. if (match == 0) /* if a, b and c are not equals to each other*/
9. if (a + b <= c)
10. printf("Not a triangle"); return 0.0; }
11. else if (b + c <= a)   {
12. printf ("Not a triangle"); return 0.0; }
13. else if (a + c <= b) {
14. printf("Not a triangle"); Return 0.0; } else {
15. double p = (a + b + c)/2.0; printf ("Scalene");
16. return sqrt (p*(p-a)*(p-b)*(p-c)); /* compute square */ }
17. else if (match == 1) /* if (a = b ≠ c) */
18. if (a + b <= c) {
19. printf ("Not a triangle"); return 0.0; }
20. double h = sqrt (pow (a, 2) - pow(c/2.0, 2)); printf ("Isosceles");
21. return (c*h)/2.0; /* compute square */ }
22. else if (match == 2) /* if (a = c ≠ b) */
23. if (a + c <= b) { 24. printf ("Not a triangle"); return 0.0; } else {
25. double h = sqrt (pow (a, 2) - pow (b/2.0, 2)); printf ("Isosceles");
26. return (b*h)/2.0; /* compute square */ }
27. else if (match == 3) /* if (b = c ≠ a) */
28. if (b + c <= a) {
29. printf ("Not a triangle."); return 0.0; } else {
30. double h = sqrt (pow (b, 2) - pow (a/2.0, 2)); printf ("Isosceles");
31. return (a*h)/2.0; /* compute square */ } else { /* if (a = b= c) */ printf("Equilateral");
32. return (sqrt (3.0)*a*a)/4.0; /* compute square*/ }

Program 1. Triangle Type Determination

Table 1. Metamorphic Relations for Program

Relation no.	Metamorphic relation	Domain Relation	Condition
Mr1	(a',b',c')=(b,a,c)	TriType(a',b',c')=TriType(a,b,c)	{(a,b,c)la=b&a+b>c}
Mr2	(a',b',c')=(a,c,b)	TriType(a',b',c')=TriType(a,b,c)	{(a,b,c)lb=c&b+c>a}
Mr3	(a',b',c')=(c, b, a)	TriType(a',b',c')=TriType(a,b,c)	{(a,b,c)la=c&a+c>b}
Mr4	(a',b',c')=(2*a, 2*b, 2*c)	TriType(a',b',c')=4*TriType(a,b,c)	Nil
Mr5	(a',b',c')=($\sqrt{2b^2+2c^2- a^2}$,b,c)	TriSquare(a',b',c')=TriSquare(a,b,c)	{(a,b,c)la^2=b^2+c^2}

Following the approach we have considered only four mutants for the original program based on two types of mutant operators: AOR (Arithmetic Operator Replacement) and DSA (Data Statement Alterations)

Mutant 1: Exchange statements 3 and 5;
Mutant 2: Replace statement 15 with "$p= (a+b+c)*2$";
Mutant 3: Replace "/2" in 21, 26 and 31 with "*2";
Mutant 4: Replace statement 32 with *return sqrt(3)*a*a/2*.

The next step after mutant creation is the use of genetic algorithm to generate and optimize test data. It includes selection, crossover, and mutation. The algorithm is iterated until the population evolves to form a solution to the problem, or until a maximum number of iterations have taken place. Table 2 and 3 show the results of first and second iteration respectively. Table 4 shows the result after final cross over and mating.
Different notations used in the tables are:

X: Test data set
F(x): corresponding fitness value calculated for each test data, by adding the weights of the path followed by it in the CFG.
Pi: probability for the corresponding data
$Pi = F(Xi) / (\Sigma (F(Xi)))$
Ci: cumulative probability
Ran: Random number generated for the test data
Ns: Test data number that has cumulative probability just greater than the corresponding random number.
Mating pool: This column contains the number of times a test data appears in the Ns column. The data obtained from the Ns values is written in binary representation
Initial population selected for testing: (0, 5, 8), (13, 10, 9), (8, 5, 6), (8, 2, 9), (8, 0, 5), (3, 4, 5), (1, 5, 0), (3, 4, 2), (3, 5, 7).

Table 2. Results after First Iteration

S. No	X	F(X)	Pi	Ci	Ran	Ns	Mating Pool
1.	(0,5,8)	58	0.090	0.090	0.934	9	0
2.	(13,10,9)	74	0.114	0.204	0.474	5	0
3.	(8,5,6)	74	0.114	0.318	0.374	4	1
4.	(8,2,9)	58	0.090	0.408	0.618	6	1
5.	(8,0,5)	66	0.102	0.510	0.979	9	2
6.	(3,4,5)	80	0.123	0.633	0.533	6	3
7.	(1,5,0)	74	0.114	0.746	0.618	6	0
8.	(3,4,2)	82	0.127	0.873	0.275	3	0
9.	(3,4,2)	82	0.127	1.000	0.487	5	2

Fitness value after first iteration i.e. $\Sigma F(X) = 648$.

Table 3. Results after Second Iteration

S. No	X	F(X)	Pi	Ci	Ran	Ns	Mating Pool
1.	(5,2,9)	58	0.085	0.085	0.487	5	0
2.	(0,3,4)	58	0.085	0.170	0.739	8	1
3.	(10,1,2)	66	0.097	0.267	0.618	7	0
4.	(9,5,3)	66	0.097	0.364	0.382	5	0
5.	(2,6,6)	102	0.150	0.514	0.890	9	3
6.	(3,3,8)	70	0.103	0.617	0.793	8	0
7.	(4,2,3)	82	0.121	0.738	0.821	8	1
8.	(5,5,5)	98	0144	0.882	0.507	5	3
9.	(3,4,5)	80	0.118	1.000	0.101	2	1

Fitness value after second iteration i.e. $\sum F(X) = 680$.

Table 4. Final Results

S. No	X	F(x)	Pi	Ci	Ran	Ns	Mating Pool
1.	(6,6,6)	98	0.138	0.138	0.321	3	1
2.	(5,4,3)	82	0.115	0.253	0.723	7	1
3.	(3,3,5)	86	0.121	0.374	0.901	9	1
4.	(3,0.1)	66	0.093	0.467	0.698	7	0
5.	(5,3,3)	102	0.144	0.611	0.501	5	2
6.	(2,3,5)	58	0.082	0.693	0.491	5	0
7.	(6,3,0)	66	0.093	0.786	0.178	2	2
8.	(1,2,9)	58	0.082	0.868	0.089	1	0
9.	(5,4,5)	94	0.132	1.000	0.881	9	2

Fitness value after third iteration, i.e. $\sum F(X) = 710$. It clearly indicates that test data is being optimized.

Table 5. Results of Metamorphic Testing

TC	P	M1	M2	M3	M4	Mutant Detected
(6,6,6)	T	T	T	T	F	M4
(5,4,3)	T	T	F	T	T	M2
(3,3,5)	T	F	T	F	T	M1,M3
(5,3,3)	T	T	T	F	T	M3
(5,4,5)	T	F	T	F	T	M1,M3
(6,3,0)	T	T	T	T	T	-

Thus, the optimized test data is (6,6,6), (5,4,3), (3,3,5), (5,3,3), (6,3,0), (5,4,5)

This test data (TC) is finally used for metamorphic testing on the original program(P) and its mutants(M1, M4). The results of applying metamorphic testing are shown in Table 5. All the four mutants are detected by testing. Hence Mutation score reaches 1, which indicates effectiveness of test data. Table 6 reports test summary of all mutants by metamorphic testing with metamorphic relation Mri in Columns 2 to 6, where "1" represents that component under test passes the test case and "0" represents its failure.

Table 6. Test Summary Report for All Mutants with Metamorphic Relations

Test Case	Mr1	Mr1	Mr3	Mr4	Mr5	Fault Detection Ratio
(6,6,6)	1	1	1	1	0	80%
(5,4,3)	0	0	0	1	1	40%
(3,3,5)	1	0	0	1	0	40%
(5,3,3)	0	1	0	1	0	40%
(5,4,5)	0	0	1	1	0	20%
(6,3,0)	0	0	0	1	0	20%

Based on the above data, the following conclusions can be drawn:

1. Metamorphic testing augmented with genetic algorithm and mutation testing approach helps in minimizing the number of test cases

2. Metamorphic Relations forth is strongest as compared to all other metamorphic relations as it can be traversed by all the test cases. Thus one can opt for strong relations based on these facts as metamorphic relation selection is crucial to metamorphic testing.

3. Test suite (6,6,6) is the best as it satisfies four different metamorphic relations in all the mutants.

5   Conclusions and Future Work

This paper investigates the effectiveness of Metamorphic testing method in conjunction with Genetic algorithm with a case study of triangle type determination program. The technique of mutation testing has been incorporated. It has been observed that the number of test cases can be reduced substantially to achieve optimization. The use of Genetic algorithm requires the choice of a set of genetic operations between many possibilities. Currently, pair-wise crossover and flip-flop mutation has been considered. Fault detection ration and Mutation score are applied to analyze the effectiveness of testing method.

The case study has shown that metamorphic testing augmented with genetic algorithm result in test suite optimization and in selection of good metamorphic relations. Though we have worked upon a small program, In future we can apply and check the same for more complex programs from different domains, which can generate different metamorphic relations. We will further investigate the role of good and weak metamorphic relations to check the correctness of the program for which no test oracle is present. The number of generations, population size, crossover and mutation probabilities will be varied to initialize the optimization process as all these parameters have great influence on the performance of genetic algorithms.

References

1. Chen, T.Y., Feng, J., Tse, T.H.: Metamorphic testing of programs on partial differential equations: a case study. In: 26th IEEE Annual International Computer Software and Applications Conference (COMPSAC), pp. 327–333. IEEE Computer Society Press, Los Alamitos (2002)
2. Gotlieb, A., Botella, B.: Automated metamorphic testing. In: The Twenty-Seventh IEEE Annual International Computer Software and Application Conference (COMPSAC 2003), pp. 34–40. IEEE Computer Society Press, Los Alamitos (2003)
3. Zhou, Z.Q., Huang, D.H., Tse, T.H., Yang, Z., Huang, H., Chen, T.Y.: Metamorphic testing and its applications. In: 8th International Symposium on Future Software Technology (ISFST 2004). Software Engineers Association, Japan (2004)
4. Chen, T.Y., Huang, D.H., Tse, T.H., Zhou, Z.Q.: Case study on the selection of useful relations in metamorphic testing. In: The Fourth Ibero-American Symposium on Software Engineering and Knowledge Engineering (JIISIC 2004), pp. 569–583. Polytechnic University of Madrid, Spain (2004)
5. Michael, C.C., McGraw, G.E., Schatz, M.A., Walton, C.C.: Genetic algorithm for dynamic test data generation. Technical Report, RSTR-003-97-11
6. Girgis, M.R.: Automatic test data generation for data flow testing using a genetic algorithm. Journal of Universal Computer Science 11(6), 898–915 (2005)
7. Dong, G., Changhai, N., Baowen, X., Lulu, W.: An effective iterative metamorphic testing algorithm based on program path analysis. In: The Seventh International Conference on Quality Software (QSIC 2007), Portland, Oregon, USA, pp. 292–297 (2007)
8. Srivastava, P.R., Kim, T.H.: Application of genetic algorithm in software testing. International Journal of Software Engineering and Its Applications 3(4) (October 2009)
9. Chan, W.K., Cheung, S.C.: A metamorphic testing approach for online testing of service-oriented software applications. International Journal of Web Services Research X (2010)

Reducing Classification Times for Email Spam Using Incremental Multiple Instance Classifiers

Teng-Sheng Moh and Nicholas Lee

Department of Computer Science, San Jose State University
San Jose, CA, U.S.A.
Teng.Moh@sjsu.edu, nicholasjlee@gmail.com

Abstract. Combating spam emails is both costly and time consuming. This paper presents a spam classification algorithm that utilizes both majority voting and multiple instance approaches to determine the resulting classification type. By utilizing multiple sub-classifiers, the classifier can be updated by replacing an individual sub-classifier. Furthermore, each sub-classifier represents a small fraction of a typical classifier, so it can be trained in less time with less data as well. The TREC 2007 spam corpus was used to conduct the experiments.

Keywords: Multiple instance classifiers; email spam; Naïve Bayes; TREC 2007 spam corpus.

1 Introduction

Unsolicited, commercial emails have become an unavoidable daily nuisance. Spam not only fills up inboxes, wastes network resources and lowers employee productivity, but can also be used for more nefarious purposes, such as collecting passwords and other personal information. As companies expand their online presence, the importance of email grows as both a marketing tool and a medium for sending essential and, more importantly, timely communications to their customers. Therefore, correctly identifying spam in a timely manner not only impacts the end recipient, but also the legitimate companies.

It has been estimated that 75% of all emails are unsolicited, i.e., spam [1]. Unlike traditional junk mail, spam can be sent with little or no cost to the sender [2]. However, the cost to the recipient can be enormous. It costs companies billions of dollars a year in terms of lost employee productivity on top of the money spent on anti-spam tools.

This paper focuses on multi-instance incremental classifiers for identifying solicited (ham) and unsolicited (spam) emails. Classifiers are trained at fixed time intervals based on a sample of emails gathered between those training times. As a new classifier is trained, the oldest classifier is discarded in order to limit the number of active classifiers at any given time. When a new email arrives, a majority vote is taken on the remaining classifiers in order to determine if the email is ham or spam.

The remainder of this paper is organized as follows. Section 2 discusses related work. Section 3 describes the proposed algorithm. Section 4 presents the experimental results. Section 5 concludes the paper.

S. Dua, S. Sahni, and D.P. Goyal (Eds.): ICISTM 2011, CCIS 141, pp. 189–197, 2011.

2 Related Work

The algorithm presented in this paper applies aspects of the concepts described in this section. The main driving force behind this algorithm is the grey list concept which is a spam focused application of multiple instance learning. Multiple instance learning algorithms reduce any bias from a specific classifier. Automation can then be achieved with learning classifiers to keep the classifier up to date.

2.1 Grey List Analysis

Islam and Zhou proposed a new email classification technique based on grey list analysis [3]. The term grey list is derived from two email filtering techniques called black listing and white listing. On one hand, white listing is a filtering technique that ensures all emails sent from any contact contained in a particular list is considered ham. On the other hand, a black list considers any email sent from an email or IP address in a separate list as spam.

The grey listing technique uses multiple independent classifiers working together to classify a single email. The number of classifiers can vary, but each type of classifier is different while the training data is the same. The type of classifier varies since each type has its own different strengths and weaknesses. The training data is fixed for all the classifiers in order to limit the attributes and ensure consistency.

While this approach brings the false positive rates down, it creates a third classification type for the end recipients to sort through. To address this issue the algorithm was updated to remove the grey listing portion from the equation and replace it with a simple majority vote [4]. The majority vote can be done as a simple majority vote or a ranked majority vote where each classifier is essentially weighted. This automates the grey listing resolution without introducing additional overhead for the sender or the recipient.

Since the classifiers work off of the same training data, the quality of the training data can be an issue. If one classifier is biased, the other classifiers can automatically correct this issue. But if the majority of the classifiers are biased, then they must all be retrained.

2.2 Multiple Instance Learning

Kang and Naphade utilized multiple instance learning to classify images [5]. Images can be classified in multiple categories. Taking different portions of an image can produce two different, yet valid classifications. For instance, an image of a deer in the woods can be classified as both an animal and scenery.

Zhou, Jorgensen and Inge took this idea and applied it to spam classification [6]. It was specifically used to combat good word attacks. Good word attacks get past traditional classifiers by increasing the number of non-spam words, therefore decreasing the weight of the spam words. The spam words can be shown in larger font or bolded in order to stand out and convey the original message. This algorithm takes an email and splits it into multiple instances. Emails can be split in half or the words can be broken down into ham, spam or neutral words. Splitting up an email into multiple instances combats good word attacks by decreasing the impact of good words.

As words transition from spam to ham and vice versa, the list of words must constantly be updated and the classifier must be retrained in order to split the emails correctly.

2.3 Learning Classifier

Sirisanyalak and Sornil implement a learning classifier [7]. A classifier should learn and evolve over time like the human immune system. As spam changes, classifiers should change as well. New spam words need to be added, and older words should also carry less weight over time until they are eventually discarded. Once they are discarded, they can be re-added if they begin to appear again. Essentially, each spam word has an expected lifetime and once the lifespan ends, it is discarded. This keeps the classifier up to date but does not weigh down the classifier with unused attributes. However, it lessens the weight of a valid attribute based solely on its lifespan and not its frequency. Therefore, a commonly used spam attribute contributes to the classification process less and less over time. Once it dies, that attribute is not accounted until it is added back, even though it could still be found in some spam email.

3 Proposed Algorithm

In this section, the proposed spam classification algorithm is described. It utilized both majority voting and multiple instance approaches. Note that in our method, the majority voting is not weighted.

At any given time, a predetermined number of sub-classifiers are active. Each sub-classifier utilizes the same type of classifier but is trained with different datasets. Each dataset represents a set interval of time with no overlapping data. For illustration, the following discussion uses three sub-classifiers with each dataset representing six months of data. The first sub-classifier is trained with data from months 0–6, the second with months 6–12 and the third with months 12–18. Each sub-classifier produces the classification of an email. Once each sub-classifier has produced its result, a majority vote is taken to determine the actual classification of the email. After six months, a new sub-classifier is trained with training data collected from months 18–24. The oldest sub-classifier is then removed and the new sub-classifier is added to the multiple-instance classifier. Fig. 1 outlines the classifier update process.

This algorithm combines aspects from the classification approaches discussed in Section 2. The first benefit is that of incremental learning. Old classifiers trained with obsolete data are removed and new classifiers with new data are added. This reduces overhead because the classifier can forget old, unused spam words. Any old spam

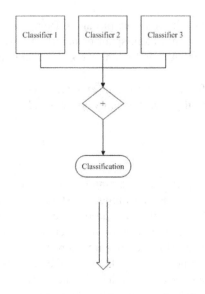

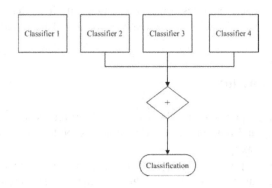

Fig. 1. Updating Classifiers

words that are still being used would still be included in the newer classifiers since they should be part of the training dataset. The second benefit is that the training is performed in stages. Classifiers do not need to be trained all over again every six months. This also creates different sets of attributes to be checked for, which essentially splits the email into multiple parts. Furthermore, if the classifier is found to be biased, an individual sub-classifier can be replaced quickly and easily.

4   Experimental Results

Naïve Bayes is a common classifier used in many spam filters. Bayesian filters achieve high spam detection rates for text emails. However, they have difficulty filtering emails that use different encoding types and non-English characters [8]. Despite these shortcomings, Naïve Bayes was used as the base classifier due to its use in spam filters such as SpamAssassin.

The following section outlines the experimental results produced by text emails contained in the TREC 2007 spam corpus [9]. Due to the limited sample size of the data, the timeframe for the datasets had to be adjusted so the full implementation of the algorithm was not possible.

Three approaches of training classifiers were used. The first type was a single classifier that was trained once and never updated. This type was only included as a baseline comparison. The second type was a normal/retrained classifier that was retrained using all pertinent data at a set interval. The third type was the incremental classifier approach that is outlined in Section 3 of this paper. All the tests were performed on a 2.53GHz Intel Core 2 Duo computer with 2GB of RAM.

4.1 False Positive Rates

Results of testing with 8 different sub-classifiers are shown in Figures 2 and 3. It can be seen from Figure 2 that the ham false positive rate for the incremental classifier type does not perform as well as the normal/retrained classifier type. However, in Figure 3 the spam false positive rates for the incremental classifier type were three times less than that of the normal classifier type, thus the incremental classifier type is

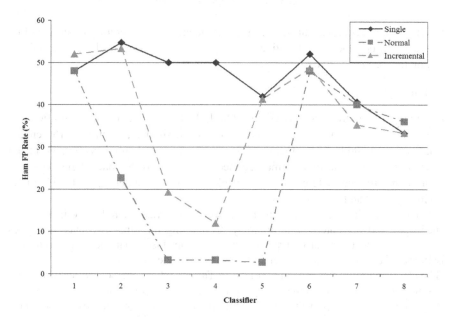

Fig. 2. Ham False Positive Rate

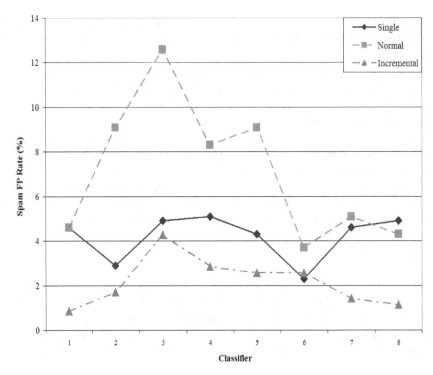

Fig. 3. Spam False Positive Rate

three times more accurate. Further analysis on the attribute generation process will be required to reduce the false positive rate. In summary, these results are promising based on this prototype experiment.

4.2 Timing Results

The training times are summarized in table 1. Each incremental sub-classifier was trained with a dataset of 50 emails and the other classifier types utilized 150 emails per dataset. It should be noted that the first three incremental sub-classifiers utilized a dataset equivalent to the first normal/retained classifier and the lone single classifier. This accounts for the different number of classifiers lists under each classifier type (also in Figures 2 and 3).

For the clarity in analysis, the single classifier type was ignored since it was only trained once and presented as an anomaly. Comparing the other two classifier types shows the incremental classifier type was over six times faster than the normal/retained classifier type as a whole and over eight times faster on the average.

The classification times are summarized in Table 2 where each classifier type using the same dataset of 500 emails. It shows that the classification times for the incremental classifier type were much higher than the other two classifier types. This was mainly caused by the fact that the incremental classifier type had to run each dataset through three different sub-classifiers and this was done serially. If this were done in parallel, the times could be reduced [10].

Table 1. Training Times in Seconds

Dataset	Incremental	Normal / retrained	Single
1	6.23	78.38	77.74
2	9.08	81.34	-
3	12.29	86.07	-
4	6.86	69.12	-
5	8.88	63.45	-
6	8.44	55.67	-
7	6.26	51.45	-
8	6.51	65.91	-
9	9.06	-	-
10	8.84	-	-
Total	82.46	551.39	77.74
Average	8.25	68.92	77.74

Table 2. Classification Times in Seconds

Dataset (Incremental)	Incremental	Normal / Retrained	Single
1	69.22	16.34	18.57
2	53.53	17.88	13.89
3	51.96	14.99	14.99
4	48.80	15.59	14.08
5	45.86	19.94	14.49
6	46.46	12.78	14.48
7	53.76	13.84	14.76
8	54.21	14.32	15.34
Total	423.81	125.66	120.59
Average (Serial)	52.98	15.71	15.07
Average (Parallel)	17.66	15.71	15.07

The classification times were still a bit higher for the incremental classifier type even when run in parallel. However, the lower spam false positive rates and large savings in training times greatly compensates for this minor increase, 3.9 ms per email, in classification time.

5 Conclusions

This paper has shown the proposed incremental multiple instance classifier as a viable spam filter. The main objective of this approach is reinforced by the much improved training times. The approach may be easily applied to any classification algorithm to make it incremental as well as modular. These modular components can be quickly and easily replaced. The low spam false positive rates demonstrate a marked improvement over the single and normal/retrained classifier approaches. Thus, the proposed algorithm is superior for classifying emails.

While these results are promising, there are areas that may be enhanced with further experimentation. The limited amount of data forced the algorithm to be adjusted since the date range for the datasets was not long enough. Large data is difficult to obtained and the data that is publicly available tends to be relatively small and dated.

The next area that can be improved upon is the attribute generation process. The evolving nature of spam also makes it difficult to maintain a somewhat stable attribute list since the content varies and is not limited to one language. Utilizing fuzzy attributes can help improve classifications accuracy and can handle misspelled words [11].

One additional area for further investigation is the base classifier. Naïve Bayes was chosen as a base classifier since it is widely used. Other classification algorithms such as AdaBoost or K-nearest neighbor may also be used instead.

Acknowledgments. Special thanks are due to Reviewers 2 and 4 for their many helpful comments.

References

1. Hoanca, B.: How good are our weapons in the spam wars? IEEE Technology and Society Magazine 25(1), 22–30 (2006)
2. Carpinter, J., Hunt, R.: Tightening the net: a review of current and next generation spam filtering tools. Computers & Security 25(8), 566–578 (2006)
3. Islam, M.R., Zhou, W.: An Innovative Analyser for email classification based on grey list analysis. In: 2007 IFIP International Conference on Network and Parallel Computing Workshops, pp. 176–182. IEEE Computer Society, Washington, DC (2007)
4. Islam, M.R., Zhou, W., Chowdhury, M.U.: MVGL Analyser for Multi-Classifier Based Spam Filtering System. In: The Eighth IEEE/ACIS International Conference on Computer and Information Science (ICIS), pp. 394–399. IEEE Computer Society, Washington, DC (2009)
5. Kang, F., Naphade, M.R.: A generalized multiple instance learning algorithm with multiple selection strategies for cross granular learning. In: 2006 IEEE International Conference on Image Processing, pp. 3213–3216. IEEE Press, New York (2006)
6. Zhou, Y., Jorgensen, Z., Inge, M.: Combating good word attacks on statistical spam filters with multiple instance learning. In: Nineteenth IEEE International Conference on Tools with Artificial Intelligence (ICTAI), pp. 298–305. IEEE Computer Society, Washington, DC (2007)
7. Sirisanyalak, B., Sornil, O.: An artificial immunity-based spam detection system. In: 2007 IEEE Congress on Evolutionary Computation (CEC), pp. 3392–3398. IEEE Press, New York (2007)

8. Yeh, C.-C., Chiang, S.-J.: Revisit Bayesian approaches for spam detection. In: Ninth International Conference for Young Computer Scientists (ICYCS), pp. 659–664. IEEE Computer Society, Washington, DC (2008)
9. SPAM Track Guidelines - TREC 2005-2007,
   `http://plg.uwaterloo.ca/~gvcormac/spam/`
10. Islam, R., Zhou, W., Xiang, Y., Mahmood, A.N.: Spam filtering for network traffic security on a multi-core environment. Concurrency and Computation: Practice and Experience 21(10), 1307–1320 (2009)
11. Tran, D., Ma, W., Sharma, D., Nguyen, T.: Possibility theory-based approach to spam email detection. In: 2007 IEEE International Conference on Granular Computing (GRC), p. 571. IEEE Computer Society, Washington, DC (2007)

Resolving Ambiguous Queries via Fuzzy String Matching and Dynamic Buffering Techniques

Olufade F.W. Onifade and Adenike O. Osofisan

Department of Computer Science, University of Ibadan,
Oyo State, Nigeria
onifadeo@loria.fr, mamosho@yahoo.co.uk

Abstract. The general means for representing user information need is through query. Obtaining desired information is therefore dependent on the ability to formulate some set of words to match the database content. The problem of non-retrieval arises when the query fails to predictably and reliably match a set of document either because of limited knowledge, wrong input and/or supposedly simple errors like words or character transposition, insertion, deletion or total substitution. The accruable risk is better imagined for a scenario where information is employed for strategic decisions. With myriad of string matching function to deal with some of these query problems, the problem has not abated because of uncertainty which engulf the process. This research proposed a fuzzy-based buffering technique to compliment a fuzzy string matching model in a bid to accommodate query matching problems that result from ambiguous query representation.

Keywords: Fuzzy string matching, fuzzy buffering, query evaluation, information retrieval.

1 Introduction

The bulk of information employed for decision purpose resides in databases or what can be referred to as information delivery systems. Access to this information is facilitated by a technique made available by the information retrieval system (IRS) put in place. The sole responsibility of IRS is in providing fast, effective and reliable content-based access to vast amount of information organized as documents (information items) [17,13]. In a bid to access the organized documents, the user is required to formulate his/her query with a set of constraints in a bid to determine the relevance of the query to the information items. The main components of IRSs are: collection of documents, a query language allowing the expression of selection criteria synthesizing the user's needs, and the matching mechanism which estimates the relevance of the documents to the query [13]. Attempt to estimate the relevance of each document with respect to a specific user need is based on a formal model which provides a formal representation of both documents and the user queries. Using the trio of documents collection, query language and the matching mechanism

S. Dua, S. Sahni, and D.P. Goyal (Eds.): ICISTM 2011, CCIS 141, pp. 198–205, 2011.

in an IRS, the input represents the user's query while the corresponding output reflects the relevance estimation of the user information need (query) and the information collection.

It can therefore be established that IRS are composed primarily to provide information solution for decision-making problems, that is: identifying the information items corresponding to the user's information preferences in terms of relevance. Most decision making are based on three main components: obtaining relevant information (from memory or external world), construction of the decision or problem space followed by attempt to fix the acquired information appropriately into the decision problem structure, and assessing the values and likelihoods of different outcomes. The above submission further buttresses the importance of information in everyday life. With myriad of information retrieval tools and techniques and the emergence of search engines as an outshoot of IR the story has not recorded much change. This is sequel to the fact that most search engines are developed from the traditional retrieval models defined several years back. A typical example is the Boolean query which forces the user to precisely express his/her information needs as a set of un-weighted keywords. The result of this is inability to express and accommodate vagueness and uncertainty as part of requirement for specifying selection criteria which tolerate imprecision [14].

This research builds from the fuzzy string matching technique described in [11, 12]. We build up from the buffering mechanism described in [1] to fortify the above fuzzy string matching technique as a robust tool for retrieval operation. Our consideration is based on the issue of relevance which kept recurring in the search for information need from information items. Relevance is contextual, dependent on individual interpretation and some other factors that cannot be adequately described, thus retrieval tools like the Boolean model cannot handle such ambiguous operation. With this notion our focus is to create a flexible means of accessing information systems and databases with focus on reducing the risk accruable from retrieval.

In the rest of this work, we review approximate string matching in section two, and present briefly our novel fuzzy string matching model in section three. We discuss the fuzzy buffering model in section four with some examples and conclude the work in section five.

2 Approximate String Matching

Traditionally, approximate string matching algorithms are classified into two categories: on-line and off-line. With on-line algorithms the pattern can be pre-processed before searching but the text cannot. In other words, on-line techniques do searching without an index. Early algorithms for on-line approximate matching were suggested by [9]. The algorithms are based on dynamic programming but solve different problems. Sellers' algorithm searches approximately for a substring in a text while the algorithm of Wagner and Fisher calculates Levenshtein distance, being appropriate for dictionary fuzzy search only [6]. On-line searching techniques have been repeatedly improved. Perhaps the most famous improvement is the bimap algorithm (also known as the shift-or and shift-and algorithm), which is very efficient for relatively short pattern strings [2]. The Bitap algorithm is the heart of the UNIX searching utility agrep. An excellent review of on-line searching algorithms was done by [9].

Although very fast on-line techniques exist, their performance on large data is grossly inadequate. Text pre-processing or indexing makes searching dramatically faster. Today, a variety of indexing algorithms have been presented. Among them are suffix trees, metric trees and n-gram methods. In computing, approximate string matching provide a means for finding approximate matches to a pattern in a string. The closeness of a match is measured in terms of the number of primitive operations necessary to convert the string into an exact match. This number is called the edit distance — also called the Levenshtein distance — between the string and the pattern. The usual primitive operations are:

    i. *Insertion* (e.g., changing *cot* to *coat*),
    ii. *Deletion* (e.g. changing *coat* to *cot*), and
    iii. *Substitution* (e.g. changing *coat* to *cost*).

Some approximate matchers also treat transposition, in which the positions of two letters in the string are swapped as a primitive operation e.g. "transposition of cost to cots". Different approximate matchers impose different constraints. Some matchers use a single global un-weighted cost, that is, the total number of primitive operations necessary to convert the match to the pattern [8]. For example, if the pattern is coil, foil differs by one substitution, coils by one insertion, oil by one deletion, and foal by two substitutions. If all operations count as a single unit of cost and the limit is set to one, foil, coils, and oil will count as matches while foal will not. Other matchers specify the number of operations of each type separately, while still others set a total cost but allow different weights to be assigned to different operations. Some matchers allow separate assignments of limits and weights to individual groups in the pattern.

Most approximate matchers used for text processing are regular expression matchers. The distance between a candidate and the pattern is therefore computed as the minimum distance between the candidate and a fixed string matching the regular expression [6]. If the pattern is co.l, using the POSIX notation in which a dot matches any single character, both coal and coil are exact matches, while soil differs by one substitution. While the above present a flexible means for string matching, the mode of implementation still does not adequately support the vagueness and uncertainty earlier discussed as a debilitating factor to ambiguous queries. In the next section, we present our proposition on this.

3 Fuzzy String Matching

Fuzzy string matching is our attempt to guide against the risk accruable form some class of dirty-data, which include strings that are miss-spelt, inconsistent entries, incomplete context different ordering and ambiguous data. Consider the strings '*onifade*' and '*onitade*'. The two strings are practically the same, but for the character 't' in the later. The problem arises when a typical matching algorithm encounter this entry, once no direct relationship can be established, it would be ignored. However, when viewed fuzzily, the two strings have a lot in common. Firstly, we can establish that the substring *'oni'* and *'ade'* are in the same position when the two strings are analyzed concurrently. Another point is that they both have the same number of character and thus the main problem is either in misspelling or transposition.

The above described scenario formed the basis for the Fuzzy string matching algorithm analyses shown in Fig. 1. In order to favourably and concurrently compare the user's string and the database contents, two dynamic buffers were created at the commencement of the operation. One holds the unmatched characters of the user sub input *'buffer1'* and the other holds the unmatched characters of the database substring *'buffer2'*.

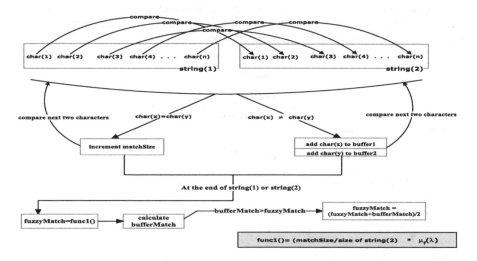

Fig. 1. Fuzzy String Matching Model [11]

The algorithm then scans the character content of the two strings concurrently. When the characters are similar, the variable indicating how many characters were matched is incremented. If the characters are dissimilar, the two characters are stored in *buffer1* and *buffer2* respectively. After all the characters might have been compared, it gets to the end of one of the strings (in the case where the size of the two strings are not the same), the fuzzy match value is calculated based on the level of containment or belongingness (via fuzzy membership function 'func1()') of the *matched character size* and the *size of the database substring* (see Fig. 3). The above operation does not do away with the unmatched characters, instead they are considered to generate some other entries to be displayed alongside the retrieved entries.

While this could generate a high volume of redundant entries, the user has the opportunity to decrease the level of fuzziness at the application level and thus reducing the number of entries. We considered the above as exigent for two reasons, extreme cases of misspelling as in the cases of dyslexia, and when the supplied query forms a subset of the database content but not a whole e.g. *'Oberman'* and *'Hoberman'*. A fuller discussion on this model and its operation can be found in [11].

In what followed, we referred to two sets of buffers that were not explicitly described in the above paper, and which constitute the main contribution of this work. This is the focus of the next section.

4 Fuzzy Buffering Model

Buffering is a technique used for facilitating synchronization between and amongst dissimilar devises. There exist different types of buffer management techniques, a list of which were presented in [1]. We however look at the static buffering, linear buffering and modulo buffering to create an environment for discussing our work. These buffering methods assume that samples are queued in the arc buffers in the arrival order and access is via the movement of buffer indices.

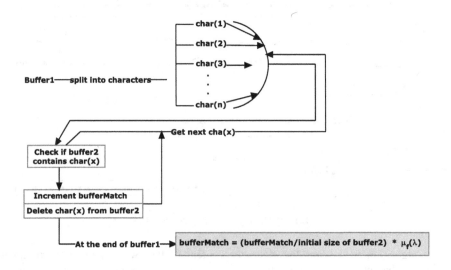

Fig. 2. FuzzyMatch Buffering Model

Static buffering has a limited area of applications, thus its only useful whenever the indices does not change at runtime. Linear and modulo buffering also have their drawbacks, linear buffering requires a large size buffer to function while modulo buffering requires a runtime computation to function appropriately. The overhead of the above methods amongst others does not support the retrieval operation we propose for our work. Thus, we propose a fuzzy based buffering model which employs the fuzzy partitioning method to determine the belongingness of any subset of a string to a particular string.

Users with disabilities such as dyslexia for example could spell a word with the same character content but in most cases, the characters are muddled up. For example a dyslexic could spell '*clement*' as 'elcmten'. In order to trap cases like these, the algorithm analyses the character content of the two strings even if their characters do not match concurrently. To do this, the unmatched characters placed in buffer1 and buffer2 described above were analysed to check for similarity. The case of dyslexics could be considered as an extreme case, but research has shown that most of failed-hit in retrieval operation are due to misspellings. Google and Yahoo search have however propose some level of fuzziness in addressing such problems.

In Fig. 2, we present the buffering operation which is another important model included in FuzzyMatch. The character splitting operation is not complex, but it is important to feed the buffering model for character analysis and fuzzy decision making. After the analysis, a **buffer Match** value is produced. This buffer value is then compared to the *fuzzy match* value. If it is larger, then the new fuzzy value becomes the average of the *fuzzy match* and the *buffer match*. However, if it is less, then the buffer value is discarded and the fuzzy value remains unchanged. When this process is applied to the strings *'clement'* and 'elcmten', a fuzzy match value of comparable to 50% is recorded which we considered fair enough considering that the strings cannot be matched concurrently and could have been otherwise discarded in other search engines.

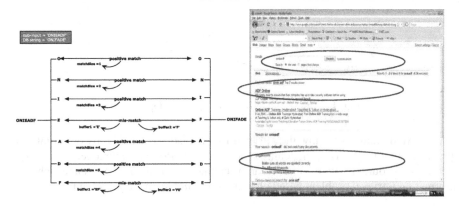

Fig. 3. (a) Buffering operation for the string   (b) Google results for the same string

While it is almost impossible to eradicate data errors at input, manipulation, and deletion, it is important to note that the effect of these errors can be mild or grave depending on the situation. The accruable risk thus is better imagined in real life scenario. With our developed tool, the focus is to reduce the risk in information retrieval processes. We are considering basically some element of the classes considered as dirty data which include: spelling error, omission, insertion and sometimes deletion. We submitted the same input *'ONIEADF'* to Google search engine and the result is interesting. Before discussing the Google result, let us look at the string comparison vis-à-vis the buffering pattern employed by our combined approach to resolving such ambiguity.

The model compares the two buffers containing the unmatched characters produced from stage 2 to check for similarity in their character content. The product of this stage is the buffer match value. Fig. 4 attempts to explicitly capture the operations involved in string splitting, matching and the buffering pattern. Once a character presence can be established in the string, the buffer content continues to be manipulated dynamically until the last entry is considered in the string. This results into the fuzzy match which is the multiplicative effect of the buffer match and the level of belongingness. The fuzzy function employed helps to determine the level of fuzziness in the pattern of arrangement of the user's input and used same to assist in possible rearrangement.

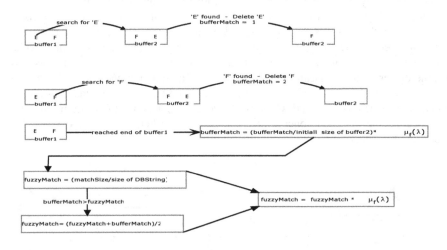

Fig. 4. Strings Comparison and Buffering Operations

We have demonstrated how the retrieval for "onieadf" is handled by our combined approach of fuzzy string matching and fuzzy buffering technique. Our result compares favourable with other search engine in dealing with some inherent information retrieval problem based on ambiguous and incoherent user entries.

5 Conclusion

Adequate definition of user information need is the only way by which information retrieval exercise can be near full success. However it has been established that on many occasions, users don't even know how to define and formulate their queries appropriately in other to result into the right information item stored up in the database. We reiterated that this factor is of grave consequence more importantly when strategic decision are involved. In a bid to ameliorate this, we proposed in this research a fuzzy buffering model to handle ambiguous queries from the definition of information need. The result showed that, issue like transposition, deletion, insertion and/or omission which hitherto could result into no-hit scenario were properly handled in our model. We also facilitate a robust manner via which user can determine the level of fuzziness of their queries and employed same to minimize redundant entries. In the future, we hope to build a complete framework representing complete retrieval tool.

References

1. Aderounmu, G.A., Ogwu, F.J., Onifade, O.F.W.: A dynamic traffic shaper technique for a scalable QoS in ATM networks. In: ICCCT, Austin, Texas, USA, August 14-17, pp. 332–337 (2004)
2. Baeza-Yates, R., Navarro, G.: Fast approximate string matching in a dictionary. In: Proc. SPIRE 1998, pp. 14–22. IEEE CS Press, Los Alamitos (1998)
3. Bordogna, G., Pasi, G.: Controlling retrieval trough a user-adaptive representation of documents. International Journal of Approximate Reasoning 12, 317–339 (1995)
4. Bordogna, G., Pasi, G.: Modelling vagueness in information retrieval. In: Agosti, M., Crestani, F., Pasi, G. (eds.) Lectures in Information Retrieval. Springer, Heidelberg (2001)
5. Crestani, F., Pasi, G.: Soft information retrieval: applications of fuzzy set theory and neural networks. In: Kasabov, N., Kozma, R. (eds.) Neuro-Fuzzy Techniques for Intelligent Information Systems, pp. 287–313. Physica-Verlag, Springer-Verlag Group (1999)
6. Dyke, N.V.: Levenshtein: Levenshtein distance metric in scheme (2006), http://www.neilvandyke.org/levenshtein-scheme/
7. Hahn, J., Chou, P.H.: Buffer optimization and dispatching scheme for embedded systems with behavioral transparency. In: Seventh ACM and IEEE International Conference on Embedded Software Salzburg, Austria, pp. 94–103 (2007)
8. Navarro, G.: A guided tour to approximate string matching. ACM Computing Surveys 33(1), 31–88 (2001), doi:10.1145/375360.375365
9. Navarro, G., Baeza-Yates, R., Sutinen, E., Tarhio, J.: Indexing methods for approximate string matching. IEEE Data Engineering Bulletin 24(4), 19–27 (2001)
10. Oh, H., Dutt, N., Ha, S.: Shift buffering technique for automatic code synthesis from synchronous dataflow graphs. In: Proceedings of the Third IEEE/ACM/IFIP International Conference on Hardware/Software Code Sign and System Synthesis, pp. 51–56 (2005)
11. Onifade, O.F.W., Thiery, O., Osofisan, A.O., Duffing, G.: Dynamic fuzzy string-matching model for information retrieval based on incongruous user queries. Paper presented at the 2010 WCE, London, pp. 283–288 (June 2010)
12. Onifade, O.F.W., Thiery, O., Osofisan., A.O., Duffing, G.: A fuzzy model for improving relevance ranking in information retrieval process. In: International Conference on Artificial Intelligence and Pattern Recognition (AIPR 2010), Florida, USA (July 2010)
13. Pasi, G.: Fuzzy sets in information retrieval: state of the art and research trends. In: Bustince, H., et al. (eds.) Fuzzy Sets and Their Extensions: Representation, Aggregation and Models, pp. 517–535. Springer, Heidelberg (2008)
14. Robins, D.: Interactive information retrieval: context and basic notions. Information Science 3(2), 57–61 (2000)
15. Rocchio, J.J.: Relevance feedback in information retrieval. Prentice Hall, Englewood Cliffs (1971)
16. Rupley, M.L.: Introduction to query processing and optimization, 2, http://www.cs.iusb.edu/technical_reports/TR-20080105-1.pdf
17. Salton, G.: Automatic text processing - the transformation. In: Analysis and Retrieval of Information by Computer. Addison Wesley Publishing Company, Reading (1989)
18. Smeaton, A.F.: Progress in the application of natural language processing to information retrieval tasks. The Computer Journal 35(3), 268–278 (1992)
19. van Rijsbergen, C.J.: Information retrieval. Butterworths, London (1979)
20. Voorhees, E.M., Harman, D.K.: Overview of the Eighth Text Retrieval Conference (TREC-8). In: Information Technology: The Eighth Text Retrieval Conference (TREC-8). NIST SP 500-246, 1-23, GPO: Washington, D.C (2000)

A Deterministic Approach for the Propagation of Computer Virus in the Framework of Linear and Sinusoidal Time Variation of Birth Rate of Virus

Nistala Suresh Rao[1] and Jamwal Deepshikha[2]

[1] Computer Science and
[2] IT, Jammu University, India
suresh_jmu@yahoo.co.in, jamwal.shivani@gmail.com

Abstract. In this research paper the variation (growth & fall) of infection in computers due to virus, in a network of computers, based on susceptible-infected-susceptible (SIS) model is investigated. This is carried out by considering linear and sinusoidal variation for birth rate of virus. The rate of growth and fall of infected nodes is found to vary smoothly and attain a saturation value asymptotically in case of linear variation. While in case of sinusoidal variation, the number of infected nodes follows a sinusoidal variation. For number of nodes greater than the threshold value, there is a damping nature for subsequent cycles. For number of nodes less than or equal to the threshold value, the number of infected nodes non- uniformly decreases. Ultimately the number of infected nodes approaches asymptotically a small value.

Keywords: SIS model, deterministic approach, linear and sinusoidal variation, epidemic threshold.

1 Introduction

Computer virus is one of the frontline areas for research in the present day scenario of computer technology. This offers a great risk to computer systems endangering both the corporation systems of all frames and personal computers used for scientific work, bank accounting and consultancy applications etc. Further, usage of internet facilities increased the number of damaging virus incidents. Since the first trials on how to combat viruses, biological analogies were mainly used as tools to understand the propagation behavior of computer viruses as they share many characteristics with each other. Local systems in a computer network can be attacked through the proliferation of a malicious code, which spreads with virus code along the network and produces network-wide disorders, exactly similar to spreading of diseases in a biological system. Thus the attacks against networks are designated as worms and viruses in analogy to biological terms.

S. Dua, S. Sahni, and D.P. Goyal (Eds.): ICISTM 2011, CCIS 141, pp. 206–213, 2011.
© Springer-Verlag Berlin Heidelberg 2011

In view of similarities between biological and computer viruses, the mathematical models [2, 5] used in biological viruses, can be adopted in computer viruses also. In the present study the same philosophy is followed. Some important parameters such as birth rate, death rate, delay time and vigilance time etc play a vital role to understand the dynamics of computer virus propagation. In the previous studies [1, 3, 4, 6, 7, 8, 9, 10, 11] on computer viruses, models such as SIS and SIR were used, keeping parameters such as birth and death rates as constant. In the present investigation, variation of birth rate with time is considered using SIS model. Further two types of variation of birth rates (β) namely i) linear and ii) sinusoidal are considered in this research work.

The subject matter in the present study is arranged in a sequential order as methodology, results and discussions, conclusions, acknowledgement and references.

2 Methodologies

The SIS model is very simple in its structure and relatively easy to analyze. In this model, in a system of computers, each of the computers is denoted as a node. A pair of nodes is connected by an edge through which virus propagates. In other words one node can infect another node. The probability per unit time that a node infects another node is known as birth rate of computer virus, β. The probability per unit time that a node gets cured is known as death rate of virus, γ. Further, the probability per unit time of the node being cured will be γ only, irrespective of it getting infected by any other node.

In the general model described above, the standard homogeneous interaction can easily be formulated by connecting all pairs of nodes and making all birth and cure rates identical. For simplicity, we assume that the infection rate along each edge to be β and the cure rate for each node to be γ. Such a model provides a reasonably accurate picture of many aspects of the dynamics in the epidemics of computer virus.

The main goal in this investigation is to study the behavior of SIS model with N nodes. Several techniques such as i) deterministic analysis, ii) probabilistic analysis, and iii) simulation analysis etc. are usually considered for the study. However, in the first instance, to have an insight into the problem, we have adopted the deterministic analysis keeping the other analysis for future exercise.

2.1 SIS Model with Birth Rate (β) as Constant (Deterministic Approach)

In a given network of computers let the number of nodes be N. Out of these at a time "t" the number of susceptible nodes be S and the number of infected nodes be I, so that N=S+I. A susceptible node can become infected at a rate βSI and an infected node can be cured and become susceptible again at a rate γI. Then the over all rate of production of infected nodes per unit time is $\beta SI - \gamma I$. Thus the deterministic differential equation describing the time evolution of I(t) is given by

$$\frac{dI}{dt} = \beta SI - \gamma I = \beta(N-I)I - \gamma I = -\beta I^2 + (\beta N - \gamma)I \tag{1}$$

The steady state solution of this equation is given by

$$I(t) = \frac{I_0(N-\rho)}{I_0 + (N-\rho-I_0)e^{-kt}} \quad \text{for } k \neq 0$$

$$I(t) = \frac{1}{\beta t + I_0^{-1}} \qquad \text{for k=0} \tag{2}$$

where $k = \beta N - \gamma$

$$\begin{aligned} \text{As } t \to \infty, \; I(t) &= N - \rho && \text{if } N > \rho \\ &= 0 && \text{if } N \leq \rho \end{aligned} \tag{3}$$

2.2 SIS Model with Specific Rate of Infection as Function of Time

The birth rate of infection β is considered to be varying with time i.e β is a function of time t. In the beginning the general expression is given. Later both the cases i.e linear and sinusoidal variations are described.

The differential Equation (1) becomes

$$\frac{dI}{dt} = (\beta(t)N - \gamma)I - \beta(t)I^2 \tag{4}$$

Put J=1/I, $\dfrac{dJ}{dt} = (-1/I^2)\dfrac{dI}{dt}$

Substituting, the above in Equation (4) we get:

$$\frac{dJ}{dt} = -(\beta(t)N - \gamma)J + \beta(t) \tag{5}$$

Equation (5) is a first order, first degree differential equation which can be solved by standard methods. The steady state solution is given by:-

$$I(t) = \frac{e^{\int (\beta(t)N - \gamma)dt}}{\int \beta(t)e^{\int(\beta(t)N-\gamma)dt} dt + I_0^{-1}} \tag{6}$$

Knowing the nature of β(t) and with the help of trapezoidal rule the numerator and denominator can be evaluated and hence I(t) can be found for various values of 't'.

2.2.1 Linear Variation of Birth Rate β with Time

In this case, birth rate is considered as $\beta(t) = \beta_1 + \beta_2 t$ and on substituting $\beta(t)$, in Equation (6), the expression becomes

$$I(t) = \frac{e^m}{\beta_1 \int_0^t e^{(m} dt + \beta_2 \int_0^t t e^m dt + I_0^{-1}} \tag{7}$$

where $m = (\beta_1 N - \gamma)t + \beta_2 N t^2 / 2$

The trapezoidal rule is used to evaluate the integrations and hence I(t) can be determined for different values of time.

2.2.2 Sinusoidal Variation of Birth Rate β with Time

In this case, birth rate $\beta(t) = \beta \sin \omega t$. On substituting for $\beta(t) = \beta \sin \omega t$ in Equation (6), the expression becomes:

$$I(t) = \frac{e^{k(1-\cos \omega t)-\gamma t}}{\displaystyle\int_0^t (\beta \sin \omega t)e^{k(1-\cos \omega t)-\gamma t}\, dt + I_0^{-1}}, \text{ where } k = \beta N / \omega \quad (8)$$

The trapezoidal rule is used to evaluate the integration in the denominator and finally I(t) is obtained for different values of time.

3   Results and Discussions

As described in the previous section the rate of infected nodes, in a network of computers, depends on the parameters such as number of initial nodes (N), initial infected nodes (I₀), the death rate of virus (γ), the birth rate of virus (β) and on the threshold value ρ. These parameters are varied in general and the variation of I(t) vs. time is enunciated by Equations (2) and (3) where β is constant in the entire duration. From this it can be seen that when N is greater than ρ, there is an epidemic. I(t) increases smoothly and asymptotically attains a saturation value, N-ρ. When N is less than or equal to ρ, I(t) decreases smoothly and ultimately approaches zero value. These are eventually epidemic and non-epidemic situations.

Whereas, if the birth rate varies with time, then ρ is no more constant. However, the initial conditions change for i) N greater than ρ and ii) N less than or equal to ρ in course of time. Out of the several combinations of parameters tried, some significant values for the parameters were finally considered. The variation of I(t) is discussed below.

3.1   Linear Variation of Birth Rate of Computer Virus (β)

The variation of birth rate of virus is chosen as $\beta(t) = \beta_1 + \beta_2 t$ .Thus initially, i.e., at t=0, the value of $\beta(0) = \beta_1$. As time increases, β(t) increases. As a result the value of the epidemic threshold ($\rho = \gamma / \beta(t)$) keeps on decreasing. Initially, if ρ is less than N, it will continue to remain less than N. When initially ρ is equal to N, then as time increases, β(t) will increase and ρ will be continuously decreasing. Similarly, if ρ is greater than N initially, the value of ρ becomes equal to N and then starts decreasing as time increases. Further it may be noted that it is always significant to compare the number of nodes, N with respect to the epidemic threshold ρ, according to Equation (3) to find out the status of the epidemic.

The results arising from the above discussion are depicted graphically in Fig. 1 for various values of N, I₀, γ, β₁, β₂. These are shown as series 1-3. In all the cases, under study, N is always greater than ρ(initial), irrespective of its value, during the entire duration of time. Hence there is an epidemic growth and, finally, I(t) attains a saturation value.

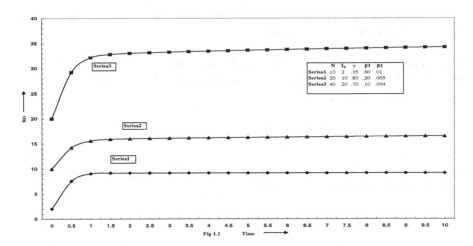

Fig. 1. Variations of Infected Nodes with Time (N > ρ at t = 0. Birth Rate Varies Linearly)

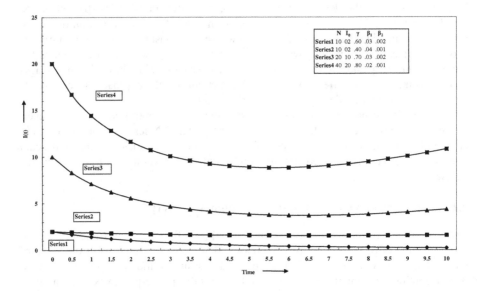

Fig. 2. Variation of Infected Nodes with Time (N ≤ ρ at t = 0. Birth Rate Varies Linearly)

In Fig. 2, results are shown in series 1-4 for N less than or equal to ρ. The values of N, I_0, γ, $β_1$, $β_2$ are shown in a rectangular box. In all the cases, N is less than or equal to ρ (initial). As time proceeds, ρ decreases and thereby N has a tendency to become greater than ρ. Thus initially, there is a decrease in the value of I(t). Gradually I(t) flattens for a while and at the instance when N is greater than ρ, there is a growing tendency in I(t). The results in Figs. 1 and 2 are consistent with the model. The crucial point to be noted is, for N greater than ρ there is an epidemic and for N less than or equal to ρ, there is no epidemic.

3.2 Sinusoidal Variation of Birth Rate of Computer Virus (β)

In sinusoidal variation, birth rate is considered as $\beta(t) = \beta\sin\omega t$. It is to be noted that the threshold value, ρ changes sinusoidally. Further in Equation (8) each of the terms $\beta\sin\omega t$ and $(1-\cos\omega t)$ are cyclic. The term $-\gamma t$ in the exponent of the exponential term becomes prominent for large values of t. As a result the sinusoidal character for the variation of I(t) vs. t is inherent in the propagation. However, it may be noted that the term, $-\gamma t$ plays a significant role in damping the values of I(t) at the position of crest $(3\pi,5\pi,\ldots)$ and trough $(4\pi,6\pi,\ldots)$ corresponding to their respective values at π and 2π in successive cycles. Eventually, if more cycles are covered in a given duration of time this effect can be noticed. Further, the presence of cycles is more pronounced if the term $k = \beta N/\omega$ is larger than 0.3. Ultimately with advancement of time, I(t) decreases exhibiting the inherent sinusoidal behavior, and asymptotically tends to zero.

The values of N, γ, β, I_0 and ω are shown in a rectangular box in Figs. 3(a), 3(b) and Fig. 4. For Figs. 3(a) and 3(b), N is greater than ρ. Similarly for Fig. 4, N is less than or equal to ρ. 'ρ' is the value corresponding to magnitude of $\rho(t)$ at first peak value. Different situations are indicated in the figures as different series. In Fig. 3(a) for the series1-3, the value of k is greater than 0.3 and for series 4, the corresponding value is less than 0.3, where $k = \beta N/\omega$. In Fig. 3(b) for series 1, k value is less than 0.3 and for series 2 and 3, k value is greater than 0.3. In Fig. 4, the value of k is less than 0.3 for all the series. The variation of I(t) is clearly represented in accordance with the discussions.

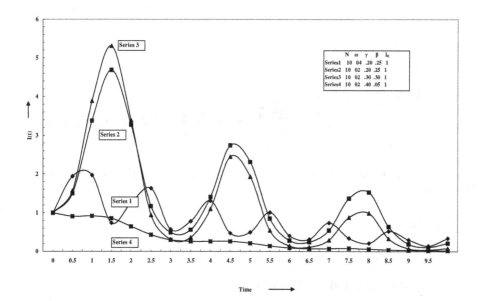

Fig. 3 (a). Variation of Infected Nodes with Time (N > ρ at $\omega t = \pi/2$. Birth Rate Varies Sinusoidally)

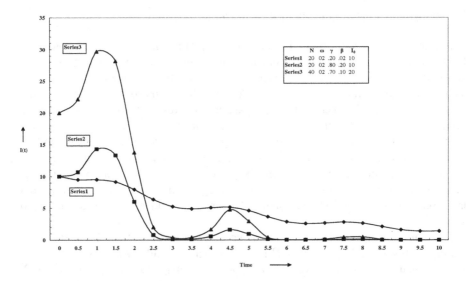

Fig. 3 (b). Variation of Infected Nodes with Time ($N > \rho$ at $\omega t = \pi/2$. Birth Rate Varies Sinusoidally)

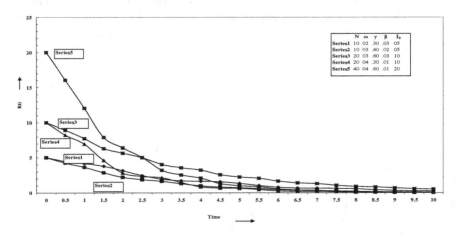

Fig. 4. Variation of Infect ed Nodes with Time ($N \leq \rho$ at $\omega t = \pi/2$. Birth Rate Varies Sinusoidally)

4 Conclusions

From Sections 2 and 3 the conclusions drawn are given below:

4.1 Linear Variation of Birth Rate of Virus

(i) The rate of growth of I(t) varies smoothly and attains asymptotically a saturation value provided N is greater than ρ ($\rho(t)$ at t=0).

(ii) The rate of fall of I(t) varies smoothly and attains asymptotically a saturation value provided N is less than or equal to ρ (ρ(t) at t=0).

(iii) The rate of rise and fall is steeper for higher values of ρ (ρ(t) at t=0).

4.2 Sinusoidal Variation of Birth Rate of Virus

(iv) For N greater than ρ(magnitude of ρ(t) at first peak), the variation of I(t) is sinusoidal and has damping nature in the peak values in the successive cycles. As time increases, the value I(t) approaches zero.

(v) For N less than or equal to ρ (magnitude of ρ (t) at first peak), there is no apparent sinusoidal behavior. However the value of I(t).

Keeps decreasing non-uniformly and finally approaches zero.

Acknowledgments

Both the author's expresses their deep sense of gratitude and thanks to Prof. N.K. Rao (Retd.) for his constant pursuation and useful discussions throughout this investigation.

References

1. Xinchu, F., Small, M., Walker, D.M., Zhang, H.: Epidemic dynamics on scale free networks with piecewise linear infectivity and immunization. Phys. Rev. E 77, 1–8 (2008)
2. Kapoor, J.N.: Mathematical models in biology and medicine, 2nd edn. East West Press, India (1999)
3. Kephart, J.O., White, S.R.: Directed-graph epidemiological models of computer viruses. In: Proceedings of the 1991 IEEE Computer Society Symposium on Research in Security and Privacy, Oakland, California, pp. 343–359 (1993)
4. Liu, Q.X., Jin, Z., Liu, M.X.: Spatial organization and evolution period of epidemic model using cellular automata. Phys. Rev. E. 74, 1–6 (2006)
5. Murray, J.D.: Mathematical models in biology and medicine, 3rd edn. Springer, New York (2002)
6. Mishra, B.K., Saini, D.K.: SEIRS epidemic model with delay for transmission of malicious objects in computer networks. Applied Mathematics and Computations 188(2), 1476–1482 (2007)
7. Piqueira, J.R.C., Navarro, B.F., Monteiro, L.H.A.: Epidemiological models applied to viruses in computer networks. Journal of Computer Science 1, 31–34 (2005)
8. Vazquez, A.: Spreading dynamics on small world networks with connectivity fluctuations and correlations. Phy. Rev. E 74, 1–7 (2006)
9. Wang, Y., Wang, C.: Modeling the effects of timing parameters on virus propagation, Worm, October 27 (2003)
10. Wang, J., Wilde, P.D.: Properties of evolving e-mail networks. Phys. Rev. E. 70(6), 1–8 (2004)
11. Xu, X.-J., Xun, Z., Mendes, J.F.F.: Impact of preference and geography on epidemic spreading. Phys. Rev. E 76, 1–4 (2007)

Pre-processed Depth First Search

Rushabh Hathi and Irfan A. Siddavatam

Department of Information Technology,
K.J. Somaiya College of Engineering Mumbai, India
rushabhhathi@gmail.com, irfansiddavatam@engg.somaiya.edu

Abstract. In this paper, we present a new algorithm for the Depth first search traversing based on Hashing. The data in the nodes are stored in a hash table and corresponding identifier or key is stored in the tree. Thus, whatever might be the data; the entries in the tree will be only integral numbers. This pre processing mark existence and location of node before searching, this causes to faster searched results. Experimental results show that proposed algorithm is simpler and faster of traversing the graph that are repeatedly queried for different goals or paths. Implementation is carried out in Java, and compared with standard DFS and BFS.

Keywords: Search algorithms, software complexity, breadth-first search, depth-first search.

1 Introduction

A graph, G, is a pair of sets (V, E), where V is a finite set of vertices and E is a subset of V x V – a set of edges. That is, each edge is a pair of vertices. Usually, the number of vertices in a graph is denoted by n = |V|, and the number of edges is denoted by m = |E|. Note that $0 \leq m \leq n^2$ if we allow self edges and $0 \leq m \leq n$ (n-1) if we don't. A walk in a graph is a finite sequence of vertices $(v_1, v_2 ... v_k)$ such that for all i between 1 and k-1, $(v_i, v_{i+1}) \in E$. That is, each pair of neighboring vertices in a walk must have an edge between them. This is called a walk from v_1 to v_k. A path is a walk that never visits the same vertex twice. The length of a path is the number of edges in it (for unweighted graphs) or the sum of edge weights (for weighted graphs). A cycle is a walk from some vertex u to u. An Euler cycle is a cycle that visits each edge exactly once. A Hamiltonian cycle is a cycle that visits each vertex exactly once. It's interesting that finding an Euler cycle in a graph can be done in O(n) time, but finding a Hamiltonian cycle is NP hard – no one knows if it can be done in polynomial time. There are several data structures suited for representing graphs. An adjacency matrix, M, is an n x n matrix of zeroes and ones, where M[i][j] is 1 if and only if the edge (i, j) is in E. It requires $O(n^2)$ memory and can answer in constant time the question, "Does G have an edge (i,j)?" An adjacency list, L, is a set of lists, one for each vertex, where L[i] is a list of all vertices j, such that we have an edge (i,j). Since there are n such lists (one per vertex) and for each edge, there is one entry in the corresponding list, the structure requires O (n+m) memory. A typical way of creating an adjacency list in java is to make a Node as Node (A), Node (B), and NodeConnect(A,B) where

S. Dua, S. Sahni, and D.P. Goyal (Eds.): ICISTM 2011, CCIS 141, pp. 214–220, 2011.

Node is class and NodeConnect is method that connect node by adjacency matrix. Finally, another common graph data structure is simply an edge list – a list (or a set) of edges. It requires O (m) space. An undirected graph is called connected if for every pair of vertices, u and v, there is a path from u to v. A sub graph of G= (V,E) is a graph G'=(V', E',) where V '⊂V and E'⊂E. A connected component of G is a maximal connected sub graph of G [1]. All graphs considered for evaluation in this paper are undirected graph [1]. It is arguable that graphs are most commonly employed for searching. For all graph searching algorithms, there exists a general outline for graph traversal. The idea is to begin at a vertex and branch out to other vertices until you reach your destination. Most searches can be done with a list and a loop. You start at the root vertex, perform an operation, add in more vertices to the list, and repeat until you have reached the destination or the list is empty. The difference in searching techniques is the order in which neighboring vertices are placed in the queue. The order is designated by $f(n) = g(n) + h(n)$, where $g(n)$ is the current cost of the edges traversed and $h(n)$ is an underestimated heuristic (an estimated cost) from the current vertex to the destination. Below is a table of different searches based on the utilization of $g(n)$ and $h(n)$. Note that $g(n) = 0$ denotes that there are no costs (weights) associated with the graph and that $h(n) = 0$ signifies that the algorithm does not utilize a heuristic. The category of proposed algorithm falls under uninformed search [2].

Table 1. Search Algorithm Type

g\h	0	nonzero
0	uninformed search	Greedy best-first search
nonzero	uniform-cost search	A* search

The uninformed algorithm assumes that the program has time to search all the way to terminal states, which is usually not practical. We propose an improvement in uninformed depth first search algorithm, where we hash actual values of node and convert give graph to new hashed tree. Thus by having known hashed tree, algorithm depth search only in the branch of terminal state. This feature improves the time complexity as verified in the result (by comparing with BFS and DFS). This paper illustrates the proposed Pre-processed DFS and evaluates their performance. Section 2 provides background by introducing the BFS and DFS algorithm, summarizing and reviewing related research literature. Section 3 outlines the algorithms we have implemented. Section 4 discusses results for sample state space. Section 5 conclusion drawn from our results and outlines ongoing directions for our research.

2   Background

In this section, we will present the search algorithms we use for this comparative study. We will also discuss how each of the algorithms works fundamentally and summaries by reviewing advantages and disadvantages.

2.1 Breadth First Search

In graph theory, breadth-first search (BFS) is a graph search algorithm that begins at the root node and explores all the neighboring nodes. Then for each of those nearest nodes, it explores their unexplored neighbor nodes, and so on, until it finds the goal. BFS is an uninformed search method that aims to expand and examine all nodes of a graph or combination of sequences by systematically searching through every solution. In other words, it exhaustively searches the entire graph or sequence without considering the goal until it finds it. It does not use a heuristic algorithm. From the standpoint of the algorithm, all child nodes obtained by expanding a node are added to a FIFO (i.e., First In, First Out) queue. In typical implementations, nodes that have not yet been examined for their neighbors are placed in some container (such as a queue or linked list) called "open" and then once examined are placed in the container "closed".

Steps in Algorithm:

1. Enqueue the root node.
2. Dequeue a node and examine it.
3. If the element sought is found in this node, quit the search and return a result.
4. Otherwise enqueue any successors (the direct child nodes) that have not yet been discovered.
5. If the queue is empty, every node on the graph has been examined – quit the search and return "not found".
6. Repeat from Step 2 [3][6][7].

2.2 Depth First Search

Depth-first search (DFS) is an algorithm for traversing or searching a tree, tree structure, or graph. One starts at the root (selecting some node as the root in the graph case) and explores as far as possible along each branch before backtracking. Formally, DFS is an uninformed search that progresses by expanding the first child node of the search tree that appears and thus going deeper and deeper until a goal node is found, or until it hits a node that has no children. Then the search backtracks, returning to the most recent node it hasn't finished exploring. In a non-recursive implementation, all freshly expanded nodes are added to a stack for exploration [4][7][8][9].

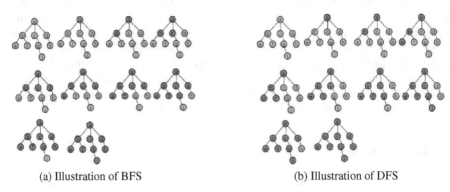

(a) Illustration of BFS   (b) Illustration of DFS

Fig. 1. Illustration of BFS and DFS

3 Algorithm

This algorithm is used to traverse through and search for a particular element a search tree. The algorithm does some pre processing on the data before construction of the tree. When a node is populated, the data actually goes in a hash table and an equivalent integral identifier is actually stored in the tree. Thus, the original tree might be consisting of alphabets as shown in Fig. 1-a. But after preprocessing, the tree look like Fig. 1-b.

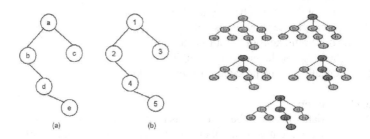

Pre-Processing of Tree

Fig. 2. Illustration of Pre-DFS

Once this is done, we can determine the number of children of each node and thus eliminate the branch which has total children lesser then the value we are searching for. Thus, the actual search is done only in the branch where probability of finding the goal node is 100 percent. The illustrations are provided in Fig. 2. The algorithm and data structure used for implementation in java are as follows:

The data structure of the tree node:

class name: Mytree

1. int data: The data of the node.
2. int c: Specifies the number of children of the node.
3. Mytree parent: Points to the parent of the node.
4. int max,min : Specifies the range of each node (data of max and min of childs)
5. static int count : Counts the number of nodes in the tree.
6. Static Mytree root : Points to the root element

Part 1:

1. The first task is to create a look up table.
2. The lookup table has only 2 entries: node number and node value.
3. Every node is given a number and mapping is done here in the table.

A Table 2 shows the table mapping actual node value and equivalent integer node value after processing

Table 2. Mapping of Node Values

NODE NUMBER	NODE VALUE
1	A
2	B
3	C
4	D

We presume that the tree is made using node number in the data

Insert function:
1. Accept the data to be inserted from the user.
2. Check whether the root is null.
3. If it is so make root point to that node.
4. Then ask for the number of children.
5. Repeat the same procedure for each child.

Search function:
1. The data of node to be searched is taken from user.
2. The equivalent node number is extracted from the lookup table.
3. This value is stored in a variable called 'x'.
4. A property of the table is that if the corresponding node is not found then it will return a number which is very large.
5. Check the status of root.
6. If root== null then display "element not found" and return..
7. If root is not null then proceed.
8. The search function will require a static variable called "nodes_scanned".
9. This variable is used to keep count.
10. We start from 'root'.
11. Set the value of nodes_scanned as 1.
12. If x > count display "Element not found." and return.

4 Results

In this section, we report on the experimental results of our study. The algorithm was tested with a sample search tree as shown in Fig. 3.

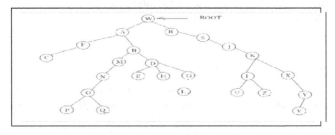

Fig. 3. Sample Search Tree

Table 3. Experimental Results

Node	Execution time in pico sec		
	BFS	**DFS**	**Pre-DFS**
V	8412487	3652131	39016
Z	6793017	3441694	36501
K	5884371	1414565	56494
B	680020	288369	38855
Q	4145666	2960631	39166
P	3272372	2220435	38970
C	4731360	796919	39755

Each of the algorithm searches the specified node and execution time was noted. Table 3 gives the comparison in terms of execution time for BFS, DFS and Pre DFS. The algorithms were implemented in JAVA and executed on Intel Core 2 DUO processor with 1 GB RAM. The graph for time taken for searching various nodes is illustrated in Fig. 3.

5 Conclusions

In this paper, we have examined the performance BSF, DSF and also presented a custom search Pre- DSF in an improved DSF algorithm. From our studies, we showed that breath-first search has weak performance in terms of execution time when it comes for search of individual node. This was obvious since for the node at lowest

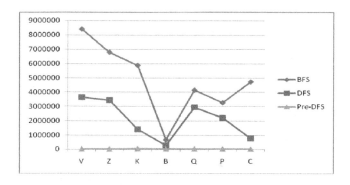

Fig. 4. Execution Time vs. Terminal Node for BFS, DFS, Pre-DFS

level whole tree needs to be traversed. In DFS nodes at the higher level requires less time as compared to leaf node. This variation in term of execution time is overcome in Pre-DFS; it provides almost constant Execution time from top to leaf node search. The constant line in graph shown in Figure 4 is justifies the same. We can say that with property of constant search time Pre-DFS will find application in areas like game theory, networking, etc.

References

1. Naverniouk, I., Chu, F.: Graph algorithms,
   http://www.cs.cornell.edu/~wdtseng/icpc/notes/graph_part1.pdf
2. Chen, J.: Converter introductory graph theory part I,
   http://activities.tjhsst.edu/sct/lectures/graph1.pdf
3. From Wikipedia, the free encyclopedia,
   http://en.wikipedia.org/wiki/Breadth-first_search
4. From Wikipedia, the free encyclopedia,
   http://en.wikipedia.org/wiki/Depth-first_search
5. From Bioinformatics Wikidot,
   http://bioinformatics.wikidot.com/graph-theory-algorithms
6. Cormen, T.H., Leiserson, C.E., Rivest, R.L., Stein, C.: Introduction to algorithms, 2nd edn., pp. 540–549. MIT Press, McGraw-Hill (2001); Section 22.3: Depth-first search
7. Knuth, D.E.: The art of computer programming 1, 3rd edn. Addison-Wesley, Boston (1997)
8. Russel, S., Norvig, P.: Artificial intelligence, a modern approach (2003)
9. Kurant, M., Markopoulou, A., Thiran, P.: On the bias of BFS (Breadth First Search). International Teletraffic Congress, ITC 22 (2010)

Soft Computing Model to Predict Average Length of Stay of Patient

Bindu Garg[1], M.M. Sufyan Beg[1], A.Q. Ansari[2], and B.M. Imran[1]

[1] Department of Computer Engineering, Jamia Millia Islamia,
New Delhi-110025, India
[2] Department of Electrical Engineering, Jamia Millia Islamia,
New Delhi-110025, India
bindugarg80@gmail.com, mmsbeg@hotmail.com,
aqansari@ieee.org, abbabeta@gmail.com

Abstract. Forecasting the average Length of Stay (LoS) of a patient is prime aspect for all hospitals to effectively determine and plan services demanded at various level. Prediction of LoS plays a vital role in strategic decision making by health care administrators. In this paper, a dynamic computational model based on time series, implemented using soft computing techniques is presented to forecast average length of stay of patient. Aim of designing proposed model is to overcome the drawbacks of the exiting approaches and derive more robust and accurate methodology to forecast LoS of patient. Subsequently, the performance of the proposed model is demonstrated by comparing the results of proposed model with some of the pre-existing forecasting methods. In general, the findings of the study are interesting and superior in terms of least Average Forecasting Error Rate (AFER) and Mean Square Error (MSE) values.

Keywords: Time series, soft computing, fuzzy logic, average length of stay (LoS), average forecasting error rate, mean square error.

1 Introduction

Traditionally, it is considered that the provision and use of health services is of great significance in meeting the health care needs of population. The demand for effective health services is ever increasing. As a result, hospitals are absorbing tremendous amount of resources to improve their services. Quality of health care services intern depends on the efficient planning and management of hospital resources. Remarkably, LoS of in-patients has been found as a proxy for measuring the consumption of hospital resources. LoS is vital not only for the overall understanding of system activity but also for improving functionality of health care institutes. OTA case studies scripted in USA acknowledged the importance and impact of LoS on health outcomes by introducing the concept of Myocardial Infarction (MI) [1]. Further in this direction, forecasting the LoS can be miraculous for patient planning, medical resource utilization and overall management of health care systems. The aim of forecasting the LoS can also be seen in terms of individual care. In addition, the advanced prediction of patient LoS in hospital can play a significant role in health

S. Dua, S. Sahni, and D.P. Goyal (Eds.): ICISTM 2011, CCIS 141, pp. 221–232, 2011.

insurance plans and reimbursement systems. Hence, forecasting accurate average LoS of patient is very crucial factor in health care domain. Evaluating and predicting LoS information is a challenging task [2]. It is revealed from past research that LoS varies and variations are not only unpredictable but also time consuming [3]. Consequently, infinitesimal research work is done in this direction. The existing methods are outdated and inaccurate to measure LoS. These are limited only to laboratory application and case studies. Few application of soft computing techniques have been made in the area of patient planning and medical resource utilization [4] but these applications could not depict the complexity of health care systems, making them impracticable for hospitals. However, LoS is essential for the operational success of hospitals. It is thus important to employ scientifically sound and robust soft computing based methods for predicting LoS.

This paper can broadly be divided into six sections. Section 1 is current section, short introduction on the significance and role of LoS in a hospital. Section 2 briefly refers about related work. Section 3 describes the new dynamic computational algorithm for forecasting in detail. Section 4 demonstrates the effectiveness of projected forecasting model to predict average length of stay of patient in health care domain. Section 5 evaluates and compares the result of proposed forecasting model with previous forecasting models. Section 6 has conclusion.

2   Related Work

Song and Chissom [5,6,7,8] used fuzzy set theory to develop models for fuzzy time series and applied them on the time series data of University of Alabama to forecast enrollments. Song and Chissom used an average auto correlation function as a measure of dependency. Chen [9,10,11] presented simplified arithmetic operations and considered high order fuzzy time series model. Hunrag [12,13], Hsu and Chen [14], Hwang and Chen [15], Lee Wang and Chen [16], Li and Kozma [17], Melike and Degtiarev [18], Melike and Konstsntin [19]; all developed number of fuzzy forecasting methods with some variations. Jilani, Burney and Ardi [20, 21] partitioned the universe of discourse into equal length interval and developed method based on frequency density partitioning. Singh [22] developed forecasting models using computational algorithm.

Aforementioned research on fuzzy time series for forecasting problems focused on obvious linguistic values; while ignoring slight, but potentially crucial clues behind those obvious ones. Hence dearth of strong standard methodology to predict LoS and its crucial-ability in health care domain motivated us to develop an effectual and consistent dynamic soft computing based time series model to forecast average length of stay of a patient. In this paper an innovative forecasting model is proposed to rectify these imperfections. Proposed method capitalizes on available information with different perspectives. Model endeavor the issue of improving forecasting accuracy by introducing the concept of dynamic event discretization function and novel approach of weighted frequency density based distributions for length of intervals.

Research in this paper utilizes a computational fuzzy based algorithm to conceptualize the technique that can accurately predict the LoS. Scope and suitability of the proposed model in health care domain is very considerable. For verification, we collected month wise data for 2 years from 2008 to 2009 of a territory care hospital and exercised our proposed technique on this data.

3 Proposed Prediction Algorithm (LoS)

In this section, we present the new dynamic computational algorithm for forecasting of time series data. Strength of algorithm lies in integrated usage of event discretization function, dynamic weighted frequency based distribution function and fuzzy logic relationship in an inventive way. Primarily it calculates event discretization function for time series in terms of RoCs and defines universe of discourse on these RoCs. Thereafter, it applies the concept of weighted frequency density based distributions on intervals so that intervals get weightage on the basis of maximum rate of change along with frequency of RoCs. As a next step, the fuzzy logical relationships is formulated in such a way that impact of previous and next value is also accounted on current time series fuzzy value F_i besides the influence of fuzzy intervals F_{i-1} and F_{i+1} on F_i. Subsequently, to utilize derived information effectively, some arithmetic operations are performed in a creative way. Finally, forecasted value is generated using predicted ROC.

3.1 Key Concepts of Algorithm

Event discretization function. The discretization operation causes a large reduction in the complexity of the data. This process is usually carried out as a first step toward making data suitable for numerical evaluation. In our problem, event discretization function can be defined in a way, so that its value at time t index correlates with the occurrence of the event at particular specified time in the future.

RoC(t) = (X(t+1) - X(t)) \ X(t), where X(t+1) is value at time t+1 index and X(t) is actual value at time t index. ROC is the rate of change of value from time t to t+1.

Frequency density based partitioning procedure
- Determine number of RoC fall in each fuzzy interval
- Identify the interval with maximum frequency of RoC.
- Search for existence of similar interval (sign ignored) having frequency of RoC with a difference of less than or equal to one, e.g., [-1 -2] and [2,1] are same interval (sign ignored).
- In case any such interval does not exist, continue with normal procedure of Frequency distribution.
- Otherwise select the interval having maximum rate of change among these two and divide it into four sub intervals.
- Repeat the same process for next two intervals of highest frequency. Further divide these intervals into three and two sub intervals, respectively. Let all subsequent intervals remain unchanged in length.frequency density procedure is demonstrated in Table 3.

Optimization of fuzzy relationship procedure. Using this procedure, more optimized F_i at particular time t would be generated. Obtain the fuzzy logical relation for $F_i{\rightarrow}F_j$ and $F_j{\rightarrow}F_k$, where F_j is fuzzy value   at time y. F_i is fuzzy value at time y-1. F_k is fuzzy value at time y+1. Get their corresponding RoCs .D is time variant parameter which is calculated as:

$$\text{Calculate } D = \| (RoCj - RoCi) | - | (RoCk - RoCj) \| \tag{1}$$

To generate nearest and optimized value, some simple arithmetic operations are performed [19]. F_{val} is defuzzified value obtained in step 6 of proposed algorithm in section 3.3. X_i, Y_i, Z_i and W_i are positive added fraction in F_{val} of D, D/2, D/4 and D/8 respectively. XX_i, YY_i, ZZ_i and WW_i are positive subtracted fraction in F_{val} of D, D/2, D/4 and D/8, respectively.

Notations used in the procedure

[*Fj] is corresponding interval uj for which membership in Fj is Supremum (i.e. 1).

L[*Fj] is the lower bound of interval uj.

U[*Fj] is the upper bound of interval uj.

l[*Fj] is the length of the interval uj whose membership in Fj is Supremum (i.e. 1).

M[*Fj] is the midvalue of the interval uj having Supremum value in Fj

Initialize Count=1 and Target = F_{val}

i.	$X_i = F_{val} + D$, if $X_i \leq L$ [*Fj] and $X_i \geq U$ [*Fj] then Target= X_i and Count=Count+1
ii.	$XX_i=F_{val} - D$, if $XX_i \leq L$ [*Fj] and $XX_i \geq U$ [*Fj] then Target=Target + XX_i and Count=Count+1
iii.	$Y_i= F_{val} + D/2$, if $Y_i \leq L$ [*Fj] and $Y_i \geq U$ [*Fj] then Target= Target + Y_i and Count=Count+1
iv.	$YY_i=F_{val} - D/2$, if $YY_i \leq L$ [*Fj] and $YY_i \geq U$ [*Fj] then Target=Target + YY_i and Count=Count+1
v.	$Z_i= F_{val} + D/4$, if $Z_i \leq L$ [*Fj] and $Z_i \geq U$ [*Fj] then Target= Target + Z_i and Count=Count+1
vi.	$ZZ_i = F_{val} - D/4$, if $ZZ_i \leq L$ [*Fj] and $ZZ_i \geq U$ [*Fj] then Target= Target + ZZ_i and Count=Count+1
vii.	$W_i=F_{val} + D/8$, if $W_i \leq L$ [*Fj] and $W_i \geq U$ [*Fj] then Target= Target + W_i and Count=Count+1
viii.	Wwi=Fval-D/8, if WWi $\leq$L[*Fj] and WWi$\geq$U[*Fj] then Target= Target+WWi and Count=Count+1
ix.	FRoCj = Target/Count (FRoC is optimized forecasted RoC at time y)
x.	Return FRoCi

3.2 Basic Definitions Used in Algorithm

Fuzzy Set: A fuzzy set is a pair (A, m) where A is a set and m:A$\rightarrow$ |0,1|

For a finite set A={x_1, ..., x_n}, the fuzzy set (A, m) is often denoted by {$(m(x_1)/x_1)$, ..., $(m(x_n) / x_n)$}. For each, x $\in$ A, m(x) is called the grade of membership of x in (A, m).

Let x $\in$ A. Then x is not included in the fuzzy set (A, m) if m(x)=0, x is fully included if m(x)=1, and x is called fuzzy member if 0 < m(x) < 1.

The set $x \in A \mid m(x)>0$ is called the support of (A, m) and the set $x \in A \mid m(x)=0$ is called its kernel.

Time Series: A series of observations made sequentially in time constitute. In time domain analysis, a time series is represented by a mathematical model $G(t) = O(t) + R(t)$, where $O(t)$ represents a systematic or ordered part and $R(t)$ represents a random part. The fact is that the two components cannot be observed separately and may involve several parameters.

Fuzzy Time Series: If fuzzy set $F(t)$ is caused by more fuzzy sets; $F(t-n)$, $F(t-n+1)$.........$F(t-1)$, the fuzzy relationship is represented by A_{i1}, A_{i2} ....$A_{in} \rightarrow A_j$, here $F(t-n)=A_{i1}$, $F(t-n+1) = A_{i2}$ and so on $F(t-1) = A_{in}$. The relationship is called nth order fuzzy time series model.

Average Forecasting Error Rate: AFER can be defined as

$$AFER = (\textstyle\sum_{t=1}^{n}(\mid A_t - F_t \mid / A_t)) / n * 100\% \tag{2}$$

Where, A_t is actual value and F_t is forecasted value of time series data at time t.

Mean Square Error: MSE can be defined as

$$MSE = \textstyle\sum_{t=1}^{n}(A_t - F_t)^2 / n \tag{3}$$

where, A_t is actual value and F_t is forecasted value of time series data at time t n is total number of time series data.

3.3   Algorithm Steps

Step 1: Event discretization function is calculated for given time series data t=1 to n:

Step 2: Define the universe of discourse on RoC as U and partition it into equal intervals say; u_1, u_2, u_3, u_4.....................u_n of equal lengths.

Step 3: Call frequency density distribution procedure.

Step 4: Define each fuzzy set F_i based on the re-divided intervals and fuzzify the time series data where fuzzy set F_i denotes a linguistic value of the RoC represented by a fuzzy set. We use a triangular membership function to define the fuzzy sets F_i [20].

Step 5: After fuzzification of historical data, establish the fuzzy logic relationships using rule:

Rule: If Fj is the fuzzy production at time period n, Fi is the fuzzify production at time period n -1 and F_k is the fuzzify production at time period n +1 then the fuzzy logical relation is denoted as Fi→Fj and Fj→F_k. Here, Fj is called current state, Fi is the previous state and F_k is next state. RoC_j is percentage change at time frame n, RoC_k is percentage change at time frame n-1 and ROC_k is percentage change at time frame n+1.

Step 6: Let us assume the fuzzify value of RoC at particular time period is F_j, calculated in step5. Approximate value of RoC can be generated at same time j using defuzzification formula [20]. In this formula f_{j-1}, f_j, f_{j+1} are the mid points of the fuzzy intervals F_{j-1}, F_j, F_{j+1} respectively. F_{val} is defuzzify value of F_j. Above formula fulfills the Axioms of Fuzzy set like monotonicity, boundary condition, continuity and idempotency.

Step 7: Optimized forecasting of data F_j for the time period n and onwards is done as follows: For y=2 to ... Y (end of time series data). Call optimization fuzzy relationship procedure

Step 8: Calculate Forecasted value as:

$$Forecast_{val} = (x(t)_y * FRoC_{y+1}) + x(t)_y \qquad (4)$$

where, y=1 to n-1. Here, Forecast$_{val}$ is forecasted value at time y+1. $x(t)_y$ is the actual value at time y and FRoC$_{y+1}$ is corresponding value obtained at step7.

4   Proposed Model Exercised to Forecast Average Length of Stay of Patient

Average length of stay of particular month is calculated as ALS=H_d/D, where H_d is total number of hospital days in a month (H_d) and D is discharges. Here total hospital days is the sum of number of days spent in the hospital by each inpatient discharged during the time period examined regardless of when the patient was admitted. Total Discharge is the number of inpatient released from the hospital during the time period examined. This figure includes deaths. Births are excluded unless the infant was transferred to the hospital neonatal intensive care unit prior to discharge.

Step 1: Histological Data used for forecasting is for the year 2008 and 2009. We calculated RoC$_{t+1}$ of every month of year 2008 and 2009. RoC of hospital days and discharges for each month is shown in Table 1.

Table 1. Calculation of RoC

Month (2008)	H_D	ROC$_{HD}$	D	ROC$_D$	ALS	Month (2009)	H_D	ROC$_{HD}$	D	ROC$_D$	ALS
01	9993		1351			01	10223	-0.47%	1502	18%	7
02	9101	-8.93%	1378	2%	7	02	9515	-6.93%	1563	4.06%	6
03	10999	20.85%	1490	8.13%	7	03	10787	13.37%	1657	6.01%	7
04	9589	-12.82%	1418	-4.83%	7	04	9926	-7.98%	1378	-16.84%	7
05	9328	-2.72%	1500	5.78%	6	05	9819	-1.08%	1604	16.4%	6
06	8654	-7.23%	1392	-7.2%	6	06	9325	-5.03%	1589	-0.94%	6
07	10789	24.67%	1534	10.2%	7	07	10985	17.8%	1697	6.8%	6
08	9517	-11.79%	1411	-8.02%	7	08	9355	-14.84%	1437	-15.32%	7
09	10231	7.5%	1253	-11.2%	8	09	9729	4%	1398	-2.71%	7
10	10789	5.45%	1320	5.35%	8	10	10213	4.97%	1389	-0.64%	7
11	9654	-10.52%	1240	-6.06%	8	11	9312	-8.82%	1278	-7.99%	7
12	10271	6.39%	1268	2.26%	8	12	9423	1.19%	1267	-0.86%	7

Step 2: Define the universe of discourse U on hospitals days and discharges. Partition it into intervals u_1, u_2 ... u_n of equal length. The RoC of hospital days ranges from -15% to 24.67% and for discharges RoC ranges from -17% to 22%. The universe of discourse for hospital days is U = [-15, 25] and for discharges it is U = [-18, 30]; partition these U into eight equal intervals.

Step 3: The weighted frequency density based distribution and partitioning of ROC is given in Table 2.

Table 2. Frequency Distribution

H_D		D	
Interval	Freq-uency	Interval	Freq-uency
{-15,-10}.	4	{-18, -12}	2
{-10,-5}.	6	{-12,-6}	5
{-5,0}	2	{-6,0}	5
{0,5}	3	{0,6}	5
{5,10}	4	{6,12}	3
{10,15}	1	{12,18}	2
{15,20}	1	{18,24}	1
{20,.25}	2	{24,30}	0

Table 3. Defining Fuzzy Set

Linguistic	Interval	Linguistic	Interval
H_1	{-15,-12.5}	D_1	{-18, -12}
H_2	{-12.5, -10}	D_2	{-12,-10.5}
H_3	{-10,-8.75}	D_3	{-10.5,-9}
H_4	{-8.75,-7.5}	D_4	{-9,-7.5}
H_5	{-7.5,-6.25}	D_5	{-7.5,-6}
H_6	{-6.25,-5}	D_6	{-6,-4}
H_7	{-5,0}	D_7	{-4,-2}
H_8	{0,5}	D_8	{-2,0}
H_9	{5,6.67}	D_9	{0,3}
H_{10}	{6.67, 8.34}	D_{10}	{3,6}
H_{11}	{8.34, 10}	D_{11}	{6,12}
H_{12}	{10,15}	D_{12}	{12,18}
H_{13}	{15,20}	D_{13}	{18,24}
H_{14}	{20,25}	D_{14}	{24,30}

Step 4: Define each fuzzy set F_i on the re-divided intervals and fuzzify the time series data where fuzzy set F_i denotes a linguistic value of the RoC represented by a fuzzy set is shown in Table 3.

Step 5: Create fuzzy logical relationship as shown in Table 4.

Table 4. Fuzzy Relationship

Month	ROC_{HD}	Fuzzy	FLR (H_d)	ROC_{HD}	Fuzzy	FLR (D_d)
2008/01						
2008/02	-8.93%	H_3	$H_3{\rightarrow}H_{14}$	2%	D_9	$D_8{\rightarrow}D_{11}$
2008/03	20.85%	H_{14}	$H_{14}{\rightarrow}H_3, H_{14}{\rightarrow}H_1$	8.13%	D_{11}	$D_{11}{\rightarrow}D_8, D_{11}{\rightarrow}D_6$
2008/04	-12.82%	H_1	$H_1{\rightarrow}H_{14}, H_1{\rightarrow}H_7$	-4.83%	D_6	$D_6{\rightarrow}D_{11}, D_6{\rightarrow}D_{10}$
2008/05	-2.72%	H_7	$H_7{\rightarrow}H_1, H_7{\rightarrow}H_5$	5.78%	D_{10}	$D_{10}{\rightarrow}D_6, D_{10}{\rightarrow}D_5$
2008/06	-7.23%	H_5	$H_5{\rightarrow}H_7, H_5{\rightarrow}H_{14}$	-7.2%	D_5	$D_5{\rightarrow}D_{10}, D_5{\rightarrow}D_{11}$
2008/07	24.67%	H_{14}	$H_{14}{\rightarrow}H_5, H_{14}{\rightarrow}H_2$	10.2%	D_{11}	$D_{11}{\rightarrow}D_5, D_{11}{\rightarrow}D_4$
2008/08	-11.79%	H_2	$H_2{\rightarrow}H_{14}, H_2{\rightarrow}H_{10}$	-8.02%	D_4	$D_4{\rightarrow}D_{11}, D_4{\rightarrow}D_2$
2008/09	7.5%	H_{10}	$H_{10}{\rightarrow}H_2, H_{10}{\rightarrow}H_9$	-11.2%	D_2	$D_2{\rightarrow}D_4, D_2{\rightarrow}D_{10}$
2008/10	5.45%	H_9	$H_9{\rightarrow}H_{10}, H_9{\rightarrow}H_2$	5.35%	D_{10}	$D_{10}{\rightarrow}D_2, D_{10}{\rightarrow}D_5$

Table 4. (*continued*)

2008/11	-10.52%	H_2	$H_2{\rightarrow}H_9$, $H_2{\rightarrow}H_9$	-6.06%	D_5	$D_5{\rightarrow}D_{10}$, $D_5{\rightarrow}D_9$
2008/12	6.39%	H_9	$H_9{\rightarrow}H_2$, $H_9{\rightarrow}H_7$	2.26%	D_9	$D_9{\rightarrow}D_5$, $D_9{\rightarrow}D_{13}$
2009/01	-0.47%	H_7	$H_7{\rightarrow}H_9$, $H_7{\rightarrow}H_5$	18.45%	D_{13}	$D_{13}{\rightarrow}D_9$, $D_{13}{\rightarrow}D_{10}$
2009/02	-6.93%	H_5	$H_5{\rightarrow}H_7$, $H_5{\rightarrow}H_{12}$	4.06%	D_{10}	$D_{10}{\rightarrow}D_{13}$, $D_{10}{\rightarrow}D_{11}$
2009/03	13.37%	H_{12}	$H_{12}{\rightarrow}H_5$, $H_{12}{\rightarrow}H_4$	6.01%	D_{11}	$D_{11}{\rightarrow}D_{10}$, $D_{11}{\rightarrow}D_1$
2009/04	-7.98%	H_4	$H_4{\rightarrow}H_{12}$, $H_4{\rightarrow}H_7$	-16.84%	D_1	$D_1{\rightarrow}D_{11}$, $D_1{\rightarrow}D_{12}$
2009/05	-1.08%	H_7	$H_7{\rightarrow}H_4$, $H_7{\rightarrow}H_6$	16.4%	D_{12}	$D_{12}{\rightarrow}D_1$, $D_{12}{\rightarrow}D_8$
2009/06	-5.03%	H_6	$H_6{\rightarrow}H_7$, $H_6{\rightarrow}H_{13}$	-0.94%	D_8	$D_8{\rightarrow}D_{12}$, $D_8{\rightarrow}D_{11}$
2009/07	17.8%	H_{13}	$H_{13}{\rightarrow}H_6$, $H_{13}{\rightarrow}H_1$	6.8%	D_{11}	$D_{11}{\rightarrow}D_8$, $D_{11}{\rightarrow}D_1$
2009/08	-14.84%	H_1	$H_1{\rightarrow}H_{13}$, $H_1{\rightarrow}H_8$	-15.32%	D_1	$D_1{\rightarrow}D_{11}$, $D_1{\rightarrow}D_7$
2009/09	4%	H_8	$H_8{\rightarrow}H_1$, $H_8{\rightarrow}H_8$	-2.71%	D_7	$D_7{\rightarrow}D_1$, $D_7{\rightarrow}D_8$
2009/10	4.97%	H_8	$H_8{\rightarrow}H_8$, $H_8{\rightarrow}H_3$	-0.64%	D_8	$D_8{\rightarrow}D_7$, $D_8{\rightarrow}D_4$
2009/11	-8.82%	H_3	$H_3{\rightarrow}H_8$, $H_3{\rightarrow}H_8$	-7.99%	D_4	$D_4{\rightarrow}D_8$, $D_4{\rightarrow}D_8$
2009/12	1.19%	H_8	$H_3{\rightarrow}H_8$,	-0.86%		D_8

Step 6: Defuzzified value of Roc is presented in *Table 5* as specified

Table 5. Defuzzification of RoC

H_D				D			
Month-08	H_{val}	**Month-09**	H_{val}	**Month-08**	D_{val}	**Month-09**	D_{val}
01		01	-2.69%	01		01	20.11
02	-9.41%	02	-6.76%	02	7.20	02	3.27
03	20.54%	03	12.26%	03	7.83	03	7.83
04	-12.8%	04	-8.03%	04	-4.54	04	-13.50
05	-2.69%	05	-2.69%	05	3.27	05	13.70
06	-6.76%	06	-4.44%	06	-6.48	06	-2.40
07	20.54%	07	16.76%	07	7.83	07	7.83
08	-11.2%	08	-12.8%	08	-8.11	08	-13.50
09	7.32%	09	7.32%	09	-11.53	09	-2.14
10	4.57%	10	7.32%	10	3.27	10	-2.40
11	-11.2%	11	-9.41%	11	-6.48	11	-8.11
12	4.57%	12	7%	12	7.20	12	-2.40

Step 7 and Step 8: Forecasted value of ALS is shown in Table 6, along with MSE and AFER.

Table 6. Calculations for Forecasting ALS, MSE, and AFER

Month	$FROC_{Hd}$	$FROC_D$	Fore-H_D	Fore-D	Fore-LoS	$(A_i - F_i)^2$	$|A_i - F_i|/A_i$
2008/01							
2008/02	-9.41	2	9052.6587	1378.02	6.56932	0.18548254	0.06152526
2008/03	22.12	7.83	11114.141	1485.8974	7.47975	0.23016014	-0.06853573

Table 6. (*continued*)

2008/04	-12.8	-5.13	9591.128	1413.563	6.78507	0.04619366	0.03070387
2008/05	-2.69	5.64	9331.0559	1497.9752	6.22911	0.05249249	-0.0381854
2008/06	-6.76	-7.03	8697.4272	1394.55	6.23673	0.05603952	-0.03945445
2008/07	24.63	7.83	10785.48	1500.9936	7.18556	0.03443267	-0.02650863
2008/08	-11.8	-8.11	9518.0558	1409.5926	6.75235	0.06133291	0.03537926
2008/09	6.86	-11.53	10169.866	1248.3117	8.1469	0.02157858	-0.01836206
2008/10	6.06	5.02	10850.999	1315.9006	8.24606	0.06054676	-0.03075781
2008/11	-11.2	-6.48	9580.632	1234.464	7.76097	0.05713769	0.02987936
2008/12	6.46	3.26	10277.648	1280.424	8.02675	0.00071577	-0.00334424
2009/01	-2.69	20.11	9994.7101	1522.9948	6.56254	0.19137366	0.06249468
2009/02	-6.86	4.05	9521.7022	1562.831	6.0926	0.00857453	-0.01543313
2009/03	12.26	7.83	10681.539	1685.3829	6.33775	0.43857229	0.09460684
2009/04	-8.03	-13.5	9920.8039	1433.305	6.92163	0.00614218	0.01119601
2009/05	-2.19	15.69	9708.6206	1594.2082	6.08993	0.00808789	-0.01498878
2009/06	-4.44	-1.2	9383.0364	1584.752	5.92082	0.00626897	0.01319613
2009/07	17.99	7.83	11002.568	1713.4187	6.42141	0.17758812	-0.07023534
2009/08	-13.7	-14.69	9483.3505	1447.7107	6.55058	0.20197477	0.06420229
2009/09	4.78	-2.8	9802.169	1396.764	7.01777	0.00031579	-0.00253862
2009/10	4.52	-1.08	10168.751	1382.9016	7.3532	0.1247496	-0.05045701
2009/11	-8.96	-8.11	9297.9152	1276.3521	7.28476	0.08108669	-0.04067961
2009/12	2.3	-1.23	9526.176	1262.2806	7.5468	0.29898744	-0.07811392

MSE=0.1022 AFER=-0.09

5 Performance Measure of Proposed Model

In this section, we evaluated the forecasting performance of our proposed model on enrollment data of University of Alabama and compared the results with previous models on same data. All previous models used same enrollment data set as benchmark [2-23].

Table 7. Comparison of MSE & AFER

	Song Chissom	Chen	Hwang Chen & Lee	Jilani, Burney & Ardi	Hur-ang	Singh	Jilani & Bur-ney	Jilani, Burney & Ardiadv	Mered-ith & John	Proposed Method
MSE	775687	321418	226611	227194	86694	90997	82269	41426	21575	*9917.17*
AFER	4.38%	3.12%	2.45%	2.39%	1.53%	1.53%	1.41%	1.02%	0.57%	*0.34%*

The forecasting accuracy is measured in terms of mean square error (MSE) and average forecasting error rate (AFER). Lower value of MSE and AFER are measure of higher forecasting accuracy. It can be observed from Table 7 that the obtained value of MSE and AFER is lowest in case of proposed model. The comparative study of MSE, AFER and the forecasted values obtained by our designed model clearly indicates its superiority over already existing soft computing time series models.

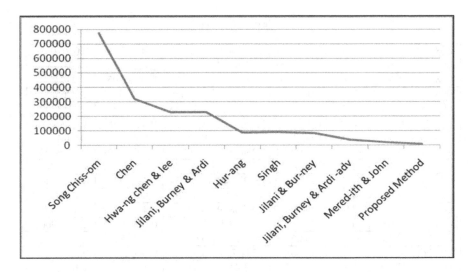

Fig. 1. Comparitive Study of MSE

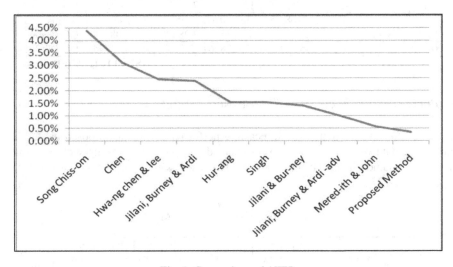

Fig. 2. Comparison of AFER

6 Conclusions

Pioneered dynamic computational method can be employed as an accurate and reliable means for estimating and predicting the LoS of patients in any territory care hospital. Presented model can also be considered as a strong standard methodology for better resource allocation, planning and management in health care domain. New technique introduced achieves the best accuracy having least mean square error among all related work in the field of forecasting till date. Further, design of proposed method can be used as base framework to develop decision support system for health care institutes. Such decision support system can significantly improve the efficiency of health care services. In future, proposed model can be extended to optimize with genetic algorithm.

References

1. Chassin, M.R.: Length of stay and outcome: myocardial infarction, variations in hospital length of stay: their relationship to health outcomes. Health Technologies Case Study 4, NTIS Order #PB84-111483 (August 1983)
2. Weissman, C.: Analyzing intensive care unit length of stay data: problems and possible solutions. Critical Care Medicine 25(9), 1594–1599 (1997)
3. Clarke, A., Rosen, R.: Length of stay, how short should hospital care be? European Journal of Public Health 11(2), 166–170 (2001)
4. Walczak, S., Scorpio, R., Pofahl, W.: Predicting hospital length of stay with neural networks. In: The Proceedings of the Eleventh International FLAIRS Conference, pp. 333–337 (1998)
5. Song, Q., Chissom, B.S.: Fuzzy time series and its models. Fuzzy Sets and Systems 54, 269–277 (1993)
6. Song, Q., Chissom, B.S.: Forecasting enrollments with fuzzy time series Part I. Fuzzy Sets and Systems 54, 1–9
7. Song, Q., Chissom, B.S.: Forecasting enrollments with fuzzy time series: Part II. Fuzzy Sets and Systems 62, 1–8 (1994)
8. Song, Q.: A note on fuzzy time series model selection with sample autocorrelation functions. Cybernetics and Systems: An International Journal 34, 93–107 (2003)
9. Chen, S.M., Hsu, C.-C.: A new method to forecasting enrollments using fuzzy time series. International Journal of Applied Science and Engineering 2(3), 234–244 (2004)
10. Chen, S.M.: Forecasting enrollments based on fuzzy time series. Fuzzy Sets and Systems 81, 311–319 (1996)
11. Chen, S.M.: Forecasting enrollments based on high-order fuzzy time series. Cybernetics and Systems: An International Journal 33, 1–16 (2002)
12. Huarng, K.: Heuristic models of fuzzy time series for forecasting. Fuzzy Sets and Systems 123, 369–386 (2002)
13. Huarng, K.: Effective lengths of intervals to improve forecasting in fuzzy time series. Fuzzy Sets and Systems 12, 387–394 (2001)
14. Hsu, C.C., Chen, S.M.: A new method for forecasting enrollments based on fuzzy time series. In: The Proceedings of Seventh Conference on Artificial Intelligence and Applications, Taichung, Taiwan, Republic of China, pp. 17–22
15. Hwang, J.R., Chen, S.M., Lee, C.H.: Handling forecasting problems using fuzzy time series. Fuzzy Sets and Systems 100, 217–228 (1998)
16. Lee, L.W., Wang, L.W., Chen, S.M.: Handling forecasting problems based on two-factors high-order time series. IEEE Transactions on Fuzzy Systems 14(3), 468–477 (2006)
17. Li, H., Kozma, R.: A dynamic neural network method for time series prediction using the KIII model. In: The Proceedings of the 2003 International Joint Conference on Neural Networks, vol. 1, pp. 347–352 (2003)
18. Melike, S., Degtiarev, K.Y.: Forecasting Enrollment Model Based on First-Order Fuzzy Time Series. In: The Proceedings of World Academy of Science, Engineering and Technology, vol. 1, pp. 1307–6884 (2005)

19. Melike, S., Konstsntin, Y.D.: Forecasting enrollment model based on first-order fuzzy time series. In: The Proceedings of the International Conference on Computational Intelligence, Istanbul, Turkey (2004)
20. Jilani, T.A., Burney, S.M.A., Ardil, C.: Fuzzy metric approach for fuzzy time series forecasting based on frequency density based partitioning. Proceedings of World Academy of Science, Engineering and Technology 23, 333–338 (2007)
21. Jilani, T.A., Burney, S.M.A., Ardil, C.: Multivariate high order fuzzy time series forecasting for car road accidents. International Journal of Computational Intelligence 4(1), 15–20 (2007)
22. Singh, S.R.: A computational method of forecasting based on fuzzy time series. International Journal of Mathematics and Computers in Simulation 79, 539–554 (2008)

Development of Three-Stage Methodology for a Strategy-Based Integrated Dynamics Model from a Value Chain Reconfiguring Perspective

Chung-Chou Tsai[1], Sununta Siengthai[1],
Vilas Wuwongse[2], and Donyaprueth Krairit[1]

[1] School of Management
[2] School of Engineering and Technology,
Asian Institute of Technology,
Km. 42, Phaholyothin Highway, Klong Luang, Thailand
smokymt@ms54.hinet.net, S.Siengthai@ait.ac.th,
vw@cs.ait.ac.th, donya@ait.ac.th

Abstract. The dynamics modeling system has forced the value chain reconfiguration to have fast-paced adaptability of clustered module structure associated with BSC-KIT (Balanced Scorecard, Knowledge-Integrated Traceability). It acts as a major driving support in developing the efficient appraisal and distinction of generic hierarchy of distribution decision factors. This paper proposes a Strategy-based Integrated Dynamics Model (SIDM) for process-driven capability of value chain reconfiguration using three-stage methodology. Its backbone is formed by the BSC-KIT of AHP-BSC structure (Analytic Hierarchy Process), ARIS-BSC platform (Architecture of Integrated Information System), hybrid ST-ARIS architecture (Systems Thinking) and a set of ARIS-EPCs architecture (Event-driven Process Chains) of decision-making strategy. SIDM development aims to supply findings of the events-and-patterns behavior of the aggregated comparison matrix, and to derive a precisely strategic mapping of causality for the industrial value chain. All these elements are taken to build cognition of a management-driven platform of the computer-aided process simulation model (CPSM).

Keywords: Balanced Scorecard (BSC), Analytic Hierarchy Process (AHP), Architecture of Integrated Information System (ARIS), Systems Thinking (ST), Value Chain, Event-driven Process Chains (EPC), Decision-Making, Modeling, Clustered Module Structure.

1 Introduction

Taiwan Publishing Market was faced with immense difficulties (from 2003 till now) which had resulted in the collapse of the industry's structure and publishing value chain. The development trend of future value-chain model is grim and raises concerns as to "How can publishers maintain a normal process model with high total value chain cost?" [1]. The answer lies in constructing a new competitive process modeling that will overcome the gap between supply and demand, transaction structure and total

S. Dua, S. Sahni, and D.P. Goyal (Eds.): ICISTM 2011, CCIS 141, pp. 233–244, 2011.

value chain costs [2]. This paper aims to propose an innovative, effective and feasible three-stage methodology for value chain reconfiguration that combines computer-aided assumptions and causality reasoning mechanism via multi-method procedures in identifying the performance of value chain, referenced with knowledge- based hierarchy and modeling process/simulation, in consideration with experts agreement on event identification phases [3].

By using qualitative and quantitative methods for conceptual building, we recognized an optimized alternative which is more commonly known as events-driven representation [4]. Mobilizing BSC-KIT presents an effective AHP and ARIS methods in identifying statistical results within the validation model [5, 6]. Navigating the sustainable development is realized via utilizing clustered process modeling which is constructed to identify the influence of relevant decision criteria on a target key performance indicators (KPI) setting [7, 8, 9, 10, 11]. The strategic framework of Strategy-based Integrated Dynamics Model (SIDM) development is then composed.

2 Perspective on Clustered Module Structure in SIDM Design

The formulation of business strategy is focused on multi-attribute activity portfolios within the process- and function-oriented strategic planning to identify the definition of strategic thrust and evaluation of specific business action, then creating the process-driven capability of value chain [12]. Gilmour (1999) developed specific multi-method procedures. From a strategic network perspective, he applies the technique of non-linear value chain to improve supply-chain performance [13]. This method asserted that a process modeling might provide new insight in evaluating the performance and capability of value-chain process, comparing the gap in literature, and providing knowledge-base for the toolset variables in consolidating the analysis and design. BSC model has proven to be a powerful tool for strategic communication and planning that aids in causality-building matrices and scenarios of strategic framework [14]. Expounding on an actual case study, Scheer and Nüttgens (2008) have shown that using ARIS-EPCs architecture in organizational projects or SME and mid-market can reduce process costs by 40% and time factors by more than 30% [15]. Hsi (2001) uses a medical center as an example to construct an ARIS process simulation model of electronic procurement that indicates the simulation results of ARIS-EPCs in the inventory control system [16]. Result shows that the empirical practice of the average time to complete electronic procurement by the contract is about 10 days as opposed to a non-contract which is about 28 to 33 days. According to the setting of generic hierarchy of distribution decision factors, the average procurement time was reduced by about 53.33%, the average turnover rate of inventory increases 14.28%, and the average inventory cost of hospitals decreases.

SIDM is a three-stage methodology for understanding how different distribution decision factors affect the value of knowledge-based hierarchy. The theory-building cognition is based on the theoretical justification variable in methodological literature on how decision-making strategy reacts to the validating core proposition of exchange value-validated model and on process-driven capability. In addition, it can replicate the activity of management-driven interface, resulting in activities expected under simulation situation. Rabelo et al. (2007) constructed a practical approach of hybrid system dynamics and discrete-event simulation (SD-DES) model with the toolset variable of AHP/SD/DES for identifying two major themes according to multi- method procedure which is recognized as a value chain process analysis associated with simulation [9]. O'Donnell (2005) indicates the system dynamics focused on how the relevant decision criteria linked together to form a holistic network with multi-causal relationships and archetypes -associated control [17]. Wu et al., (2007) applied the integrated dynamics decision making (IDDM) method to construct a project planning system to improve the efficiency and synergy of supply chain collaborative operations [18]. IDDM method integrates AHP and Quality House methods as a decision support system to determine the related comparison weighting of multi-attributes in decision-making contribution and to confirm business priority by the score distribution of quality function deployment. Loebbecke and Huyskens (2009) provided a five-stage methodology and a specific software application to develop a model-based decision support system [19]. Based on performance- evaluated view, Han et al. (2009) develop a simulation model, capable of identifying two major themes: macro- and micro-process analysis [20]. The former, macro- process, is a process-based performance framework perspective which deals with major business process operation. The later, micro, uses simulation mechanism contributes. Process-driven simulation is a BPR model in itself [20][21]. It seems multi-method procedure brings maximized outcome.

3   Case Study Research Methodology for Conceptual Building

3.1   Defining the Knowledge-Integrated Traceability (KIT): Achieving an Empirical Inquiry

The knowledge-base and inference/reasoning mechanisms emphasize common characteristics and the traceable value of integrated feasibility as two major themes of dynamic behavior and knowledge foci. This common characteristics are the priority of weighting to explore elements including (1) the critical success factors (CSFs) of real-situation context in publishing industry, (2) hierarchy to construct the consistency and traceability of AHP-BSC structure in systems thinking archetypes (Fig. 1), (3) the target KPIs (key performance indicators) setting to identify events, (4) behavioral patterns of aggregated comparison matrix and results sustainability, (5) process decomposition and clustered module process to define the causality between AHP technique and ARIS method, and (6) the performance improvement of validation model to achieve the research purpose in total cost saving and throughput time of value chain reconfiguration. The common characteristics are a virtual necessity in three-stage method which delivers the traceable value of integrated feasibility for the statistical description of strategic value chain (Table 1). This view is focusing on the research design, theory-building, and empirical inquiry on knowledge foci which are the central feature of value chain reconfiguration.

3.2   Facilitating Perspectives on Three-Stage Methodology with Process Technology

BSC-KIT is a validation model that identifies accessibility assessment hierarchy criteria with critical success factors (CSFs). Relevant literature review on multi- method procedures described process technology as quantifiable decision context and it was consisted of the combination of outcome- and process-driven evaluation. To fulfill it, three stages are involved in designing SIDM.

Stage 1 aims to construct clustered module structure through CSFs acquisition in generically hierarchical structure of distribution decision factors (Fig.1). Three themes are identified in the structure. Theme one is a design of "methodology" view of how decision factor uncertainties and multi-criteria decision-making operate [4]. Theme two is a simplification of "hierarchy" view of how each component of the clustered module structure contributes [9]. Theme three is a process-driven "independence" viewpoint. Process-driven modeling is used to conceptualize the feasibility within systems thinking archetypes. Knowledge-integrated traceability (KIT) is then used as a statistical application for theory-building matrix while aiming at the combination of AHP technique and ARIS method (Table 1). Stage 2 highlights important advantages of system dynamics modeling: linking process-driven modeling with clustered process modeling to construct a strategic mapping of clustered module structure via causality feedback loop diagram and systems thinking archetype [22]. Thus, it changes a conceptualizing layout of systems thinking archetype into a process simulation model of STELLA-systems thinking [23, 24]. Stage 3, simulation results are used to demonstrate the process-driven capability of value chain to replicate management-driven interface activity qualitatively. To carry out the influence of relevant decision criteria on a target KPI test quantitatively [6], STELLA-systems thinking of causality diagram is converted into a simulation model using Excel spreadsheet [24]. Then, cross-examination phase of validity and reliability of research are conducted.

4   The Application of Process-Driven Modeling in SIDM

4.1   Validity and Reliability

SIDM is a three-stage methodology, therefore, validity and reliability is discussed in three stages.

The first stage is to apply Balanced Scorecard (BSC) and Knowledge-integrated traceability (KIT) in two related perspectives on strategic activities that are significant to most industrial value chain structure and to all levels of generically hierarchical structure [4][25]. These perspective are: (a) constructing decision factors of value chain function deployment from a review of relevant methodological literature and experts-consensus cognition, and then followed by transformation design of BSC-KIT with real-world data; and (b) the transferability and feasibility between knowledge-base and reconfiguration mapping tool is tested by applying an in-depth interview to AHP-BSC structure (Table 1).

The validity of the second-stage, research design, is considered under threat due to the combining use of integrated toolsets, validation of self-designed study, practicability, and limitations of literature review on relevant themes (Table 1). This stage consists of the causality of the collapse and reconfiguration of three building blocks in graphic value chain.

The first building block, macro hierarchy criteria analysis, is conducted to identify the measurement of pair-wise comparison weighting and the consistency of reciprocal judgment matrix as the prototype to model a generic pattern of theory-building matrix.

Meanwhile, the statistical results of AHP-BSC are used in constructing ST architecture which is a diagram of feedback loops and strategic mapping of relevant decision criteria.

The second building block, a decision-making base characterized by subjective judgment, is conducted to model the ARIS House in ARIS-BSC platform. It explores the relationships among decision factors which are gathered from events and patterns of behaviors in ARIS-EPCs architecture. The third building block, micro hierarchy criteria analysis using hybrid ST-ARIS process simulation model, is conducted to predict the performance improvement.

The third-stage methodology is to improve the validity in designing the simulation model during SIDM development (Table 1). First, a pilot study is executed to build a systematization layout of clustered process modeling from the theory-building cognition. Secondly, the in-depth interview and questionnaire approach is adopted as a form of mathematical triangulation that includes a group of experts, questionnaire interview and valid questionnaire response, and quantitative measures. The validation model of SIDM is based on a specific cross-examination format, which were results of simulation and can be replicated in the management-driven platform activity.

Reliability means the same results can be repeated in data-collection procedure and dependability can be expected under traceable situation. To lower interviewees' absent-minded answers in questionnaires, all interviews were carried out by the first author face-to-face to ensure the reliability.

Table 1. Lineament of Descriptive Techniques: The Communication of Toolset

Lineaments	AHP Method	ARIS Method	BSC Method	Value Chain Deployment	System Thinking
Scaling	Ratio / Priority.	Ratio / Priority	Ratio / Priority	Ratio / Squeeze	Ratio / Time Delay
Performance Elicitation	Pair-Wise Comparison	Customization Software / Cost and Time Superiority / Process Optimization / Simulation.	Budget Controlling / Action Planning	CSFs / Strategic Alignment / Integrated Resource / Extending Business / Customer Value	Butterfly Effect / Beer Game / Dynamic System
Weighting	Normalized Ratio via Eigenvalues	Real-Environmental Data via KPI.	The Benchmark and Method of Evaluation / KPI	Business Process Outsourcing / KPI	Events and Patterns of Behavior / KPI
Synthesis	Additive / Eigenrectors.	Event-to-Event / Unique Address Space.	Four Perspectives / Strategic Mapping	Business Process Model	Archetypes and Interventions
Structure	Hierarchical	ARIS Platform / Hierarchical / Process Modeling	Core Strategic Organization / Juxtaposition	Organization-P etal Platform / Hierarchical	Learning Organization / Feedback Loop
Set-in Reaction	Synthesis / Consistency / Measure / Technique Produced Weightings	BPM / Process Decomposition / Measure / Continuous Improvement	Strategic Toolset / Performance Measurement / Cost Saving & Throughput Time	Horizontal Coordination	Dynamic Complexity / Mental Change

Source: The data collation and identification of Qualitative Research with Knowledge-Integrated Strategy and Literature Review by Author in this research. (2008)

4.2 Case Study: Trials and Results of Simulation

In this research, a group of four experts was chosen to identify the CSFs selection of industrial value chain. These experts' consensus was aligned with the event identification phase. Simultaneously, their consensus was used to construct AHP-BSC, which is a necessary and important structure of the research. In composing decision elements and designing two questionnaires: AHP and AHP in-depth interview. The statistical results are presented in Fig.1 and the ensuing paragraph summarizes the findings.

There are five entities of participants. The first entity was 31 subjects, their average score served as the standard. The 2nd to the 4th entities were publishers A, B and C. The 5th entity was a group of experts. These four entities were targeted population and their performance was compared to the standard. Table 2 shows the comparison of participants' performance in currently empirical practice and in CPSM. It reveals that in currently empirical practice, process-driven capability of value chain, cost structure budge capability and throughput time controlling are all too weak, which are the reasons why Taiwan publishing market collapsed. After simulating with computer-aided process simulation model, results revealed that all the above variables improved, for example, process cost savings reduced 26.6% and throughput time controlling increased 23.8%. The entity of experts performed very well, just as expected. It proves that SIDM is a well- designed and feasible model.

Table 2. The Comparison of Participants' Performance in Currently Empirical Practice and CPSM

Research Participants	Currently Empirical Practice of Publishing Value Chain				Computer-aided Process Simulation Model (CPSM)			
	(1)	(2)%	(3)	(4)%	(5)	(6)%	(7)	(8)%
Publishing Industry (The Average of 31 publishers)	149.1		198.3		121.9	18.2	144.6	27.1
A publisher	224.8	50.8	249.8	26.0	175.3	22.0	190.5	23.7
B publisher	98.0	(34.3)	86.5	(56.4)	78.5	19.9	69.5	19.7
C publisher	232.8	56.1	235.0	18.5	185.0	20.5	185.5	21.1
A Group of Experts (The Average of 4 Experts)	184.4	23.7	193.3	(2.5)	113.8	38.3	113.3	41.4
The Average of CPSM in (6) and (8)						26.6		23.8

Source: Data were collected by the 1st Author (2008)
Note: (1) to (8) represents participants' performance in various categories.
(1) ~(4) Currently Empirical Modeling of Publishing Value Chain
    (1) Process Cost Savings
    (2) Management Business Review (%) in Process Cost Savings: Comparing the standard group with other four groups
    (3) Throughput Time Controlling
    (4) Management Business Review (%) in Throughput Time Controlling
(5)~(8) CPSM
    (5) Process Cost Savings
    (6) Improvement (%) in Process Cost Savings
    (7) Throughput Time Controlling
    (8) Improvement (%) in Throughput Time Controlling

4.3 The Strategic Framework for SIDM Development

The traditional mechanistic approach is unable to cope with simulation-based construction, as these are often related to complex, and process-driven "independence" view of the individual model itself [25]. The overall structure of SIDM is a process that is designed to create a real-accuracy simulation-based design, moving through a series of holistically simplified, consistent and accurate hierarchy modeling phases, and consists of three sub-modules: AHP-BSC structure, strategic mapping of causality diagram of ST archetypes, and hybrid ST-ARIS simulation model. The detailed description is below.

Two strategic frameworks are implemented in this paper (Fig. 1 and Fig. 2). These strategic frameworks are virtually essential in system dynamics. Fig. 1 differs from an operational framework. It is designed to provide a conclusive management-driven platform to support managers to strengthen decision-making quality while dealing with real situation and resources integration. In order to implement value chain reconfiguration, BSC-KIT was used to manage strategic activities to support research methodology. Consequently, these strategic frameworks propose a computer-aided process simulation model (CPSM) and establish a management-driven platform, based on accessibility assessment hierarchy criteria with AHP-BSC structure. To define the validating core compositions among all variables in knowledge-generated and model-based methodology, the consistency of comparison matrix and a target KPI set are used as measures of contribution indices in simulation-based design. Results are depicted in Fig. 1, by means of the causality feedback loop diagram composed by 9 archetypes of system thinking [22].

There is as follows (Fig. 1):

- Limits to Growth (No.1) are the initial growth in the element of the architecture which is eventually limited by action constraint affecting.
- Shifting the Burden (No.2) is an action based on short-term symptom and ability to act based on the long term.
- Eroding Goals (No.3) is also known as the "boiled frog syndrome."
- Escalation (No.4) is an action that increases performance but results in decreased performance over the long-term running.
- Success to the Successful (No.5) is a myth that "the rich get richer to get more resources and a greater ability to improve."
- Tragedy of the Commons (No.6) is also known as the "bullwhip effect."
- Balancing Loop and Reinforcing Loop (No.7) is growth or decline of the element of the architecture.
- Fixes That Fail (No.8) is typically a result of manifest problem symptom rather than root-cause data.
- The Attractiveness Principle (No.9) is addressing one limit that run into multiple "Limits to Growth."

The dynamic process reengineering of Fig. 1 results in STELLA-systems thinking in Fig. 2 (Fisher & Potash, 2008; STELLA–Systems, 2008). Excel spreadsheet, being capable of mathematical computation, is substituted for the inaccessible commercial software, for the purpose.

4.3.1 Strategic Activity in Archetypes of Systems Thinking

Success of causality feedback loop diagram consists of context-connected systems thinking archetypes, which is a scientific and strategic activity between causal imagery and maximized outcome [22]. Various representation techniques of inter- relationships are the heart or core decision factors of Fig. 1. The most widely used technique focuses on using the sensitivity test, the KPIs identification of ARIS-BSC platform and simulation results of ARIS-EPCs architecture within the STELLA- systems thinking [24].

First, the behavioral factors of financial impetus and strategic outsourcing motivate the publisher to model the triad-power relationship. This brings better performance evaluation on short-term symptomatic solution but it reduces the adapted ability in transaction structure collapse. It in turn results in the incapability of improving financial impetus for long-term fundamental strategic objectives and measures (No.1 and 2 archetypes).

Secondly, the thick-dark arrows indicate influential linkages by different limiting factors. Their linkages are expanded in key success feedback loops, critical success archetypes and interventions. For example, the transaction structure promoted the changeability of the transaction manner portfolio (No.5 archetype); transaction platform encouraged the accounts payable and accounts receivable to create the efficiency of cash-to-cash (No.9 and 3 archetype), strategic outsourcing reduced the process cost saving in editing and publishing operation (No.3 and 6 archetype). Therefore, the behavioral factors, by the distributor are the accounts payable and accounts receivable to improve the cash return cycle time of transaction manner and cash-to-cash results in eliminating the whole value chain system of process cost saving and throughput time controlling, and also eliminates the producing costs controlling of pre-editing operation to perform as the strategic outsourcing is eliminated.

Thirdly, the thick-light arrows show a better fulfillment of commercialization of knowledge for contributing value-oriented archetypes and interventions. For instance, the financial impetus advanced managing process in budget execution (No.4 and 8 archetype) and strategic outsourcing promoted business clusters in risk sharing. That means the organizational infrastructure produces the requested horizontal cooperation in the short-term strategic activity, but makes the problem worse of the triad-power relationship in the long-term strategic planning, and typically a result of addressing the competitive advantage symptoms rather than collaborative mechanism. In fact, the requesting horizontal cooperation in the short-term strategic activity need be taken again and again because feasibility and effectiveness easily wears off. So, to better understand the long-term strategic planning, it is necessary to take the strategic network and tie the 1st and the 2nd suppliers and customers as the business drivers of system dynamics modeling and simulation reality.

4.3.2 Decision-Making Strategy in STELLA-Systems Thinking

Fig. 1 depicts the clustered process modeling which includes distribution decision factors, one-way causality feedback loop, and clustered module structures. Each clustered module structure is developed in the programming of system dynamics together with the functionality and boundary of a module means in EPCs, as shown in Fig. 1 Afterwards. Setting the language of STELLA-systems thinking with stock, flow, convert, and connector is based on Table 1. This work is an attempt towards building transforming prototype of a programming of systems thinking change for Fig. 1, and exploring the efficiency of SIDM in the interrelationships of events and patterns of behavioral factors of ARIS-EPCs architecture, which is the language of STELLA-systems thinking as an activity of decision-oriented interface to transform Fig. 1 into Fig. 2 and unfold the stocks in the clustered module structure and form a formula that represents the total stocks of a conclusive management-oriented interface in order to depict conclusions of decision-making strategy.

ARIS-BSC Platform is shown in the Fig. 2. It is a useful management-driven platform to make a hybrid ST-ARIS simulation model from generic hierarchy of distribution decision factors of AHP-BSC structure, which is associated with ARIS-EPCs architecture in ARIS House. Fisher and Potash (2008) have found that the

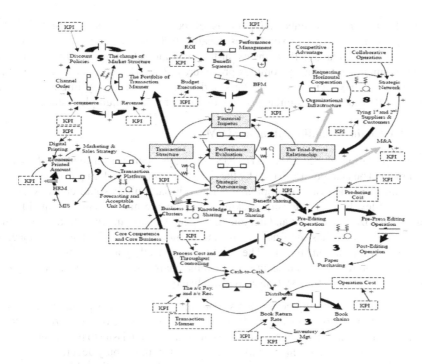

Fig. 1. The Strategic Mapping of Causality in Systems Thinking

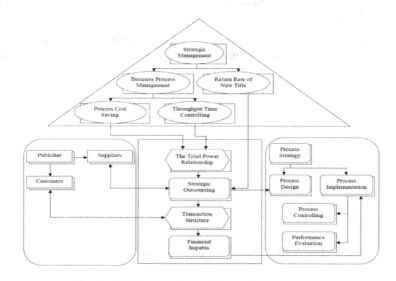

Fig. 2. ARIS House: Hybrid ST-ARIS Simulation Model

ARIS-BSC Platform is associated the interrelationship of functional view of ARIS House with a series of decision-driven events [23]. Events are not only triggering functions, but also showing the value of functions themselves. More importantly, Fig. 2

depicts the feature of events and patterns of behavior associated with ST-ARIS implementation. The Event-driven Process Chains (EPCs) are extended to represent the distribution decision factors flow, the event-to-event and pattern-to-pattern of behavior and to accommodate variously complex procedures [23][24]. Thus, defining the status or situation under which a process starts and ends. An event may initiate multiple functions at the same time; similarly, a function may result in multiple events.

4.4   Value Chain Reconfiguration in SIDM

The KIT validation of BSC/AHP/ST/ARIS is always changing in dynamic business environment. The changes are driven by the clustered process modeling in objectives of decision-making, and results of implementation. It is particularly significant at superior management-levels where decision-making strategies are mostly based on qualitative factors, drawn from judgment and computation. The decision-making strategy of value chain reconfiguration has the finer details and accuracies of value chain components and elements of variation (Fig. 1). The SIDM is enhanced by a four-clustered module structure to capture details and accuracies of value chain components and an optimized alternative is expected (Fig. 3). Causalities are listed:

a. value chain reconfiguring components represent the cash-to-cash issue to portfolio of benefit, transaction manner, and financial flow. High turnover rates are expected.
b. value chain reconfiguring elements represent collaboration issue to publisher's view, distributor and channel. This would increase the demand-supply accuracy.
c. value chain reconfiguration aims to demonstrate the capability of the process in exchanging the value-validated model. If the goal is accomplished, SIDM would be manifested as a good design (Fig. 3).

This implementation enables managers to utilize decision-making strategy and reasoning mechanism and SIDM would increase the managerial value in maximizing outcome (Fig. 3).

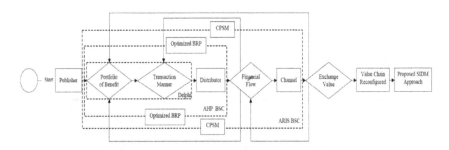

Fig. 3. The Value Chain Reconfiguration in SIDM

5   Conclusions

SIDM is a dynamics modeling system which applies a three-stage methodology and attempts to deploy the value-chain function. The modeling adopts the KIT and toolset variable of BSC/AHP/ST/ARIS in examining the virtual necessity through a stepwise clustered module structure. SIDM provides the following insights. First, the case study has demonstrated SIDM is capable of strategic activities of value chain reconfiguration. Secondly, the clustered module structure provides recognition of decision requirement in strategic framework. Thirdly, SIDM allows decision makers to overcome the knowledge regarding the outcome between computational and judgmental application in process simulation models.

To make SIDM the trends of management science, we need to do more researches, for example: implementing SIDM as a stand-alone decision support system in various businesses. It is recommended that the method outlined in this paper be replicated to test its feasibility and effectiveness. Future research should focus on building more comprehensive decision-making hierarchy with a larger number of decision factors and the consistency analysis of smaller parts along with genetic algorithm (GA), which will enable the strategic planning of SIDM to generate better quality than the approximation method.

References

1. Su, P.Z.: What's wrong of Taiwan's Publishing Industry Structure-Part I III. As-If Publishing House (2007)
2. Yang, D.H., Kim, S.C., Nam, C.G., Min, J.W.: Developing a Decision Model for Business Process Outsourcing. Computers & Operations Research 34, 3769–3778 (2007)
3. Adamides, E.D., Karacapilidis, N.: A Knowledge Centered Framework for Collaborative Business Process Modeling. J. Business Process Management 12(5), 557–575 (2006)
4. Coviello, N.E.: Integrating Qualitative and Quantitative Techniques in Network Analysis. J. Qualitative Market Research 8(1), 39–60 (2005)
5. Salmeron, J.L., Herrero, I.: An AHP-Based Methodology to Rank Critical Success Factors of Executive Information Systems. Computer Standards & Interfaces 28, 1–12 (2005)
6. IDS SCHEER: ARIS Design, ARIS Simulation. White Paper (2007), http://www.ids-scheer.com
7. Yurdakul, M.: AHP as a Strategic Decision-making Tool to Justify Machine Tool Selection. J. Materials Processing Technology 146, 365–376 (2004)
8. Wong, K.W., Li, H.: Application of the Analytic Hierarchy Process (AHP) in Multi-criteria Analysis of the Selection of Intelligent Building Systems. Building and Environment 43, 108–125 (2008)
9. Rabelo, L., Eskandari, H., Shaalan, T., Helal, M.: Value Chain Analysis Using Hybrid Simulation and AHP. International J. Production Economics 105, 536–547 (2007)
10. IDS SCHEER: Strategic Corporate Planning using Key Performance Indicators, ARIS BSC, IDS SCHEER Platform (2006), http://www.ids-scheer.com
11. Ugwu, O.O., Haupt, T.C.: Key Performance Indicators and Assessment Methods for Infrastructure Sustainability: A South African Construction Industry Perspective. Building and Environment 42, 665–680 (2007)
12. Papazoglou, M.P., Ribbers, P., Tsalgatidou, A.: Integrated Value Chains and Their Implications from a Business and Technology Standpoint. Decision Support System 29, 323–342 (2000)
13. Gilmour, P.: Benchmarking Supply Chain Operations. International J. Physical Distribution & Logistics Management 29(4), 259–266 (1999)
14. Huang, H.C.: Designing a Knowledge-based System for Strategic Planning: A Balanced Scorecard Perspective. Expert Systems with Applications (2007)

15. Scheer, A.W., Nüttgens, M.: ARIS Architecture and Reference Models for Business Process Management. J. Computer Science, White Paper (2008), http://www.iwi.uni-sb.de

16. Hsi, C.H.: Constructing a Process Reference Model of Electronic Procurement: Using a Medical Center as an Example. Master Research, Non-Publish, Chang Gung University, Taipei (2001)

17. O'Donnell, E.: Enterprise Risk Management: A Systems-thinking Framework for the Event Identification Phase. International J. Accounting Information Systems 6, 177–195 (2005)

18. Wu, S., Lee, A., Tah, J.H.M., Aouad, G.: The Use of a Multi-Attribute Tool for Evaluating Accessibility in Buildings: the AHP Approach - Facilities, vol. 25(9/10), pp. 375–389 (2007)

19. Loebbecke, C., Huyskens, C.: Development of a Model-based Netsourcing Design Support System Using a Five-stage Methodology. Europ. J. Oper. Res. 195, 653–661 (2009)

20. Han, K.H., Kang, J.G., Song, M.: Two-stage Process Analysis Using the Process-based Performance Measurement Framework and Business Process Simulation. Expert Systems with Applications 36, 7080–7086 (2009)

21. Jawahar, N., Balaji, A.N.: A Genetic Algorithm for the Two-stage Supply Chain Distribution Problem Associated with a Fixed Charge. European J. Operational Research 194, 496–537 (2009)

22. Senge, P., Ross, R., Smith, B., Roberts, C., Kleiner, A.: The Fifth Discipline: Strategies and Tools for Building a Learning Organization. Doubleday, New York (1994)

23. Fisher, D.M., Potash, P.J.: The Promise of System Dynamics Modeling in K-12 Education (2008)

24. STELLA–Systems Thinking in an Experiential Learning Lab with Animation (2008), http://www.iseesystems.com

25. Bertrand, J., Will, M., Fransoo, J.C.: Modeling and Simulation: Operations Management Research Methodologies using Quantitative Modeling. International J. Operation & Production Management 22(2), 241–264 (2002)

26. Mansar, S.L., Reijers, H.A., Ounnar, F.: Development of a Decision-making Strategy to Improve the Efficiency of BPR. Expert Systems with Applications 36, 3248–3262 (2009)

Hybrid Temporal Mining for Finding Out Frequent Itemsets in Temporal Databases Using Clustering and Bit Vector Methods

M. Krishnamurthy[1], A. Kannan[2], R. Baskaran[3], and G. Bhuvaneswari[4]

[1] Department of Computer Applications, Sri Venkateswara College of Engineering,
Sriperumbudur, Tamil Nadu, India, 602105
[2] Department of Information Science and Technology, Anna University,
Chennai, India, 600025
[3] Department of Computer Science and Engineering Anna University,
Chennai, India, 600025
[4] Sri Venkateswara College of Engineering, Sriperumbudur,
Tamil Nadu, India, 602105
`mkrish@svce.ac.in, akannan@annauniv.edu,`
`{baskaran.ramachandran,twinklebhuvana}@gmail.com`

Abstract. Hybrid Temporal Pattern Mining was designed to address the problem of discovering frequent patterns of point and interval-based events or both and it is essential in many applications, including market analysis, decision support and business management. Such methodology cannot deal with Clustering, Bit Vector and Variable Threshold. In this paper, we propose a new algorithm called RHTPM (Revised Hybrid Temporal Pattern Mining) to find the frequent temporal pattern based on Clustering, Bit Vector and Variable Threshold. The experiments demonstrate that the proposed algorithm is capable of mining frequent hybrid temporal pattern for effective decision making and has been proved to be significantly good.

Keywords: Hybrid temporal pattern, point and interval-based events, clustering, bit vector, variable threshold.

1 Introduction

1.1 Data Mining

Data mining can be defined as an activity that extracts some new nontrivial information contained in large databases [1]. The goal is to discover hidden patterns, unexpected trends or other subtle relationships in the data using a combination of techniques from machine learning, statistics and database technologies. This new discipline today finds application in a wide and diverse range of business, scientific and engineering scenarios.. In this work we propose a new algorithm: Revised Hybrid Temporal Pattern Mining (RHTPM) which is a novel data mining technique that can be used to help make decisions in a variety of applications that deal with

S. Dua, S. Sahni, and D.P. Goyal (Eds.): ICISTM 2011, CCIS 141, pp. 245–255, 2011.

time-related data from Temporal Databases, where a Temporal Database is a set of events or items with associated dates and durations.

1.2 Frequent Item Sets and Clustering

Frequent item sets play an essential role in many data mining. A frequent item set is an item set whose support is greater than some user-specified minimum support [3]. In this paper, the support threshold is not the same for all the cases and it is varied. Therefore, a varying threshold is more realistic, considering the real time transactions in practice. Clustering is the division of data into groups of similar objects.

1.3 Proposed Work

In this paper we propose an algorithm called Revised Hybrid Temporal Pattern Mining (RHTPM) that can be used to discover all frequent hybrid temporal pattern from a set of hybrid event sequences based on clustering, bit vector and variable threshold. This algorithm incorporates clustering, bit vector and variable threshold during mining and in doing so, it accomplishes two goals: it saves a lot of computations in the post-processing phase and provides more flexibility for effective decision making in many applications.

1.4 Paper Organization

The remainder of this paper is organized as follows: Section 2 focuses on Related Works, Section 3, explains the proposed algorithm. Section 4, provides the Experimental results while conclusions are given in section 5.

2 Works on Temporal Mining

2.1 Related Works

Agrawal et al [1] introduced association rules for discovering regularities between products in large scale transaction data recorded by point-of-sale systems in supermarkets. A temporal association rule is an association rule that holds during specific time intervals. Using temporal association rule mining one may discover different association rules regarding different time intervals [5].Claudio [4] discusses on association rules that may hold during some intervals but not during others. Keshari Verma[7]uses calendar schemas to discover temporal association rules. It is of the form(year, month, day). Recent researches in the field of temporal association rule mining are using Apriori based approach. Nevertheless, these proposed approaches still encounter some difficulties for different datasets such as sparse or dense dataset. This approach has two limitations. First, huge space is required to perform the mining in Apriori based temporal association rule [6]. It generates a huge number of candidates in case of a dataset, which is dense or sparse. Many algorithms have been proposed for mining frequent pattern like apriori [2] and it involves complex candidate generation and takes many scans of the database for frequency checking. Jiewai Han [5] proposed FP-tree algorithm which involves mining of frequent pattern without candidate generation. FP-tree works well when the data is dense. Jian Pei [8] invented a new technique called H-mine for frequent pattern mining which works well when the data is sparse dataset. H-mine is a frequent pattern mining algorithm that takes advantage of a hyper-linked H-struct data structure and mines for frequent patterns. The great advantage of this method is it takes only 2 scans of the database to build the H-struct and then mines. H-Mine partitions the database and then mines each partitions in memory to find the frequent patterns. Bowo Prasetyo [3] proposed an algorithm called Hmine-rev to mine frequent patterns. It is a revised algorithm of H-mine that does not need any adjustment of H-struct links. Mining frequent itemsets using Matrix [9] reduces scanning cost and execution times, but the algorithm works only for nine transactions. There has been some previous research on hybrid event sequences. Discovering hybrid temporal patterns from sequences consisting of point and interval based events proposed by Shin-Yi wu and Yen-Liang Chen [10] introduced an algorithm called HTPM where event sequences may contain both point-based and interval-based events called hybrid event sequences.

2.2 Literature Limitation

In the existing work frequent items are mined using many algorithms. Though many algorithms were proposed to reduce space and efficiency, a considerable limitations were addressed. Most of the algorithms occupied more space; generate many candidate item sets, a fixed minimum support threshold which cannot be changed during the mining and increase work in post-processing phase. In this paper, the proposed algorithm addresses all the above issues.

3 Proposed Algorithm RHTPM

3.1 Algorithm RHTPM (Revised Hybrid Temporal Pattern Mining)

We propose a new and efficient algorithm, RHTPM (Revised Hybrid Temporal Pattern Mining) discovers frequent pattern of point-based events, interval-based events or both using clustering, bit vector and variable threshold. Once the data's are stored, we form a cluster based on the similarities. Then, we apply temporal relationship with respect to the data. If the pattern satisfies the temporal relationship then we represent it as 1 else 0. This type of representation is called as bit vector. In addition to that the value of the threshold is variable. The process starts by scanning the temporal database D to generate frequent pattern. We have conducted extensive experiments to evaluate the algorithm. Here we report the results for the frequent tea importing countries for developed and developing countries, obtained using RHTPM algorithm. The experimental result is presented in figure 4 and 5. The proposed algorithm is described below.

Proposed Algorithm: RHTPM (Revised Hybrid Temporal Pattern Mining)

Input: Temporal Database, TD
Output: Frequent Pattern
begin
>   Create Cluster table [2]; Developed country[n]; Developing country[n];
>   if(Cluster table[1])
>   Compare Developed country[n] for both Tea and Coffee values;
>> If it satisfies the temporal pattern: Assign 1 into the table;
>> end; else Assign 0 into the table; end; end

A: calculate the support count for each comparison
>   begin
>   Introduce i loop varying from 1 to N
>   Introduce j loop varying from 1 to N
>   if pattern bit=1
>   Increment the count of pattern bit i.
>   end
>   Support count=count (pattern bit for i) * Total no. of comparison
>   Min_sup_thres= Sum of all the support count / Total no. of comparison
>   if (Support count>Min_sup_thres) then
>>> Output the temporal pattern as frequent.
>   end;
>   end ; else if (Cluster table [2])
>   Compare developing country[n] for both Tea and Coffee values;
>> If it satisfies the temporal pattern: Assign 1 into the table;
>> end; else: Assign 0 into the table; end; end
> Calculate the support count for each comparison.
>   goto A:
end.

4   Experimental Results on Real Data

The experimental results examines the reasons for using the clustering, bit vector and changeable threshold for mining the frequent pattern over the given time interval in the field of international trading. We all know that buying and selling of commodities or services is called trading. Let us take Export and Import Trading in our example. Tea and Coffee was the major source of export earnings in India. This paper explains how RHTPM (Revised Hybrid Temporal Pattern Mining) is applied here and how it is useful for decision making.

Considering two tables (shown below), table1: tea shows the India's Tea Export details from the year 2001 to 2009 in terms of Kilograms and table2: coffee shows the India's Coffee Export details from the year 2001 to 2009 in terms of Kilograms. We have taken only the five major countries Japan, Afghanistan, Srilanka, China and America. In table1: tea and table2: coffee (shown below) starting from 2001-2002 to 2008 -2009 are all interval-based patterns but the last row of both the tables is 2009 which is point-based pattern.

4.1 Formation of Cluster

From the below two table's table1: tea and table2: coffee, we form developed countries as one cluster and developing countries as another cluster. Based on this grouping, we build cluster table. We maintain separate cluster table for developed countries like Japan, China and America. And another cluster table for developing countries like Afghanistan and Srilanka.

Table 1. Annual sales of Tea for different countries in different time period

Year / Country	JAPAN	AFHANISTAN	SRILANKA	CHINA	AMERICA
2001-2002	680	270	401	1270	292
2002-2003	616	180	313	304	312
2003-2004	654	1195	1112	258	307
2004-2005	706	143	801	1122	309
2005-2006	787	201	785	609	347
2006-2007	681	188	531	654	309
2007-2008	641	800	157	472	348
2008-2009	546	520	165	295	399
2009	700	150	700	330	325

Table 2. Annual sales of Coffee for different countries in different time period

Year / Country	JAPAN	AFHANISTAN	SRILANKA	CHINA	AMERICA
2001-2002	160	559	338	68	278
2002-2003	23	468	139	93	270
2003-2004	11	633	208	212	418
2004-2005	34	181	337	968	378
2005-2006	17	173	267	1998	351
2006-2007	229	444	39	852	343
2007-2008	2264	586	25	322	334
2008-2009	139	771	394	261	312
2009	440	560	168	378	301

4.2 Temporal Relationship with Respect to the Values and Bit Vector

In this work we are going to compare one developed country with all other remaining developed countries with respect to their values. Say for e.g.: Japan is compared with China and America. Here, we are going to compare to find the most frequent importers of tea and coffee from Indian country. And based on the comparison and the result we can come to a decision that whether we may or may not to continue to export our products to the other countries in future. If the temporal pattern satisfies the temporal relationship then we represent it by bit 1 else by bit 0.

Support Threshold for each comparison: Support threshold for each comparison = (Number of 1's) * (Total Number of comparisons)

Finding the minimum support threshold:

Min_sup_thres = Sum of all the threshold / Total no. of comparison

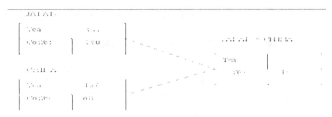

Fig. 1. Comparing Japan 'J' with China 'C' for the year 2001-2002

In figure 1, we have taken the Japan's and China's Tea and Coffee value. Since Japan's tea value is not greater than China's tea value, so we represent J>C as 0 in table 3 (Cluster table with respect to Tea values for the developed countries), since it does not satisfies the condition.

But in table 4 (Cluster table   with respect to Coffee values for the developed countries) we set J>C as 1 it means that Japan's Coffee value is greater than China's Coffee value.

Fig. 2. Comparing Japan 'J' with America 'AM' for the year 2001-2002

In figure 2, Japan's Tea value is greater than America's Tea value, so we represent J>AM as 1 in table 3 and it satisfies the condition. But in table 4 we set J>AM as 0 means that Japan's Coffee value is not greater than America's Coffee value.

Similarly, we proceeded the same steps for C>J, C>AM, AM>J and AM>C and the results are shown in table 3 and table 4.

In table 3, Bit 1 represents that one country's tea value is greater than the other country's tea value. Else it is represented by Bit 0.

Table 3. Cluster table and temporal relationship with respect to Tea values for developed countries

Year / Country	J > C	J > AM	C > J	C > AM	AM > J	AM > C
2001-2002	0	1	1	1	0	0
2002-2003	1	1	0	0	0	1
2003-2004	1	1	0	0	0	1
2004-2005	0	1	1	1	0	0
2005-2006	1	1	0	1	0	0
2006-2007	1	1	0	1	0	0
2007-2008	1	1	0	1	0	0
2008-2009	1	1	0	0	0	1
2009	1	1	0	1	0	0

Support threshold for each comparison = (Number of 1's) * (Total Number of comparisons)

BV(J>C)=01101111=(7*6) =42%   BV(J>AM)=111111111=(9*6) =54%
BV(C>J)=100100000=(2*6)=12%   BV(C>AM)=100111101=(6*6) =36%
BV(AM>J)=000000000=(0*6)=0%   BV(AM>C)=011000010=(3*6) =18%

Finding the minimum support threshold for developed countries:
Min_sup_thres=Sum of all the threshold / Total No. of comparison
Therefore Min_sup-thres = (42+54+12+36+0+18) / 6 = 162 / 6 = 27%
Frequent Temporal Pattern after pruning:

BV(J>C)=011011111=(7*6)=42%   BV(J>AM)=111111111= (9*6) =54%
BV (C>AM) = 100111101 = (6*6) =36%

The frequent temporal patterns are patterns that satisfy the Min_sup_thres. And from frequent temporal patterns we observe that Japan is the leading tea importers with threshold 42%, 54% and India is exporting tea frequently to Japan when compared with China and America.

Table 4. Cluster table and temporal relationship with respect to coffee values for developed countries

Year / Country	J > C	J > AM	C > J	C > AM	AM > J	AM > C
2001-2002	1	0	0	0	1	1
2002-2003	0	0	1	0	1	1
2003-2004	0	0	1	0	1	1
2004-2005	0	0	1	1	1	0
2005-2006	0	0	1	1	1	0
2006-2007	0	0	1	1	1	0
2007-2008	1	1	0	0	0	1
2008-2009	0	0	1	0	1	1
2009	1	1	0	0	0	1

In table 4, Bit 1 represents that one country's coffee value is greater than the other country's coffee value. Else it is represented by Bit 0.

Support threshold for each comparison = (Number of 1's) * (Total Number of comparisons)

BV(J>C)=100000101=(3*6)=18%   BV(J>AM)=000000101=(2*6)=12%
BV(C>J)=011111010=(6*6)=36%   BV(C>AM)=000111000=(3*6)=18%
BV(AM>J)=111111010=(7*6)=42%   BV(AM>C)=111000111=(6*6)=36%

Finding the minimum support threshold for developed countries:
Min_sup_thres=Sum of all the threshold / Total no. of comparison
Therefore Min_sup-thres = (18+12+36+18+42+36) / 6 = 162 / 6 =27%

Frequent Temporal Pattern after pruning:

BV(C>J)=011111010=(6*6)=36%   BV(AM>J)=111111010=(7*6)=42%
BV (AM>C) = 111000111 = (6*6) =36%

Thus America is the leading Coffee importers and India is exporting Coffee frequently to America when compared with China and Japan.

Fig. 3. Comparing Afghanistan 'A' with Srilanka 'S' for the year 2001-2002

In figure 3, Afghanistan's tea value is not greater than Srilanka's tea value, so we represent A>S as 0 in table 5 since it does not satisfies the condition. But in table 6 we set A>S as 1 meaning Afghanistan's Coffee value is greater than Srilanka's Coffee value.

In table 5, Bit 1 represents that one country's Tea value is greater than the other country's Tea value. Else it is represented by Bit 0.

Support threshold for each comparison = (Number of 1's) * (Total Number of comparisons)

BV(A>S) = 001001010 = (3*2) =6%   BV(S>A)= 110110101 = (6*2) =12%

Finding the minimum support threshold for developing countries:
Min_sup_thres=Sum of all the threshold / Total No. of comparison
Therefore Min_sup-thres   = (6+12) /2= 18/2=9%

Frequent Temporal Pattern after pruning:

BV (S>A)= 110110101 = (6*2) =12%

Thus our country is exporting tea frequently to Srilanka.

Table 5. Cluster table and temporal relationship with respect to Tea values for developing countries

Year / Country	A > S	S > A
2001-2002	0	1
2002-2003	0	1
2003-2004	1	0
2004-2005	0	1
2005-2006	0	1
2006-2007	1	0
2007-2008	0	1
2008-2009	1	0
2009	0	1

Table 6. Cluster table and temporal relationship with respect to Coffee values for developing countries

Year / Country	A > S	S > A
2001-2002	1	0
2002-2003	1	0
2003-2004	1	0
2004-2005	0	1
2005-2006	0	1
2006-2007	1	0
2007-2008	1	0
2008-2009	1	0
2009	1	0

In the above table 6, Bit 1 represents that one country's coffee value is greater than the other country's tea value. Else it is represented by Bit 0.

Support threshold for each comparison = (Number of 1's) * (Total Number of comparisons)

$$BV\ (A>S)= 111001111 = (7*2) =14\% \quad BV\ (S>A)= 000110000 = (2*2) =4\%$$

Finding the minimum support threshold for developing countries:
Min_sup_thres=Sum of all the threshold / Total No. of comparison
Therefore Min_sup-thres   = (4+14) / 2 = 18/2 = 9%
Frequent Temporal Pattern:

$$BV\ (A>S)= 111001111 = (7*2) =14\%$$

Thus our country is exporting coffee frequently to Afghanistan.

From the entire cluster table 3 to 6, the support threshold is not fixed for all the cases. The support threshold for developed countries is different from developing countries. Therefore a changeable threshold is more realistic, considering the real time transactions in practice. This provides more flexibility for effective decision making in many applications.

4.3 Experiment Results

Figure 4 and 5, we can observe that Japan is the leading tea importing countries among developed countries and Afghanistan is the leading coffee importing countries. Thus, we come to a decision that if India starts exporting tea to Japan and coffee to Afghanistan frequently in future then we can make profit out of it. Thus, by finding out the frequent importing or exporting countries and their products for the past years we can come to a decision how the future will be and how well we can make profits out of it.

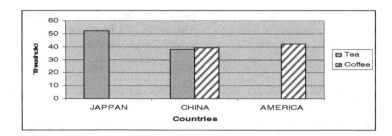

Fig. 4. Experimental Results showing frequent temporal pattern for developed countries

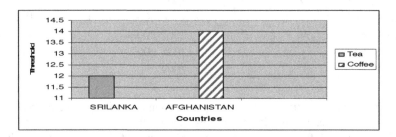

Fig. 5. Experimental Results showing frequent temporal pattern for developing countries

5 Conclusions

The exiting HTPM uses more space and consumes time. However in some applications it is necessary to handle legacy data. In such situation our new algorithm, called RHTPM, provides better performance in terms of time and space complexity for discovering patterns from hybrid event sequences. To verify the efficiency of the algorithm, we performed experiments on real time data. The experimental results show that the algorithm is efficient and better than that of previous approaches in effective decision making. Thus RHTPM can reduce the number of database scans, yield better performance, allow flexibility and reduce post computational works.

References

1. Agrawal, R., Imielinski, T., Swami, A.: Mining Association Rules Between Sets of Items in Large Databases. In: Proceedings of the SIGMOD Conference, pp. 207–216 (1993)
2. Agrawal, R., Srikant, R.: Fast Algorithms for Mining Association Rules. In: Proceedings of the International Conference of Very Large Data Bases, pp. 487–499 (1994)
3. Prasetyo, B., Pramudiono, I., Kitsuregawa, M.: Hmine-rev: Toward H-mine Parallelization on Mining Frequent Patterns in Large Databases. In: Proceedings of the IEIC Technical Report (Institute of Electronics, Information and Communication Engineers), pp. 49–54 (2005)
4. Bettini, C., Jajodia, S.G., Wang, S.X.: Time Granularities in Databases, Data Mining, and Temporal reasoning, p. 230. Springer, New York (2000) ISBN 3-540-66997-3

5. Han, J., Pei, J., Yin, Y.: Mining Frequent Pattern without Candidate Generation. In: Proceedings of the ACM SIGMOD Conference on Management of Data, pp. 487–499 (2000)

6. Ale, J.M., Rossi, G.H.: An Approach to Discovering Temporal Association Rules. In: Proceedings of the 2000 ACM Symposium on Applied Computing, vol. 1, pp. 294–300 (2002)

7. Verma, K., Vyas, O.P.: Efficient Calendar Based Temporal Association Rule. Proceedings of the SIGMOD record 34(3), 63–70 (2005)

8. Pei, J., Han, J., Lu, H., Nishio, S., Tang, S., Yang, D.: H-Mine: Hyper-Structure Mining of Frequent Patterns in Large Databases. In: Proceedings of the International Conference on Data Mining (ICDM 2001), pp. 441–448 (2001)

9. Liu, C., An, J.: Fast Mining and Updating Frequent Itemsets. In: Proceedings of the 2008 International Colloquium on Computing, Communication, Control and Management (ISECS 2008), pp. 365–368 (2008)

10. Wu, S.-Y., Chen, Y.-L.: Discovering hybrid temporal patterns from sequences consisting of point- and interval-based events. In: Proceedings of the Conference on Privacy in Statistical Databases, vol. 68(11), pp. 1309–1330 (2009)

Biztool: A Flexible Cloud-Centric Framework for Business Analytics

Amarnath Ayyadurai and Chitra Babu

SSN College of Engineering, Kalavakkam, Tamil Nadu, India
amarnatha.86@gmail.com, chitra@ssn.edu.in

Abstract. Business analytics provide large enterprises an edge over their competitors, depending upon the size of the data analyzed and the time needed to generate business models. This requires an infrastructure model that meets these huge demands on large scale data processing. Cloud computing provides low cost storage space of virtually any size on demand which can host the data perpetually. Similarly, the processing power can also be commissioned as and when needed. Enterprises are constantly in search of simple and inexpensive systems which transform the available raw data to useful information. In this context, we propose a new framework named Biztool, where a collection of data analytic operators based on Gridbatch is provided as web services that process the data remotely. Biztool is flexible and customizable through user-defined functions. Since various clients can reuse the existing operators for their needs, the development and maintenance cost are reduced.

Keywords: Business Analytics, Data Analysis, Cloud Computing.

1 Introduction

Business analytics, is a continuous iterative exploration and investigation of past business performance to gain insight and drive business planning and it is a complex task. Since this often involves "mixed" data sets that contain nominal and metric attributes, even statistical approaches can be used only up to a certain degree. As a result, the gap between the generation and the understanding of data volumes is widening.

The amount of data being collected for analysis purpose is growing at an exponential rate. Many applications are already taking hours or even days to process the data today, and it might take even longer as the data volume grows. Such long processing delays result in competitive disadvantages for the Enterprises since they cannot react fast enough.

The shift of computer processing, storage, and software delivery from desktop and local servers, to the Internet, and into next-generation data centres has led to new opportunities in data management. Consequently, there is an increasing trend to analyze the data in the emerging Cloud Computing infrastructure because of its strong value propositions.

Cloud architecture comprises a collection of commodity systems. The cost of such systems is less compared to a monolithic high performance system, which is set up specifically for running the data mining applications of an organization. The fault tolerance provided by the cloud architecture is better when compared to the single

S. Dua, S. Sahni, and D.P. Goyal (Eds.): ICISTM 2011, CCIS 141, pp. 256–264, 2011.

server architecture. Cloud computing provides flexible infrastructure by obtaining storage and processing capacity on demand.

The data analytic application must be fault-tolerant, able to exploit parallelism while reducing the development overhead. The application can be fault tolerant by distributing the processing across multiple-independent systems, and it can be done in parallel when the current computation is not dependent on previous results. This translates into communication, coordination and synchronization between multiple parallel processes which can be a difficult job for developers. Google's MapReduce methodology eliminates such difficulties of distributed parallel processing applications.

MapReduce[1] is a programming model for processing and generating large data sets. Users specify a map function that processes a key/value pair to generate a set of intermediate key/value pairs, and a reduce function that merges all intermediate values associated with the same intermediate key. Hadoop[2] is an opensource Mapreduce implementation developed by Yahoo under apache projects.

Hadoop consists of, Hadoop Distributed File System, a method of aggregating the storage space available in all the nodes in a cluster into one logical storage medium which stores data redundantly to ensure high availability and a Job Tracker, which schedules the map and reduce functions simultaneously in all the node processors in the cluster. Due to the inadequacy of the operators provided by Hadoop to address the needs of business analytic applications, additional operators are required.

GridBatch[3] is a programming model and associated library that hides the complexity of parallel programming, while giving the users complete control on how data is partitioned and how computation is distributed so that applications can have the highest performance possible. GridBatch currently consists of the following operators: Map, Distribute, Recurse, Join, Cartesian and Neighbor. The operators handle the details of distributing the work to multiple machines, thus relieving the user from the burdens of parallel programming. The limitation of Gridbatch is that it is designed for a closed system. Clients need to obtain the library of Gridbatch operators to develop their applications. This will incur higher cost in terms of development and maintenance of the system and cause difficulties in expanding the application to suit future requirements. The current trend of business applications' is migration towards Software as a Service.

From this perspective, the proposed work, Biztool is a framework of operators similar to the Gridbatch Model, which utilizes the cloud computing and the Hadoop implementation of MapReduce to provide a simple and powerful interface that enables parallelization and distribution of large-scale computations. The key differentiating feature of Biztool from Gridbatch is that the operators are provided as Web services which are well-defined, self contained functions that can be invoked via the internet.

The ability to integrate these services with an existing system is the main feature of the proposed Biztool framework model which enables its wide usage across varied domains. Providing the underlying operators as a web service will enable other users to adapt this model to suit their needs by adding more functionality. Service based data analysis can in general, be implemented using data mining application suites, large scale databases or through parallel algorithms. However, the proposed system is based on Gridbatch Programming model for the following reasons:

- It scales well and can handle very large datasets. The programming model can also adapt to new datasets which are added frequently to the existing data.

- Gridbatch is based upon Google's MapReduce algorithm which is proven to perform well in Google's private cloud environment.

The remainder of this paper is organized as follows. Section 2 discusses related work in the literature. Section 3 provides an overview of the Gridbatch Programming Model. Section 4 describes the proposed Biztool framework. Section 5 discusses the implementation details. Section 6 explains a sample case study for this framework and Section 7 concludes and provides future directions.

2 Related Work

"Deployable Suite of Data Mining Web Services for Online Science Data Repositories" described in [4] is similar to the proposed system. However, it differs from the proposed system in two aspects: It is used to access remote sensing data scattered across vast geographical network through a common portal. The system is developed using a grid computing environment with dedicated monolithic systems or a collection of dedicated workstations. On the other hand, the proposed system is developed over a cloud architecture which is prevalent in a business environment. Most of the data is distributed across nodes in a LAN or WAN.

The system discussed under Grid-based Data Mining in Real-life Business Scenario [5] is built upon a grid computing environment. The compute node, and data are gathered from distributed workstations across the internet and applied to the task scheduler to process the resources. Grid-computing systems are more widely used in scientific research. However, since the proposed system targets large scale data processing in business enterprises, it utilizes the emerging cloud computing paradigm to meet the infrastructure needs.

The key differences between distributed data mining algorithms such as Gridbatch and Parallel Data Mining (PDM) algorithms are the scale of dataset, network bandwidth usage and underlying hardware architecture as listed in Table 1.

Data analytic applications developed using query languages such as Pig are similar to our proposed system. Pig presents to the users, a higher level programming language, making it easy to write analytics applications. However, since the users are shielded from the underlying system, they cannot optimize its performance to the fullest extent.

Table 1. Comparison of Parallel Data Mining and Gridbatch

FEATURES	PDM	Gridbatch
Scale	Data is processed in Gigabytes range.	Terabyte Datasets are processed.
Interference	Inter process comm. delays execution.	Processes are distributed independently.
Load Balancing	Predefined operation load, homogenous environment.	Transient load handling, heterogeneous environment.
Communication	Uses more bandwidth between processes.	Low bandwidth is used for completion.

Gridbatch is a programming model which provides user defined operators to work on the dataset. The developer can use self defined functions to process the datasets. It also isolates the developer from low level system functions of distributed systems.

In the proposed Biztool framework, the operators have been developed as Web services. The model envisaged is originally developed for the cloud computing environment. However, it can also be used in commodity systems.

3 Gridbatch System

The Gridbatch is a programming model which extends Google's MapReduce algorithm; it does not attempt to help a programmer reason the best approach to program an application. Instead, it aims to provide a set of commonly used primitives, called operators, which the programmer can use to reduce the programming effort. It provides the following group of additional operators for working with a dataset:

- *Distribute Operator:* The rearranging of dataset or deriving subset from existing tables, or grouping similar records into a single memory chunk for processing.
- *Cartesian Operator:* It is used to match every record in one table with the record in another table irrespective of semantic constraints.
- *Join Operator:* This operator is used to merge two tables when the index value of corresponding table matches.
- *Recurse Operator:* The operator is provided to parallelize the reduce operator by applying the reduce operation to different part of the key-value data and obtain a faster result.

These operators use two data structures, Table and Indexed Table (similar to database tools). Hence, the data is stored in table format using which the operations are executed. In our system, the concept of providing developers a set of operators for developing a system to analyze data is retained, but the structure of data is not restricted to a Table or Indexed Table. Instead, the structure of data can be determined by the developers.

4 Biztool Framework

Biztool framework is a collection of data analysis operators provided for large scale data analysis. The operators are a logical grouping of several analytical operations that may be performed on the dataset. Since the framework deals with large scale data analysis, the cost of setting up and maintaining such a system can be high. This can be eliminated through the use of Cloud computing. It is a low cost solution for maintaining a large number of systems with minimal capital investment. The Biztool proposes a cost sharing model such that every user augments a specific amount of processing and storage capacity to the framework. Then when a client submits a job, it utilizes the entire capacity based on other clients demand. Similarly, the user-defined functions under the operators can be reused by other clients for their own needs. The salient aspects of the proposed Biztool framework are:

- Low cost owing to the use of cloud computing.
- Addition of new systems and storage is simpler compared to conventional systems.

- Functionality of the data analysis of framework can be customized and expanded to suit various clients.
- Structure of data is not restricted to tabular / any predefined format
- Availability of Webservice makes it easier to integrate with the existing systems.

Currently Biztool framework defines four operators: View, Process, Convert and Recurse. The framework is flexible because these operators can accommodate different user specific operations. In addition, the data to be analyzed can be of any format the user desires (text, xml, comma separated values, images, etc).

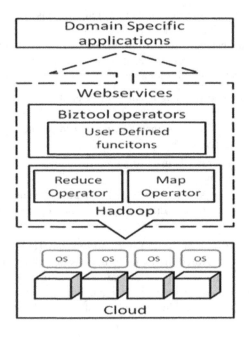

Fig. 1. Biztool Framework

4.1 Operators' Interface

The operators under Biztool framework are utilized by invoking the operator methods with the required parameters. The Webservice consists of four methods for each operator which in turn has four parameters of type *String*. The syntax of these operators is:

Operator(user_func,input,output,func_param)

The input parameter is the location of the dataset that needs to be analyzed by the user. Similarly, the output parameter specifies the location where the output will be stored. It is important to note that since the data analysis system is developed using Hadoop, the location of both the input and the output parameters of the Webservice are in terms of Hadoop Distributed File System (HDFS) paths.

These values are decided by the developer and the data analysis system administrator in advance and remain static throughout the lifecycle of the application. This Biztool framework permits multiple-tenants (Clients), such that different clients can deploy their data into the system and utilize the group of operators provided to perform their own analysis independent of each other. The main advantage of providing this interface as a Webservice is easier integration into existing enterprise application and reduce performance load at the client end.

The client side application obtains a query or user generated event and translates it into the Webservice call for this operator, with all the necessary parameters. This call submits an equivalent job to the Hadoop backend and returns a job id. This job id can be submitted to another method *progress* in the Webservice for status report. On completion of this request, the output of the analysis operation is available at the output directory. Each method in the Webservice is independent of each method and can be called asynchronously.

4.2 Hadoop Jobs

The other part of the framework is the Hadoop job defined by the first parameter *<user_func>* in the Webservice method call. This Hadoop job consists of two overridden methods *Map* and *Reduce*. It is in fact a java program written with two methods map and reduce of the Google's MapReduce paradigm. The map method generates an intermediate key value store based on some conditions. In the reduce method, it collects the key value pair produced by map and transforms into the final output defined by the user. As shown in Fig. 1, the Biztool framework consists of user defined functions. These functions are invoked through a SOAP request to the Webservice of *user_func* jobs under each operator depending upon the domain and analysis performed by the users. The last parameter *func_param* is used to obtain additional parameters if required by any of these user defined Hadoop jobs.

Currently, the following four operators have been defined under the Biztool framework:

- View Operator: It is a collection of functions which produces a view of the analysis generated in a form suitable for the user.
- Process Operator: This can be used to extract information from the dataset to generate subsets and other templates.
- Recurse Operator: Generates nonlinear, order independent analysis for any input dataset.
- Convert Operator: It is used to transform a raw input dataset to an efficient format so that the analysis can be faster and compact.

The difference between the Operators provided in the Gridbatch Model and Biztool framework, is that the input and output data structure in Gridbatch is tabular while it is customizable by the user through the user defined function in the Biztool framework. Whenever the provided functions or operators are inadequate, new functions can be written in Hadoop map and reduce functions and can be posted into the framework. Further, since the framework is provided as a Webservice, it enables easy integration with existing applications.

5   Implementation

The system consists of a private cloud built using 15 commodity workstations and EUCALYPTUS[6] – an open source software framework for cloud computing that implements what is commonly referred to as Infrastructure as a Service.

EUCALYPTUS is simple, flexible and modular with a hierarchical design reflecting common resource environments. In essence, the system allows the users to start, control, access, and terminate entire virtual machines using an emulation of Amazon EC2's SOAP and "Query" interfaces.

These virtual machine instances are used to create a virtual cluster of Hadoop. This virtual cluster consists of one jobtracker, namenode and 15 tasktrackers and datanodes. Each virtual machine created in the private cloud has one processor and 1GB RAM and 2GB of disk space. The Webservice for the operator is hosted through Glassfish server in the same virtual instance as the jobtracker/namenode.

A sample application is developed to access the data analysis system in a user-friendly manner. The client application gets a query easily understandable by the end-user and maps the request to corresponding parameters of the Webservice. The job status is displayed in the client application. Once the process is completed, the output can be viewed by the user. Since the Webservice and the Hadoop cluster are isolated, they can also be located in two different machines.

6   Case Study

The System is tested with a dataset from Wikipedia, which is an English language extraction of the articles present in the Wikipedia sites. This dataset is chosen because it is considerably large and several kind of analysis could be performed on it and is also freely available. The structure of this dataset is in xml format.

- The convert operator provides a function called *record* which converts a tree styled xml Wikipedia dataset into a linear text record which is much more compact and can be used by the *filter* function.

```
8 2007-12-22T11:17:58Z Closedmouth 1233 Ashmore and
Cartier Islands/Military

9 2005-05-19T14:50:00Z  MacRusgail 8647 Dozenal Society
of Great Britain

10 2009-11-26T14:42:55Z Magnius 8648 Daffynition

11 2009-02-06T17:18:48Z Alansohn 1234 Acoustic theory
```

Here, the columns are row id, timestamp, author name, article id and title of the article. These are extracted from the Wikipedia xml dataset.

- The *filter* function under the *view* operator is used to extract the required records from the entire dataset based on some specific condition.
- *Process* operator is used to analyze the articles text present in the Wikipedia data. For example, the *cite* function can be used to identify the citation references made in the articles.

```
1000 {{cite encyclopedia |year=2000 |author=Bunson,
Matthew |title=Hastings, Captain Arthur, O.B.E.
|encyclopedia=The Complete Christie: An Agatha Christie
Encyclopedia |publisher=Pocket Books |location=New
York}}
```

```
1000 {{cite web|
url=http://www.litencyc.com/php/speople.php?rec=true&UI
D=5054 |title= Agatha Christie (1890-1976)
|accessdate=2006-09-06 |author=Chris Willis, London
Metropolitan University}}
```

```
10005 {{cite web |last=Peterson |first=Robert |year=1998
|url=http://www.scoutingmagazine.org/issues/9810/d-
wwas.html |title=The BSA\'s \'forgotten\' founding
father |work=Scouting Magazine |publisher=Boy Scouts of
America |accessdate=2006-03-10}}
```

In this extraction, the output is *article id* and the cite references from the text of the article. Analogously, any pattern can be extracted using the *find* function provided under the framework.

- Recurse operator, consists of *listarticle* function for providing the title of all the articles in the dataset and the *pagecount* function to count the number of pages in the article.

```
1     <title>AccessibleComputing</title>

2     <title>CSS Virginia</title>

3     <title>Jacob Abbott</title>

4     <title>Wisent</title>

5     <title>Angband</title>

6     <title>November 29</title>
```

The *listarticle* function will provide the output of titles of all the articles listed in the xml dataset. Similarly, the *pagecount* will return the number of articles in the dataset.

```
<page> 51052
```

The case study uses Wikipedia dataset in two formats the xml and text files. It is interesting to note that the size of the text file is small compared to the xml file and hence the computation is faster. However, the system also supports the original XML format of data as well.

7 Conclusions

This paper has proposed a framework called Biztool that models the various operators needed for building a data analytic application, as web services. The software system is built over the cloud infrastructure and uses the Gridbatch programming model. The framework is deployed as a Webservice over a virtual Hadoop cluster created in a private cloud developed using Eucalyptus Toolkit. A client application which receives a query from the user and submits the *userfunc* jobs to the Biztool framework webservice is developed for testing. In the future, we intend to develop domain specific analytical Web services using the Biztool Framework. These Webservices will provide generic operations specific to a domain which can be customized by the client to suit their needs. Also, multitenancy issues need to be explored further.

References

1. Dean, J., Ghemawat, S.: Mapreduce: Simplified data processing on large clusters. In: Proceedings of Sixth Symposium on Operating System Design and Implementation, pp. 137–150 (2004)
2. Borthakur, D.: The Hadoop Distributed File System: Architecture and Design (2007), Retrieved from http://lucene.apache.org/hadoop
3. Liu, H., Orban, D.: GridBatch: Cloud Computing for Large-Scale Data-Intensive Batch Applications. In: Proceedings of the 8th IEEE International Symposium on Cluster Computing and the Grid, May 19-22, pp. 295–305 (2008)
4. Graves, S., Ramachandran, R., Lynnes, C., Maskey, M., Keiser, K., Pham, L.: Mining Scientific Data using the Internet as the Computer. In: Proceedings of IEEE International Conference on Geoscience and Remote Sensing Symposium (IGARSS), July 7-11, pp. 283–286 (2008)
5. Li, T., Bollinger, T., Breuer, N., Wehle, H.-D.: Grid-based Data Mining in Real-life Business Scenario. In: Proceedings of IEEE/WIC/ACM International Conference on Web Intelligence (WI 2004), September 20-24, pp. 611–614 (2004)
6. Nurmi, D., Wolski, R., Grzegorczyk, C., Obertelli, G., Soman, S., Youseff, L., Zagorodnov, D.: The Eucalyptus Open-source Cloud-computing System. In: Proceedings of 9th IEEE/ACM International Symposium on Cluster Computing and the Grid, May 18-21, pp. 124–131 (2009)

Service Availability Driven Re-configurable Embedded System Design

Anil Kumar[1] and Shampa Chakarverty[2]

[1] Patni Computer Systems, 142 E&F, Block 'B', NSEZ, Noida, India
[2] Division of Computer Engineering, NSIT, New Delhi, India
anilnbsingh@yahoo.com, apmahs@rediffmail.com

Abstract. Introduction of re-configurable hardware into embedded systems has given a new direction to fault tolerant computing. It is now feasible to satisfy the reliability and performance constraints on demanding applications while reducing the overall cost of the system. In order to evaluate the system's performance metrics, the application's specifications are mapped and scheduled to the computing and communication resources and their dynamic re-configuration capabilities are exploited. Our automated architecture design algorithm explores the design space and selects the optimal architecture which reconfigures itself into partial functional states in such a way that the most important services as perceived by the user are always available. The system availability is evaluated on the basis of Continuous Time Markov model and user's importance of service availability in partly and fully functional states. Two multi-objective genetic algorithms have been employed for architecture optimization.

Keywords: Fault Tolerance, Graceful degradation, Architecture Optimization, Multi-Objective Genetic Algorithm and CTMC.

1 Introduction

With the advance of technology, embedded systems are becoming more pervasive. Given the pervasiveness of these systems; it is necessary to consider the user's expectations about the system's Reliability and Availability (R&A) in the early stages of its design. The R&A of such systems can be improved through fault avoidance, fault tolerance, and fault forecasting. Fault avoidance relies on the use of high-reliability components and rigorous design processes to avoid the intrusion of faults or their occurrence during operation. Given the complex nature of embedded systems, all faults cannot be avoided either by fault prevention or by fault removal; fault tolerance is an alternative solution of this problem. The system built with fault tolerance capabilities continues to operate even in the presence of faults. Fault-tolerance techniques rely on component redundancy. In general, hardware redundancy in distributed embedded systems is avoided due to limited resource availability, weight and power consumption. Software redundancy is therefore, more commonly used to increase fault-tolerance. The software redundancy causes

S. Dua, S. Sahni, and D.P. Goyal (Eds.): ICISTM 2011, CCIS 141, pp. 265–276, 2011.

overall performance degradation, and it may result in missed deadlines in a real-time system. Due to cost constraints, such measures can only be applied to systems on which humans lives or large sums of money are at stake.

Fault forecasting involves the estimation of fault presence, corresponding failures and the ensuing gracefully degraded state of operation. In the graceful degraded state, the system continues to work with reduced functionality or performance. As the individual components fail or are repaired, the system makes transitions between the different states, each representing a certain level of functionality [1] [6]. It requires flexible architecture that re-configures itself on the fly to achieve maximum R&A.

System re-configuration in a faulty state can eliminate the need for redundant hardware and software, and it can also be used to increase the R&A without introducing additional redundancy [3]. Although the re-configuration does not always guarantee the full system recovery, the system re-configuration can be tailored to move the system into the graceful degrading state on individual component failure. R&A must therefore, be considered in the context of systems that can deliver multiple grades of service. Systems should be designed in such a way so that the services available in gracefully degraded states are those that are most valuable to the end user.

In this paper, we build a *Computer Aided Design* (CAD) tool for designing re-configurable multiprocessor architectures in a manner such that the system continues to provide the services which the user perceives as most important under different functional states. The tool tackles the problem of capturing the user-determined availability requirements for a system with degradable performance and uses them to construct heterogeneous architectures with high levels of performance and availability.

The design space exploration itself is a *NP-Complete*[1] problem [4]. Quest for an architecture that reconfigures itself in the event of any failure and continues to provide user-perceived important services makes the design space exploration more complicated. In order to deal with such complexity, we have used two level optimizations:

1. *Outer level optimization*: Design Exploration for selecting the best architecture.
2. *Inner level optimization*: Design Exploration for selecting the best re-configuration of the selected architecture in the event of a failure.

The outer level optimization uses a genetic algorithm to optimize cost, performance, re-configurability and system availability. The Inner level uses a genetic algorithm to optimize service availability, performance and Task Migration in each partly functional state by selecting the best architecture re-configuration achievable.

The rest of the paper is organized as follows. Section 2 describes the application's task model, the technology environment and the architectural model. Section 3 presents the proposed re-configuration model for availability. Section 4 describes the dual GA-driven design optimization. Section 5 presents the salient experimental results. We conclude the paper in Section 6.

---

[1] Nondeterministic Polynomial Complete Problems are computational problems for which no efficient solution algorithm has been found.

2 Design Environment

A typical real-time application repeatedly executes a set of tasks in a periodic or aperiodic fashion, and has a time-bound for executing the entire set of tasks. There is a specific order in which tasks need to be executed. We model the application by an acyclic, partially ordered *task precedence graph* TPG. A TPG is a pair (V, E); where V is a set of vertices, and E is a set of edges between the vertices $E \subseteq \{(v_i, v_j) \mid (v_i, v_j \in V)\}$. The set of nodes V represent the computation sub-tasks and the set of edge E represent the communication sub-tasks that carry data from source to sink tasks and also indicate the precedence constraints between the tasks. Each task has a granularity that embodies a tightly coupled major sub-function of an application that can be implemented on a single processing unit. The set $T= V \mathrel{\dot{E}} E$ contains all the sub-tasks. A set element is denoted in small case $\{v_i, v_{j....}\}$. All edges have data volume weights. The sets PI and PO represent the set of primary input tasks and primary output tasks respectively. A representative application graph TPG_{18} is shown in Fig. 1. TPG_{18} is used as a running example in this paper.

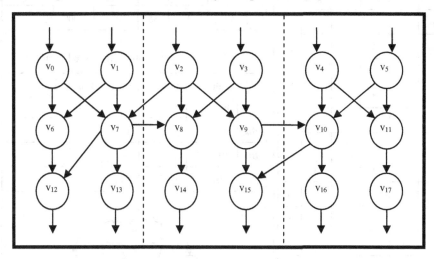

Fig. 1. TPG_{18} (Timing and Data Volume Parameters are Eliminated for Clarity)

Table 1. specifies worst case execution time of TPG_{18} tasks. A '*' entry indicates that it is not feasible to execute the task on the corresponding processing unit. The resources library consists of a database of Processing Element PE types and Bus Bt types. Table 2. Illustrates a part of the technical and cost specifications of PE types available for TPG_{18}. Table 3 illustrates a part of the technical and cost specifications of Bt types available for TPG_{18}.

3 Re-configuration for Service Availability

Re-configurable Computing is evolving as an emerging alternative for fault tolerance [3] [5] and [8]. *Field Programmable Transistor Arrays* and *Complex Programmable*

Logic Devices are the key candidate in re-configurable computing. Design exploration for reliable embedded systems using re-configurable components is becoming more and more popular, especially in service oriented architectures [9].

Table 1. Feasible Resources Library

	P_0	P_1	P_2	P_3	P_4
v_0	8	4	*	*	3
v_1	5	5	*	*	*
v_2	3	*	4	*	2
v_3	5	5	*	*	1
v_4	12	10	*	3	8
v_5	15	18	7	5	*
v_6	8	4	*	*	3
v_7	5	5	*	*	*
v_8	3	*	4	*	2
v_9	5	5	*	*	1
v_{10}	12	10	*	3	8
v_{11}	15	18	7	5	*
v_{12}	8	4	*	*	3
v_{13}	5	5	*	*	*
v_{14}	3	*	4	*	2
v_{15}	5	5	*	*	1
v_{16}	12	10	*	3	8
v_{17}	15	18	7	5	*

Table 2. Processing Element Resource Library

	Failure Rate	Repair Rate	Cost	Memory (MB)	Speed (MIPS)	Busses supported
P_0	8×10^{-5}	2	100	2048	2000	$Bt_0\ Bt_1\ Bt_2$
P_1	9×10^{-4}	1	60	1024	256	Bt_1
P_2	$4. \times 10^{-3}$	.2	40	256	2000	$Bt_0\ Bt_2$
P_3	3×10^{-3}	.8	35	512	512	$Bt_1\ Bt_2$
P_4	2×10^{-3}	1	35	1024	1000	Bt_2

Table 3. Communication Resource Library

	Failure Rate	Repair Rate	Cost	Data Transfer Rate
Bt_0	.1	4	10	50
Bt_1	.01	2	20	8
Bt_2	.001	1	30	80

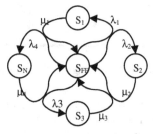

Fig. 2. CTMC Model

The design exploration for fault tolerance faces challenges like:

1. Finding the best feasible and cost efficient architecture configuration in the absence of any fault. It means the architecture should utilize the re-configurable components in improving the performance and implementing the core features. We employ a multi-objective genetic algorithm GA_1 to achieve these objectives.

2. The reconfigured architecture should meet the performance and availability goals that are established in the absence of faults. To ensure this, we employ a multi-objective genetic algorithm GA_2 to locate an optimal re-configuration.
3. Limiting consequences of faults on performance and availability. When a component becomes faulty, the reconfigured system will operate with less computing or commutation capacity, which may result in performance and availability degradation. The GA_2 explores the whole architecture space so that even with reduced capacity, the impact on system performance and availability can be mitigated and minimized.

With large numbers of devices per system, it is possible to take advantage of the multiplicity of identical and/ or re-configurable processing elements to improve the service availability and performance. In order to achieve this, it becomes necessary to efficiently map applications onto hardware or software by partitioning them onto re-configurable architecture. We develop an approach, utilizing task migration to enable the re-mapping of failing applications in environments with identical and/ or re-configurable resources.

As the system transitions from one state to another state, the importance of service availability also changes. If the system is not in a fully functional state, it should degrade the services according to the user's preferences. A flexible fuzzy rule base is defined to capture the user's availability requirement. The task graph is initially mapped and scheduled to the computing and communication resources. In the event of failure, it is then remaped depending on the application's changed requirements in the reduced functional state as guided by user preferences.

We assume that the system is brought into a known safe state prior to the re-configuration event trigger. The safe state may involve using manual or automatic steps to perform the basic operating system functions needed to keep the system stable. After all necessary components are brought to the safe state and all communication links are frozen, a system re-configuration event is fired. Since the state of the system was in a known safe state prior to re-configuration, there is little overhead to bring the system in the normal state. This can be done by sending a synchronization request to all resources.

3.1 Rule Base for Service Availability Importance

The user may desire a different level of availability for the same service under different situations. In fact, the user must be given the freedom to envision various hypothesized situations when he may have to work with only a subset of services and yet derive the maximum benefit from each available service.

To illustrate, consider a commercial *Heating Ventilation Air-Conditioning* (HVAC) unit. A chiller failure should affect the least occupied area of the building, and the most occupied area should not be affected. The occupancy at any time can be determined by the sensors installed in the building. For a given fully or partly functional state, either the user can specify or the system can use preset rules to evaluate the importance of availability of chilling services allocated for different regions. We use a Fuzzy rule based framework [7] to capture the importance-of-availability of various services under different circumstances.

The importance-of-availability weight $IoA_a(o_x)$ of a service available at a primary output o_x indicates the relative importance of that service in system state S_a . The *Output Quality Level* of a service o_x is denoted by $OQL(o_x)$. The IoA_a values of all outputs are prior determined by the designer in collaboration with the user using a fuzzy rule base framework. The rule has an antecedent **If** that specifies the output conditions when the rule is fired and a consequent **Then** the part that assigns a new importance-of-availability weight to the output. As an example, a rule can be specified as:

If [(OQL(o_n)==LOW .AND. OQL(o_{n-1})==MEDIUM) .OR. .... (OQL(o_l)==VERY HIGH) .AND. ...OQL(o_0)==REMOTE)] **Then** *[IoA(o_n)=HIGH, IoA(o_{n-1})=VERY HIGH, ... IoA(o_l)= HIGH,...., IoA(o_0)=REMOTE]*

The total availability importance in S_a is obtained by adding the IoA_a values of all the services.

$$IoA_a(Total) = \sum_x IoA_a(o_x) \tag{1}$$

The total availability importance is normalized by dividing with the total number of services or number or primary outputs.

$$IoA_a(Normalized) = \frac{IoA_a(Total)}{|PO|} \tag{2}$$

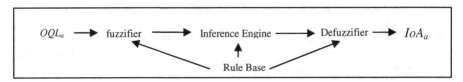

Fig. 3. Rule Base Engine

Procedure Find_IoA(OQL_a)
1. *Invoke the* Rule Base Engine *for OQL_a.*
2. *Invoke the Fuzzy Inference Engine to activate the   rules*
3. *Invoke the de-fuzzier to obtain the crisp values of availability importance weight IoA_a(o_x) for all o_x*
4. *Calculate the overall availability importance for state S_a using Equation (1).*
5. *Calculate the IoA_a(Normalized) using Equation (2).*
End Find_IoA(OQL_a)

Pseudo-code 1: *Find_IoA(**OQL_a**)*

Normal IoA_{FF} of fully functional state is unity because in a fully functional state all services may be equally important to the user. The flow of the rule base engine is shown in Fig. 3.. The vector OQL_a indicates the output quality level of each service in state S_a. OQL_a goes through a fuzzifier. The inference engine works with attribute values, which are attached to memberships function. They are created from real-valued attributes, which have been partitioned into individual fuzzy sets {*LOW*, *MEDIUM*........}. The inference engine then provides fuzzy outputs and those fuzzy outputs are de-fuzzified to obtain crisp values that represent Importance of service availability in state S_a. Pseudo-code 1 describes the steps to calculate $IoA_a(Normalized)$ for a given OQL vector.

3.2 System Availability

Assume a candidate architectural solution for the given task graph, containing a set of resources $R=PE \cup Bt$ and a task allocation function: $\psi : V \rightarrow PE, E \rightarrow Bt$. Faults in the system can occur in the processing units, in bus links or in individual software tasks. All faults are assumed to be permanent and there are no transient faults. A repaired component is as good as a new one. We assume single faults; *i.e.*, only one processor/bus can fail at a time. Moreover, the faults are reasonably spaced apart in time to allow for timely repairs. The duration of successive operating intervals as well as the repair times for all components is *Independent and identically distributed*, exponential random variables with constant failure rates.

The multiprocessor system includes hardware processing units that are vulnerable to hardware faults as well as instruction set processors that can develop faults in the processor hardware or software faults in the code being executed. We use the *Continuous Time Markov Model* (CTMC) shown in Fig. 2. to model the fail-repair mechanism for the multiprocessor system. There are $N_r=N_p+N_b$ partly functional states, corresponding to the failure of any one of N_r processors and N_b busses. After the failed processor/bus is repaired/ replaced, the system returns to its initial fully functional state. The time varying fail-repair process is described by the following differential equations:

$$\frac{dP_{FF}(t)}{dt} = -(\lambda_1 + \lambda_2 \cdots \lambda_{N_r})P_{FF}(t) + \mu_1 P_1(t) + \cdots + \mu_{N_r} P_{N_r}(t) \quad (3)$$

For state S_i: $\frac{dP_i(t)}{dt} = \lambda_i P_{FF}(t) - \mu_i P_i(t)$; $P_{FF}(0) = 1, \forall i : P_i(0) = 0$;

$$P_{FF}(t) + P_1(t) + \cdots + P_{N_r}(0) = 1$$

The steady-state time invariant state probabilities can be derived easily by setting the time differentials equal to zero and solving the set of balance equations.

$$P_{FF} = \prod_{i=0}^{N_r-1} \mu_i \left/ \sum_{j=0}^{N_r-1} \left(\lambda_j * \prod_{i=0,i \neq j}^{N_r-1} \mu_i \right) + \prod_{i=0}^{N_r-1} \mu_i \right. \quad (4)$$

The steady-state availability P_a of the ath partly functional state is given by:

$$P_a = \lambda_a * \prod_{i=0,i \neq a}^{N_r-1} \mu_i \left/ \sum_{j=0}^{N_r-1} \left(\lambda_j * \prod_{i=0,i \neq j}^{N_r-1} \mu_i \right) + \prod_{i=0}^{N_r-1} \mu_i \right. \quad (5)$$

Finally, we put it all together to evaluate the system availability. The system availability is the summation of availability of each functional state multiplied by its $IoA_a(Normalized)$. Pseudo-code 2 describes the steps for calculating system availability of architecture.

Procedure SysA(Chromosome ψ)
1. For each state S_a:

    1.1 Evaluate the OQL_a vector using the task graph and the architecture ψ with failed resource r_a.

    1.2 Calculate the value of IoA_a $(Normalized)$ in S_a using **Pseudo-code 1**

Find_IoA(OQL_a)

    1.3 Calculate P_a using equation (5)

      1.4 Calculate the availability A(a) in state S_a:

    $A(a) = P_a * IoA_a(Normalized)$

2. Calculate system availability A_{sys} considering all states.

    $A_{sys} = \sum_a A(a) + P_{FF}$

End SysA()

Pseudo-code 2: SysA (Chromosome ψ)

4 Design Optimization

Genetic Algorithms GA_1 and GA_2 are used for architecture optimization. The chromosome Ch, represents (a) processor selections (b) node allocation, i.e., computational task to processor mappings (c) bus selections (d) edge allocation i.e., the communication task to bus mappings, for a given task graph. The GA_1 algorithm starts with the randomly generated population of chromosomes and evolves towards the optimization of an objective function. The evolution occurs in four phases: processor selection, node-to-processor allocation, bus selection and edge-to-bus allocation. The objective includes Availability (Architectural as well as Service Availability Importance), Re-configurability, Cost, and Performance. The equation (9) evaluates GA_1 chromosome's fitness. The GA_1 chromosome's fitness has four aspects:

- *Cost fitness:* The cost feasibility for the given allocation ψ is defined as:

$$CF (\psi) = \frac{Ct_{max} - Ct (\psi)}{Ct_{max}} \times U (Ct_{max} - Ct (\psi)) \qquad (6)$$

Ct_{max} is a budget allocated to the system and $CF (\psi)$ is cost of architecture ψ .

- *Performance Fitness:* We employ a non-preemptive task-scheduling algorithm to determine the task schedules. We adopt an ALAP-ASAP list-scheduling [11] algorithm with a priority scheme based on the *timeliness* of the tasks. This algorithm assigns start and completion times to the tasks that can be executed. We follow an *(m, N_{po})* constrained performance criterion, *i.e.,* m^2 out of N_{po} primary outputs must meet their deadline. If this constraint is fulfilled, the system *Deadline*

---

[2] For our experiments, we take the value of m as half of N_{po}.

Meeting Ratio (DMR) is defined as the ratio of the number of tasks that are able to meet their deadlines to the total number of outputs. Thus if δ is the number of output tasks that are able to meet their deadlines then DMR is defined as:

$$DMR\ (\psi) = \delta < m\ ?\ 0\ :\ \delta/N_{po} \tag{7}$$

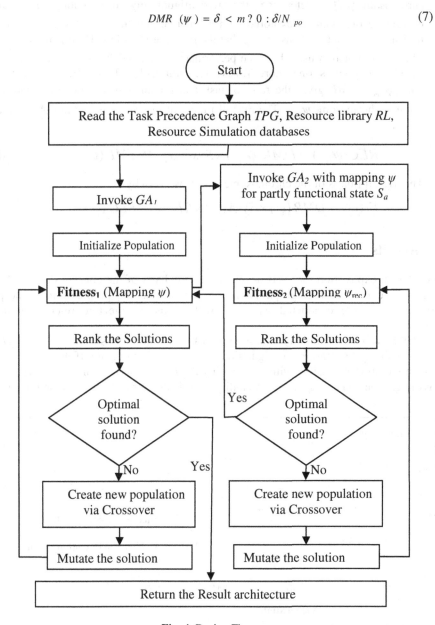

Fig. 4. Design Flow

- *Availability fitness:* The availability fitness parameter $SysA(\psi)$ is derived from the Pseudo-code 2.
- *Re-configurability fitness: REC (ψ)* is the re-configuration capability of the chromosome ψ .To determine the re-configurability of solution ψ in the occurrence of a fault is explored with a separate genetic algorithm GA_2. GA_2 is initiated with solution ψ assuming that the resource r has failed and is unavailable. The fitness function uses the same performance and availability functions as used by GA_1. GA_2 uses one more fitness function called *Task Migration Fitness TMF(ψ_{rec})*. *TMF* gives the ratio of task and a path is being migrated from the original chromosome ψ_{rec} . The overall fitness of the chromosome in GA_2 is given by:

$$REC(\psi_{rec}) = DMR(\psi_{rec}) \times SysA(\psi_{rec}) \times TMF(\psi_{rec}) \qquad (8)$$

The overall fitness of the chromosome in GA_1 is given by:

$$Ft(\psi) = DMR(\psi) \times SysA(\psi) \times CF(\psi) \times REC(\psi) \qquad (9)$$

5 Results

The design exploration tool was developed on the basis of the concept described on Fig. 4. We use the task graph TPG_{18} to demonstrate the proposed the concept. Experiments were conducted with and without the architecture re-configuration options.

The growth of availability fitness with GA_1 generations is shown in Fig. 5. GA_2 explores the architecture re-configuration possibilities in fail state. If the architecture re-configuration is feasible without violating the *DMR* criteria then the performance, task migration and user service availability is evaluated for re-configured architecture.

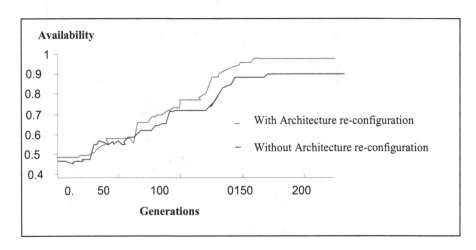

Fig. 5. Availability Fitness vs. Generation

Initially, both experiments started with the same setup. However, after only a few generations the experiment with re-configuration option achieved high availability fitness as a comparison to experiment without re-configuration. The number of successful fault re-configuration increases the probability of jumping to a fully functional state and hence the service availability is increased. During initial few generations, performance fitness was low in GA_2 due to more deadlines slips. However, after around 100 generations the *DMR* has stabilized and was comparable to the *DMR* of fully functional state.

As we can see in Fig. 6 solution S_1 (with architecture re-configuration) is more efficient than solution S_2 (without architecture re-configuration) in terms of availability as well as cost. The best solution obtained (with architecture re-configuration) is having *DMR* .8182, cost .54 units and .981843 system availability. Although from performance point of view solution S_2 is better than the solution S_1.

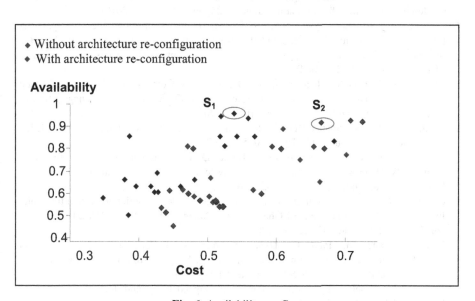

Fig. 6. Availability vs. Cost

6 Conclusion

To conclude, we have proposed an innovative mechanism to utilize the architecture re-configuration capability to meet the user's availability expectations into a partly functional state. This paper shows close relations between user's availability expectations and architecture re-configuration. An inner level optimization algorithm GA_2 is employed to explore the best re-configuration in terms of user service availability and performance in the degraded state. The main drawback of the proposed approach is the assumption that the system will be restored to its last running state after architecture re-configuration. And the restoration efforts are ignored during the architecture exploration. The proposed approach can be considered for designing applications, such as HVAC systems, which can afford the loss of internal state and can activate the whole process from scratch after architecture reconfiguration.

References

1. Glaß, M., Lukasiewycz, M., Haubelt, C., Teich, J.: Incorporating Graceful Degradation into Embedded System Design. In: Proceedings of Design, Automation and Test in Europe (DATE 2009), March 8-12, pp. 320–323 (2009)
2. Lu, S., Halan, W.A.: Incorporating Fault Tolerance into Component-based Architectures for Embedded Systems. Journal of Automation, Mobile Robotics & Intelligent System 3(1), 46–51 (2009)
3. Staroswiecki, M.: On Reconfiguration-Based Fault Tolerance. In: 18th Mediterranean Conference on Control & Automation, June 23-25, pp. 1681–1691 (2010)
4. Kianzad, V., Bhattacharyya, S.S.: CHARMED: A Multi-Objective Co-Synthesis Framework for Multi-Mode Embedded Systems. In: Proceedings of the 15th IEEE International Conference on Application-Specific Systems, Architectures and Processors (ASAP 2004), pp. 28–40 (2004)
5. Kubalik, P., Dobias, R., Kubatova, H.: Dependability Computation for Fault Tolerant Reconfigurable Duplex System. In: Proceedings of IEEE Design and Diagnostics of Electronic Circuits and Systems, pp. 98–100 (2006)
6. Beaudry, M.D.: Performance-Related Reliability Measures for Computing Systems. IEEE Transactions on Computers C-27(6), 540–547 (1978)
7. Kumar, A., Chakarverty, S.: A Fuzzy-Based Design Exploration Scheme for High Availability Heterogeneous Multiprocessor Systems. eMinds: International Journal on Human-Computer Interaction 1(4), 1–22 (2008)
8. Cheung, P.Y.K.: Process Variability and Degradation: New Frontier for Reconfigurable. In: Sirisuk, P., Morgan, F., El-Ghazawi, T., Amano, H. (eds.) ARC 2010. LNCS, vol. 5992, p. 2. Springer, Heidelberg (2010)
9. John, J., Wang, M., Pahl, C.: Dynamic Architectural Constraints Monitoring and Reconfiguration in Service Architectures. In: Babar, M.A., Gorton, I. (eds.) ECSA 2010. LNCS, vol. 6285, pp. 311–318. Springer, Heidelberg (2010)
10. Proenza, J., Almeida, L.: Position Paper on Dependability and Reconfigurability in Distributed Embedded Systems. In: Proceedings of 6th International workshop on Real Time Networks (RTN 2007), July 4-6, pp. 1–6 (2007)
11. Memik, S.O., Bozorgzadeh, E., Kastner, R., Sarrafzadeh, M.: A Super-Scheduler for Embedded Reconfigurable Systems. In: Proceeding of the International Conference on Computer Aided Design, pp. 1–4 (2001)

Comparative Analysis of Impact of Various Global Stock Markets and Determinants on Indian Stock Market Performance - A Case Study Using Multiple Linear Regression and Neural Networks

Avinash Pokhriyal, Lavneet Singh, and Savleen Singh

Management & Computer Applications, R.B.S College, Agra
pokhariyal@hotmail.com, Lavneet_agra@yahoo.co.in,
savleenagra@gmail.com

Abstract. Globalization and technological advancement has created a highly competitive market in the stock and share market industry. Performance of the industry depends heavily on the accuracy of the decisions made at performance level. The stock market is one of the most popular investing places because of its expected high profit. For prediction, technical analysis approach, that predicts stock prices based on historical prices and volume, basic concepts of trends, price patterns and oscillators, is commonly used by stock investors to aid investment decisions. In recent years, most of the researchers have been concentrating their research work on the future prediction of share market prices by using Statistical & Quantitative tools. But, in this paper we newly propose a methodology in which the Multiple Linear Regression and neural networks is applied to the investor's financial decision making to invest all type of shares irrespective of the high / low index value of the scripts, in a continuous time frame work. The proposed network has been tested with stock data obtained from the Asian Stock Market Database. Finally, the design, implementation and performance of the proposed multiple linear regression and model of simulated neural network are described.

Keywords: Stock Market Performance, Multiple Linear Regression, NIFTY 50, Artificial Neural Networks.

1 Introduction

Recently forecasting stock market return is gaining more attention, maybe because of the fact that if the direction of the market is successfully predicted the investors may be better guided and also monetary rewards will be substantial. If any system which can consistently predict the trends of the dynamic stock market be developed, would make the owner of the system wealthy. Another motivation for research in this field is that it possesses many theoretical and experimental challenges. The most important of these is the efficient market hypothesis which proposes that profit from price movement is very difficult and unlikely. In an efficient market, stock market prices fully reflect available information about the market and its constituents and thus any

S. Dua, S. Sahni, and D.P. Goyal (Eds.): ICISTM 2011, CCIS 141, pp. 277–286, 2011.

opportunity of earning excess profit ceases to exist any longer. So it is ascertain that no system is expected to outperform the market predictably and consistently. There has been a lot of debate about the validity of the EMH and many researchers have attempted to use neural networks to give a contradictory view to the Efficient Market Hypothesis. Moreover, many researchers claim that the stock market is a chaos system. Chaos is a non linear deterministic process which only appears random because it is not easily expressed. These systems are dynamic, a periodic, complicated and difficult to deal with normal analytical methods but with neural network's ability to deal with nonlinear, chaotic system, it may be possible to forecast the trends of the market. This may eventually question the traditional financial theory of efficient market. There is not much evidence that the stock market returns are perfectly linear for the very reason that the residual variance between the predicted return and the actual is quite high. The existence of the nonlinearity of the financial market is propounded by many researchers and financial analyst. Some parametric nonlinear models such as Autoregressive Conditional Heteroskedasticity and General Autoregressive Conditional Heteroskedasticity have been in use for financial forecasting as in [1, 2, 3]. But most of the non forecasting of Indian Stock Market Index Using Artificial Neural Network. Linear statistical techniques require that the non linear model must be specified before the estimation of the parameters is done. During last few years there has been much advancement in the application of neural network in stock market indices forecasting with a hope that market patterns can be extracted as in [4, 5, 6]. The novelty of the ANN lies in their ability to discover nonlinear relationship in the input data set without a prior assumption of the knowledge of relation between the input and the output. They independently learn the relationship inherent in the variables. From statistical point of view neural networks are analogous to nonparametric, nonlinear, regression model. So, neural network suits better than other models in predicting the stock market returns as in [7, 8]. This paper presents the use of artificial neural network and linear regression as a forecasting tool for predicting the direction of the market. The study also seeks to document the self similarity characteristic of the stock market. A phenomenon that is self similar, or invariance against changes in scale or size, looks the same or behaves the same when viewed at different degrees of magnification of different scales on a dimension. The dimension can be space, length, width or time. In case of a stock market the dimension is time. If a stock market possesses self similarity characteristics then the long term and the short term characteristics will be same. The rise and the falls in the market index over a year will reassemble the rise and the fall trend in a monthly scale and the same pattern can also be seen on an intraday scale. So, it can be assumed that the trend of the index price on a day is likely to be followed in a monthly trend, yearly trend and over a period of years. So, this study can help the regulators to design the market checks in advance to avoid steep fluctuation in the market.

1.1 Artificial Neural Network

Before the age of computers, people traded stocks and commodities primarily on intuition. As the level of investing and trading grew, people searched for tools and methods that would increase their gains while minimizing their risk. Statistics, technical analysis, fundamental analysis, Time series analysis, chaos theory and linear regression are all used to attempt to predict and benefit from the market's direction. None of these techniques has been proved to be the consistently correct prediction tool that is desired, and many analysts argue about the usefulness of many of the approaches. However, these methods are presented as they are commonly used in practice and represent a base-level standard for which neural networks should outperform. Also, many of these techniques are used to preprocess raw data inputs, and their results are fed into neural networks as input.

A neural network is a computer program that recognizes patterns and is designed to take a pattern of data and generalize from it. An essential feature of this technology is that it improves its performance on a particular task by gradually learning a mapping between inputs and outputs. There are no set rules or sequence of steps to follow in generalizing patterns of data. The network is designed to learn a nonlinear mapping between the input and output data. Generalization is used to predict the possible outcome for a particular task. This process involves two phases known as the training phase (learning) and the testing phase (prediction).

Regression models have been traditionally used to model the changes in the stock markets. Multiple regression analysis is the process of finding the least squares prediction equation, testing the adequacy of the model, and conducting tests about estimating the values of the model parameters as in [9]. However, these models can predict linear patterns only. The stock market returns change in a nonlinear pattern such that neural networks are more appropriate to model these changes. If stock market return fluctuations are affected by their recent historic behavior as in [10], neural networks which can model such temporal information along with spatial information in the stock market changes can prove to be better predictors. The changes in a stock market can then be learned better using networks which employ a feedback mechanism to cause sequence learning.

Recurrent networks use the back propagation learning methodology. The main difference between a feed forward back propagation network and a recurrent network is the existence of a feedback mechanism in the nodes of the recurrent network. This feedback mechanism facilitates the process of using the information from the previous pattern along with the present inputs. Copy-back/Context units are used to integrate the previous pattern into the following or a later input pattern as in [11]. This ability of recurrent networks in learning spatiotemporal patterns makes them suitable for the stock market return prediction problem.

The prediction accuracy of a network along with additional information available from recent history of a stock market can be used to make effective stock market portfolio recommendations as in [12]. To achieve these objectives, it is essential that stock market have to monitor, maintain and manage their indexes and portfolios in a systematic manner taking into account the various risks involved in these areas as in [13, 14, 15].

The objective of the study is to present the use of artificial neural network as a forecasting tool for predicting the direction of the stock market. The neural network is employed to use the homogeneous input data set which in this case is the daily returns of S&P CNX Nifty 50 Index. The study also seeks to document the self similarity characteristic of the stock market. Accuracy of the performance of the neural network is compared using various out of sample performance measures.

1.2 Multiple Linear Regression

This study uses multiple linear regression technique to define a model consisting of determinants in predicting stock market performance. The sample of this study includes data of stock market exchanges of Asia Pacific and global region for the financial period from 2005-2009 which is used in the study. It used day wise data for the financial year from 3rd Jan 2005 to 12th Dec 2009 and the data were obtained from Asian CEIC database, which is Centre Database for Asia & emerging markets and Ecostats Database. The index of the following stock markets is taken as closing index of a particular day. This study employs cross-section data multiple linear regression model. Nifty 50 Index was used as a measure of stock market performance, and hence is a dependent variable for the multiple linear regression. Twelve variables including BSE index, Nikkei 225 index, Shenzhen A China, KOSPI 11 Korea, Hang Seng Hong Kong, Jakarta CMP Indonesia, KLSE Comp Malaysia, Karachi 100 Pakistan, Straits Times Singapore, Taiwan WTD Taiwan, Down Jones, NASDAQ & Foreign Exchange Rate with respect to Gold & Silver rates were used as independent variables. The study concludes that multiple linear regression is the powerful tool in predicting stock market performance.

Thus, in this stock market prediction model, NIFTY 50 market index is considering as dependent variable and other 14 variables are considered as independent variables. This study also tested the assumptions of the linear multiple regression model, viz., multicollinerity and homoscedasticity. None of the two independent variables are highly correlated and hence, there is no multicollinearity problem exists. Further, the residuals are identically distributed with mean zero and equal variances and hence, the model does not face a problem of heteroscedasticity as in [16].

2 Data and Methodology

The data of stock market exchanges of Asia Pacific and global region for the financial period from 2005-2009 which is used in the study. It used day wise data for the financial year from 3rd Jan 2005 to 12th Dec 2009 and the data were obtained from Asian CEIC database, which is Centre Database for Asia & emerging markets and Ecostats Database. The Daily return of the index is calculated from the daily closing prices of Nifty 50 Index. The whole data set is divided in two parts. One for training the network and second for network validation. Here, different network structure will be designed having different numbers of Neurons in the input and the hidden layer. The output layer has only one neuron which gives the forecasted index value. The input to the 1st node will be today's daily return value, the 2nd node will have yesterday's return value as the input and likewise. Whereas the output will be tomorrow's forecasted daily return. A back propagation neural network learning methodology will be used to obtain the output. A subset of available daily stock return data will be used to construct the neural network. The network so constructed will be trained using the training data set. Training of a back propagation network involves obtaining optimal values for the learning rate, the momentum of learning, estimating the number of hidden layers and the number of nodes in each layer. The training of the network will be done using different combinations of learning algorithm and transfer functions. The overall error is tracked until a minima is obtained by altering the fore mentioned parameters. The net so obtained with minimum error is saved and this trained network can then be used in predicting future stock market returns. Accuracy of the performance of the neural network is compared against a traditional forecasting method.

3 Experiment Results

3.1 Linear Regression

The present paper shows the model summary of the regression for the sample stock markets. The R-Square of the model is equal to 98.9%. This means that 98.9% of the changes in the dependent variable (NIFTY 50) are due to the variations of the independent variables used in this model besides supporting the appropriate selection of proxies. The difference among the R-Square value of these studies was elucidated by the different period of times and the type of regression, viz., cross sectional or time series or both. Though, this study found a high R-Square value, few other factors which have influence on the NIFTY 50 market index were not included. The table of ANOVA shows that by using the analysis of variance, it is found that F test of the model is equal to 40402.637. This F value is largely higher than the critical value at 1% level of significance for degrees of freedom of 14. Thus, it can be concluded that among all 14 variables, 10 variables have significant effect on the dependent variable (NIFTY 50).

Table 1. Model Summary

Model	R	R Square	Adjusted R Square	Std. Error of the Estimate
1	.974[a]	.989	.989	40.505

a. Predictors: (Constant), NASDAQ, Gold, Karachi, China, Nikkei, Taiwan, Korea, Bombay, DowZones, Malaysia, Silver, HongKong, Jakarta, Singapore

Table 2. ANOVA[b]

Model		Sum of Squares	Df	Mean Square	F	Sig.
1	Regression	9.280E8	14	6.629E7	40402.637	.000[a]
	Residual	1304336.929	795	1640.675		
	Total	9.293E8	809			

a. Predictors: (Constant), NASDAQ, Gold, Karachi, China, Nikkei, Taiwan, Korea, Bombay, DowZones, Malaysia, Silver, HongKong, Jakarta, Singapore

b. Dependent Variable: Nifty

This study led to the conclusion that the NIFTY 50 stock market index has equal dependency on other Asian stock market exchange indexes. As far with results of the study, the elasticity factors are not too large but have an equal impact over index rates.

Table 3. Coefficients[a]

Model	Unstandardized Coefficients		Standardized Coefficients	t	Sig.
	B	Std. Error	Beta		
(Constant)	102.952	45.605		2.257	.024
Bombay	.260	.002	.901	106.099	.000
HongKong	.014	.002	.058	6.694	.000
Nikkei	-.013	.002	-.037	-5.381	.000
China	-.077	.014	-.030	-5.572	.000
Korea	.135	.025	.037	5.296	.000
Jakarta	.091	.022	.046	4.150	.000
Malaysia	.094	.050	.017	1.880	.060
Karachi	.018	.002	.044	10.662	.000
Singapore	-.009	.026	-.005	-.354	.723
Taiwan	-.035	.005	-.043	-6.681	.000
Gold	.158	.055	.028	2.880	.004
Silver	-6.553	2.373	-.021	-2.761	.006
DowZones	-.125	.013	-.064	-9.663	.000
NASDAQ	.187	.025	.056	7.356	.000

a. Dependent Variable: Nifty

This study also concludes that the trend of NIFTY 50 index prices doesn't have any significance with foreign exchange rates define by Gold & Silver Rates. Thus this study predict the stock market performance with respect to NIFTY 50 and other global market exchanges index prices and shows the significant tool to predict future performance analysis and impact of financial wave over Indian stock exchange(NIFTY 50).

But the Gold rate is inversely affected by the global stock market exchanges rate index. If we calculate the linear regression putting gold rate as dependent variable as other as independent variable, we find a surprising fact that Gold rate has inversely effect in world market prices with effect to the low and high index prices of global market exchanges.

Table 4. Model Summary

Model	R	R Square	Adjusted R Square	Std. Error of the Estimate
1	.963[a]	.927	.926	51.622

a. Predictors: (Constant), NASDAQ, China, Karachi, Nikkei, Bombay, Taiwan, Korea, DowZones, Malaysia, HongKong, Jakarta, Singapore, Nifty

3.2 Artificial Neural Network

Thus, the data collected in the study were randomly assigned into three different sub-samples as given in Table 4. A large number of data is needed for the training data set. Only 15% of the data are used for testing and validation purposes, due to the limited number of available sample. Feed forward neural network or multilayer perceptron with one hidden layer and seven inputs, corresponding to the seven variables are suggested to be used in the study. Experiments are done to determine the best number of neurons in the hidden layer and to evaluate the performance of the neural network in predicting bank performance. Under supervised learning, the desired output for each input is given to the network.

Table 5. Neural Network Model Summary

Neurons	Training	MSE	RSE	Validation	MSE	RSE
2	902	7.54	0.05	194	1.01	0.03
4	902	6.91	0.02	194	8.89	0.06
8	902	2.31	0.08	194	4.12	0.03
15	902	1.85	0.04	194	1.14	0.07
20	902	1.57	0.09	194	1.83	0.01
25	902	2.10	0.01	194	4.07	0.04
30	902	1.50	0.05	194	2.84	0.02

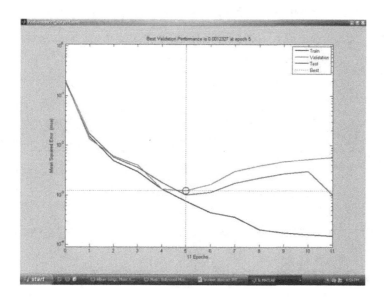

Fig. 1. Training and validation performance with respect to MSE

The network will then adjust weights in the hidden neuron so as to minimize the error obtained between the network's output and the desired output. Results for different numbers of neurons are presented in Table 4. The number of neurons with the lowest mean square error (MSE) for the testing data is chosen to be the best number of hidden nodes for the seven inputs. Neural network with 20 hidden nodes are identified to perform at its best during the training and testing. The lowest mean square error (MSE) of 1.01 is obtained. R^2 value shows a satisfactory value at 0.66868. The performance of the network during the training phase is very high for different numbers of neurons. The testing data set gives the lowest mean square error (MSE) value when the network contains 20 hidden neurons. The testing data gives a low R^2 value of about 0.669. The value indicates the proportion of total variation explained by the results is about 66.9%. Results are said to be satisfactory if the R^2 value is more than 0.80. Nevertheless, results obtained from the network has a higher R^2 value than the multiple linear regression model.

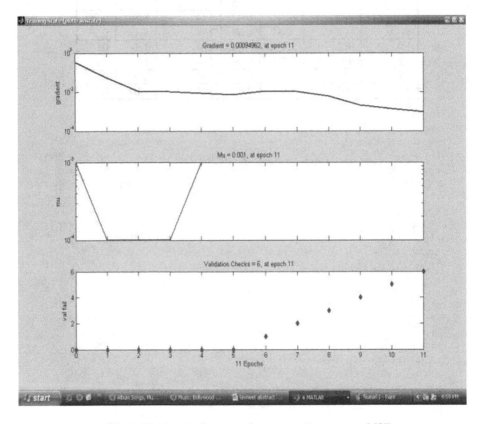

Fig. 2. Plotting the fitness performance with respect to MSE

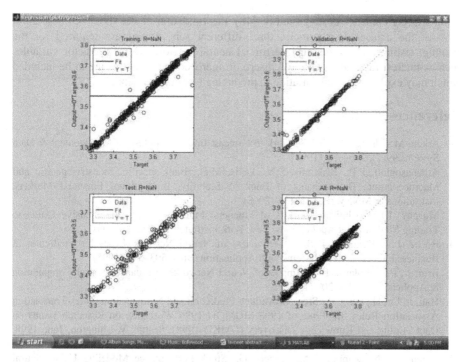

Fig. 3. Plotting of linear regression with respect to MSE

4 Conclusions

From the study, we may conclude that multiple linear regression can be used as a simple tool to study the linear relationship between the dependent variable and independent variables. The method provides two significant explanatory variables to bank performance and explains the effect of the contributing factors in a simple, understood manner. The method somehow has its limitations for its underlying assumptions are always violated when using real data. The presence of outliers also produces biased estimators of the parameters. Violations of the underlying assumptions are often accompanied by remedial measures. Data transformation, robust regression and ridge regression are among the remedial measures to be taken. Thus, this requires the needs to understand further statistical techniques, which is out of the scope of this study. The artificial neural network gives highly accurate results from the inputs. The method increases its performance with a large number of examples. An optimal number of neurons also need to be determined because the network tends to memorize with too many neurons but it can hardly make any generalization if too few are used. The method does not require any distributional assumptions and it is robust to outliers and unexpected data in the inputs. The artificial neural network outperformed the multiple regression in predicting bank performance but somehow, the method gives no explanation on the estimation of the parameters as in [7, 8, 17 and 18]. Decision makers are provided with the information on the estimated parameters from the results of multiple linear regression. The prediction of the method is only made on the mean performance and thus gives a higher MSPR value. A similar study can be performed using a larger dataset. As suggested by [18], the validation model should consist about 30% of the dataset. Furthermore, three different sub-samples are required for the training, testing and validation in artificial neural network. Other predictor variables such as Interest rate, GDP, CPI and macro or microeconomic factors are to be included which may explain the total variation in predicting bank performance.

References

1. Aiken, M.: Using a Neural Network to Forecast Inflation. Industrial Management & Data Systems 99(7), 296–301 (1999)
2. Athanasoglou, P.P., Brissimis, S.N., Delis, M.D.: Bank-specific, Industry-specific and Macroeconomic Determinants of Bank Profitability. International Financial Markets, Institutions & Money 18, 121–136 (2008)
3. Murphy, J.J.: Technical Analysis of Financial Markets – A Comprehensive Guide to Trading Methods and Applications. New York Institute of Finance (1999)
4. Boritz, J.E., Kennedy, D.B.: Effectiveness of Neural Network Types for Prediction of Business Failure. Expert Systems with Applications 9(4), 503–512 (1995)
5. Brunell, P.R., Folarin, B.O.: Impact of Neural Networks in Finance. Neural Computation & Application 6, 193–200 (1997)
6. Han, J., Lu, H., Feng, L.: Stock Movement Prediction and N dimensional Inter-Transaction Association Rules. In: Proc. of 1998 SIGMOD 1996 Workshop on Research Issues on Data Mining and Knowledge Discovery (DMKD 1998), Seattle, Washington, June 1998, pp. 12:1-12:7 (1998)
7. Refenes, A.N., Zapranis, A., Francis, G.: Stock Performance Modeling Using Neural Networks: A Comparative Study with Regression Models. Neural Networks 7(2), 375–388 (1994)
8. Chokmani, K.T., Quarda, J.V., Hamilton, S., Hosni, G.M., Hugo, G.: Comparison of Ice-Affected Streamflow Estimates Computed Using Artificial Neural Networks and Multiple Regression Techniques. Journal of Hydrology 349, 383–396 (2008)
9. Mender, B.: Introduction to Probability and Statistics, 9th edn. International Thomson Publishing (1994)
10. Tang, Z., Almeida, C., Fishwick, P.A.: Simulation: Time series forecasting using neural networks vs. Box-Jenkins methodology, pp. 303–310 (1991)
11. Morgan, S.: Neural Networks and Speech Processing. Kluwer Academic Publishers, Dordrecht (1991)
12. Roman, J., Jameel, A.: Backpropagation and Recurrent Neural networks in Financial Analysis of MultipleStock Market Returns. In: Proceedings of the 29th Annual Hawaii International Conference on System Sciences (1996)
13. Rumelhart, D.E., McClelland, J.L.: PDP Research @OUP: Parallel Distributed Processing Volume: Foundations, The Massachusetts Institute of Technology (1988)
14. Wicirow, Rumelhart: L&R: Journal of Communications of the ACM. Neural Networks: Applications in Industry, Business and Science 37(3), 93–105 (1994)
15. Gately, E.: Neural Networks for Financial Forecasting. Wiley, New York (1996)
16. Haron, S.: Determinants of Islamic Bank Profitability. Global Journal of Finance & Economics 1(1), 11–33 (2005)
17. Leshno, M., Spector, Y.: Neural Network Prediction Analysis: The Bankruptcy Case. Neurocomputing 10, 125–147 (1996)
18. Nguyen, N., Cripps, A.: Predicting Housing Value: A Comparison of Multiple Linear Regression Analysis and Artificial Neural Networks. Journal of Real Estate Research 22(3), 313–336 (2001)

Statistical Analysis for Human Authentication Using ECG Waves

Chetana Hegde[1], H. Rahul Prabhu[2], D.S. Sagar[2], P. Deepa Shenoy[2], K.R. Venugopal[2], and L.M. Patnaik[3]

[1] Research Scholar, Bangalore University, Bangalore 560 001
[2] Department of CSE, UVCE, Bangalore University, Bangalore 560 001
[3] Vice Chancellor, DIAT, Pune, India
chetanahegde@yahoo.co.in

Abstract. Automated security is one of the major concerns of modern times. Secure and reliable authentication of a person is in great demand. A biometric trait like the electrocardiogram (ECG) of a person is unique and secure. In this paper we propose an authentication system based on ECG by using statistical features like mean and variance of ECG waves. Statistical tests like $Z-$test, $t-$test and χ^2-tests are used for checking the authenticity of an individual. Then confusion matrix is generated to find False Acceptance Ratio (FAR) and False Rejection Ratio (FRR). This methodology of authentication is tested on data set of 200 waves prepared from ECG samples of 40 individuals taken from Physionet QT Database. The proposed authentication system is found to have FAR of about 2.56% and FRR of about 0.13%. The overall accuracy of the system is found to be 99.81%.

1 Introduction

Security is an emerging paradigm for understanding global vulnerabilities whose proponents challenge the traditional notion of national security by arguing that there is no proper authentication for individuals. Security is ever-challenging problem, when it comes to any of scientific, educational or government organizations. Human authentication has become mandatory in this threatening era.

Biometrics is the application of statistical analysis to biological data. It is a science of identifying a person using his physiological and/or behavioral characteristics. Biometric traits are difficult to counterfeit and hence result in higher accuracy when compared to other methods such as using passwords and ID cards. Human physiological and/or behavioral characteristic can be used as a biometric characteristic when it satisfies the requirements like Universality, Distinctiveness, Permanence and Collectability. However, in a practical biometric system, one needs to focus on other issues too like Performance, Acceptability and Circumvention[1]. Keeping all these requirements in mind, biometric traits like fingerprints, hand geometry[2], handwritten signatures[3], retinal patterns[4], facial images[5], ear[6], dental records, voice[7], DNA, ECG etc. are used extensively in areas which require security access[1].

S. Dua, S. Sahni, and D.P. Goyal (Eds.): ICISTM 2011, CCIS 141, pp. 287–298, 2011.

Most of the biometric traits mentioned above, have certain disadvantages which threaten its level of security. Some of the traits can easily be forged to create false identities. Some may be altered to hide the identity of an individual. And few other traits can be used even in the absence of the person and even if he is dead. Certain biometric traits like hand-vein patterns may fail in case of hand-injury. Though it is possible to authenticate a person with a damaged hand-vein pattern[8], it is impossible to achieve 100% accuracy as there is a limitation of threshold. Such problems can be solved using heartbeat as the biometric trait.

The heartbeat of a person cannot be copied to fake identity and it cannot be altered to hide identity. It possess an inherent liveness detection as the heartbeat can be used only if the person is alive. Also, the possible variation in heartbeat due to variation in the emotion of a person can be normalized to get the standard wave form. The heartbeat of a person is collected in the form of an electrocardiogram recording. The ECG of a person varies from person to person due to change in size, position and anatomy of the heart, chest configuration and various other factors. Electrocardiogram is a transthoracic interpretation of the electrical activity of the heart over time captured and externally recorded by skin electrodes. The different waves that comprise the ECG represent the sequence of depolarizing and repolarizing of the atria and the ventricles.

This paper is organized as follows: Section 2 deals with the related work and Section 3 presents the architecture and model. Section 4 gives problem definition. Section 5 describes the implementation of the proposed algorithm and the performance analysis. Section 6 contains the conclusions.

2   Related Work

A brief survey of the related work in the area of ECG waves and identification using ECG is presented in this section. ECG analysis is not only a very useful diagnostic tool for clinical purposes but also may be a good biometric trait for human identification. The ECG varies from person to person due to the differences in position, size and anatomy of the heart, age, sex, relative body weight, chest configuration and various other factors[9]. Biel et al.[10] showed that it is possible to identify individuals based on an ECG signal. The concern of identification system on ECG is the changes of ECG dominant fiducials during aging. Study shows that these changes are only seen up to the age of adolescence (approximately upto 14 years)[11]. After adolescence the ECG dominant features are relatively consistent. This may be a minor concern because biometric applications are mainly employed to identify adults.

The initial work on heartbeat biometric recognition used a standard 12-lead electrocardiogram for recording the data. Later a single-lead ECG was being used. It is one-dimensional, low-frequency signal that can be recorded with three electrodes (two active electrodes and a ground electrode). Biel et al.[10] used the equipment SIEMENS Megacart for the measurement of ECG. The information from SIEMENS ECG is transferred and converted to a usable format. For each person, 30 features like P wave onset, P wave duration (ms), QRS wave onset, QRS wave duration (ms) etc. were extracted. To reduce the amount of information, the correlation matrix was used. This had reduced the number of features to 12. Then the method SIMCA(Soft Independent Modeling of Class Analogy)[12] was used to classify persons. This approach proved 95% of accuracy in human identification.

The other approach proposed by Shen et al.[13] uses template matching and a Decision-Based Neural Network (DBNN). In template matching, two signals are correlated if their wave shapes are similar to one another. The value of the correlation coefficient is a measure of the similarity of the shapes of two signals. The correlation coefficient varies between +1 and −1 depending on the degree of similarity of the shapes of the two signals being compared and is not affected by the magnitude of amplitude differences in the signals[9]. A DBNN is a member of a supervised learning family, which uses both reinforced and anti-reinforced learning rules. That is, the system adjusts the weight vector either in the direction of gradient of the discriminant function (i.e., reinforced learning) or opposite to that direction (i.e. anti-reinforced learning). The DBNN adopts nonlinear discriminant functions[9,14]. Totally, seven features were extracted from the QRS complex and QT. The DBNN training was done with the features extracted from the set of waveforms of 20 individuals. The ECG template waveform for each individual subject is selected from a different temporal location in the signal. The template matching was used as a prescreening tool. Correlation coefficient was used for matching the ECG template. Even this approach showed the accuracy of 95% in identifying the individual.

Feature extraction, the prime focus of the researchers, has led to different methods for acquiring it from the wave form. Delineation is a major step in identification of the features. The QRS complex delineator is implemented using the technique proposed by Pan and Tompkins[15]. In order to determine P wave and its end fiducials, a search window is set prior to the beginning of QRS complex. The search window that approximately contained P wave is set heuristically and extended from QRS_{onset} to the beginning of heartbeat. The next promising technique for human identification using ECG was from Singh et al.[16]. In their approach, the QRS complex delineator was implemented using the technique proposed by Pan et al.[15] with some improvements. This technique employs digital analysis of slope, amplitude and width information of ECG waveforms. The identification model was designed on pattern matching and adaptive thresholding technique. This model was evaluated on the basis of correlation between corresponding attributes of the feature vectors. The accuracy of the system proposed was found to be 98.99%.

As one can observe, all the techniques proposed so far were focused on human identification but not on authentication. These techniques for human identification using ECG have used the geometrical features. Geometric features tend to be error-prone, as a minute change in the features like angle might have been ignored during approximation and/or normalization. Moreover, the computational complexity of geometric features is very high. In this paper, we propose a technique for human authentication by analyzing the nature of statistical features. The proposed technique suggests to extract statistical features like mean and variance of an ECG wave. Then the authentication is done based on the values of test-statistics obtained by applying $Z-$test, $t-$test and χ^2-test.

3   Architecture and Modeling

Initally, an identification number (ID) is given to every individual in an organization. Having the database of ECG wave signals of all the people in an organization consumes more space and the complexity of the system will increase. So, we suggest to store only the calculated features of ECG against every ID. First, we will acquire ECG waves from a 10-second sample from every person. Each of the subwaves P, QRS and T are separated from every single wave present in a sample. Now, all P subwaves are combined and normalized. The statistical features mean and variance are calculated for this normalized P subwave in a 10-second sample. The same procedure is done for both QRS and T subwaves. The features calculated in this manner are termed as *Population Features* and are stored in the database. During authentication, the person has to provide his ID and his ECG is captured. Now, the features are extracted from the subwaves P, QRS and T of a single wave. These features are called as *Sample Features*. Statistical tests are applied on population features and sample features to check the authenticity of a person.

The architectural diagram for the proposed algorithm is shown in Fig. 2. The various steps involved are explained here under.

3.1   Sampling and Data Acquisition

ECG wave formats are taken from Physionet QT database [17]. One-lead ECG is taken for analysis. The database provides a tool for sampling. Sampling is the method of selecting an unbiased or random subset of individual observations from a large population to yield some knowledge about the population. The major purpose of sampling is predictions based on statistical inference. The tool provided by Physionet database reads signal files for the specified record and writes the samples as decimal numbers on the standard output. By default, each line of output contains the sample number and samples from each signal, beginning with channel 0, separated by tabs. Fig. 3 shows the ECG wave sample of 10 seconds for one of the subjects.

3.2   Pre-processing

Preprocessing involves series of operations. The first step is to separate individual wave from a 10-second sample. Initially, the maximum amplitude is identified

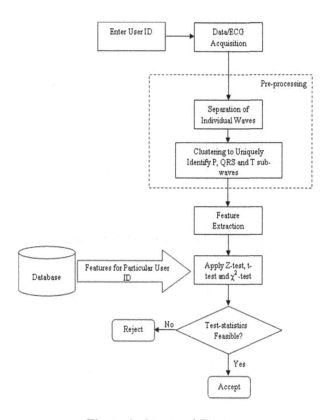

Fig. 1. Architectural Diagram

among all R waves. Then, for each wave in the sample, local maxima is calculated, which is above the certain minimum peak height. A curve represented by a function $f(x)$ is said to have a local maxima at the point x^*, if there exists some $\epsilon > 0$ such that

$$f(x^*) \geq f(x), when \ |x - x^*| < \epsilon \tag{1}$$

Local maxima for each wave is nothing but peak of the subwave R. Thus, after identifying all $R-$wave peaks, a boundary is set in the middle of two peak values to extract one ECG wave as shown in Fig. 4.

The next step is windowing using clustering. We use $k-$means clustering, which is a method of cluster analysis that aims to partition n observations into k clusters in which each observation belongs to the cluster with the nearest mean. Given a set of observations $(x_1, x_2, ..., x_n)$ where each observation is a $d-$dimensional real vector, the $k-$means clustering aims to partition the n observations into k sets $(k < n)$, $S = \{S_1, S_2, ..., S_k\}$ so as to minimize the within-cluster sum of squares (WCSS):

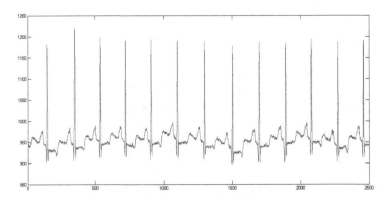

Fig. 2. ECG Wave Sample of 10 Seconds

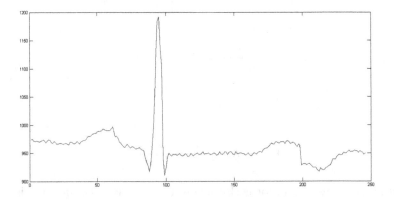

Fig. 3. One Wave Extracted from a Sample of 10 Seconds

$$argmin_{S} \sum_{i=1}^{k} \sum_{x_j \in S_i} \| x_j - \mu_i \|^2 \tag{2}$$

where μ_i is the mean of the points in S_i. The most common algorithm for $k-$means clustering uses an iterative refinement technique. Given an initial set of $k-$means $m_1^{(1)}, ..., m_k^{(1)}$, which may be specified randomly or by some heuristic method, the algorithm proceeds by alternating between two steps:

Assignment step: Assign each observation to the cluster with the closest mean-

$$S_i^{(t)} = \left\{ x_j : \| x_j - m_i^{(t)} \| \leq \| x_j - m_{i^*}^{(t)} \| \ \forall i^* = 1, ..., k \right\} \tag{3}$$

Update step: Calculate the new means to be the centroid of the observations in the cluster-

$$m_i^{(t+1)} = \frac{1}{\left|S_i^{(t)}\right|} \sum_{x_j \in S_i^{(t)}} x_j \tag{4}$$

The above approach of clustering is applied on each of the n waves in a 10-second sample. The clustering results into three clusters each consisting of P wave, QRS wave and T wave respectively. In the first cluster, there will be signal values of P wave and the other signals corresponding to QRS and T waves are set to zero. In the second cluster, only QRS signals are set and the remaining values are zero. Similarly, the third cluster represents the values for T wave and the other values are zero. Once we classify the waves in this manner, the feature extraction is possible for every subwave.

3.3 Feature Extraction

To calculate population features, first we need to combine all the P waves in a 10-second sample. For this combined set of ECG amplitude values, we calculate mean(μ_P) and variance(σ_P^2). Similarly, we compute mean and variances for both QRS and T waves, which are represented as μ_Q, μ_T, σ_Q^2 and σ_T^2 respectively.

To compute sample features during authentication process, we will consider only one wave. The sample mean$(\overline{x}_P, \overline{x}_Q, \overline{x}_T)$ and variance(s_P^2, s_Q^2, s_T^2) are calculated for P, QRS and T subwaves in that single wave.

3.4 Statistical Tests for Feature Matching

We are using three statistical tests for matching the features for authentication. The number of amplitude values for P and T subwaves are more. That is, they are large samples. Hence, we use Z−test based on mean and χ^2−test based on variance for these waves. The QRS subwave is a small sample as it consists of less number of amplitudes. So, t−test based on mean is applied on QRS wave.

$Z-$ Test

This test is used to check whether a sample $X_i \, (i = 1 \ldots n)$ is drawn from the specified population. To test this null hypothesis, the parameters used are sample mean, population mean and population variance. The test-statistic is given by -

$$Z = \frac{(\overline{x} - \mu)}{\left(\frac{\sigma}{\sqrt{n}}\right)} \tag{5}$$

Now, Z is said to follow standard normal distribution. If the calculated value of $|Z|$ is lesser than the Z-table value for a given level of significance α, then we can accept the null hypothesis and say that the sample is drawn from the population.

χ^2- Test

This test can be used to check whether a sample X_i $(i = 1 \ldots n)$ is drawn from the specified population. To test this null hypothesis, the parameters used are sample variance and population variance. The test-statistic is given by -

$$\chi^2 = \frac{(n-1)\,s^2}{\sigma^2} \tag{6}$$

The χ^2 is said to follow χ^2- distribution with $(n-1)$ degrees of freedom. If the calculated value of χ^2 is lesser than the χ^2-table value for a given level of significance α, then we can accept the null hypothesis and say that the sample is drawn from the population.

$t-$ Test

The $t-$test uses sample mean, population mean and sample variance as parameters and is applicable for small-size samples. This test is also used to test the null hypothesis that whether a sample is drawn from the specified population. The t-test statistic is given by -

$$t = \frac{(\bar{x} - \mu)}{\left(\frac{s}{\sqrt{n}}\right)} \tag{7}$$

The value t is said to follow $t-$distribution with $(n-1)$ degrees of freedom. If the calculated value of t is smaller than the $t-$table value for a given level of significance α, then we can agree that the sample is drawn from the same population.

The $Z-$test and χ^2-tests are carried out for P and T waves, whereas, the $t-$test is implemented for QRS waves. If all the tests result into the acceptance of null hypothesis that the sample is drawn from the specified population, then we can authenticate a person.

3.5   Authentication

This process involves a series of steps to authenticate a person.

1. The person who undergoes the authentication process should first enter his ID and his ECG is captured.
2. The ECG wave should be pre-processed and then features are extracted. These newly extracted features are *sample features*.
3. Now, the features stored in the database for that ID are retrieved. These features are *population features*.
4. To check whether the new features match those in the database, initally, we set the *Null Hypothesis, H_0* and *Alternative Hypothesis, H_1* as:
   H_0: The newly acquired features match those in the database OR The new sample is drawn from the specified population.
   H_1: The newly acquired features do not match those in the database OR The new sample is significantly different from the specified population.

5. To test the above null hypothesis, we apply Z−test and χ^2−test on both P and T subwaves. The t−test is applied on QRS wave.
6. If the calculated values of all these test-statistics are lesser than the corresponding values of the respective tables, then the null hypothesis can be accepted and thus we can authenticate a person. Otherwise, he will be rejected.

All the tests are carried out for the *level of significance* $\alpha = 0.01$ and $\alpha = 0.001$. That is, authenticity is tested at both of the *confidence levels* 99% and 99.9%.

4 Algorithm

4.1 Problem Definition

Given an ECG wave signal, the objectives are:

(1) To find local maxima above certain minimum peak height to separate each wave in a 10-second sample wave.
(2) To apply k−means clustering to identify subwaves P, QRS and T from every single wave.
(3) To calculate statistical features for each of the P, QRS and T subwaves and
(4) To apply statistical sampling techniques for computing the values of test-statistics Z, χ^2 and t and then for identifying authenticity of a person.

4.2 Algorithm

Four major functions are involved in the proposed technique. The first function is for separating individual waves from a 10-second wave sample. The next function is to perform windowing using clustering. The algorithm for k−means clustering is given in Table 1. The third function is to calculate statistical features. The last function is to carryout statistical sampling tests for authenticating a person.

5 Implementation and Performance Analysis

The implementation of proposed technique is developed using MatLab 7.5. The proposed algorithm is tested on ECG signals of 40 individuals. Various research works based on ECG waves proposed earlier like[10] and[16] were focused on Human identification but not authentication. Those techniques used geometrical features which are error-prone. Some promising techniques like[16] have used 19 geometrical features and the delineation is done separately for each wave. This technique is found to be time consuming and more complex in the view of real-time implementation. But, our technique uses the k−means clustering and hence each subwave is extracted at a time. So, the problems with repeated delineation are suppressed. The proposed technique uses only 2 statistical features for each subwave and in total 6 features for one wave. Computational complexity

Table 1. Algorithm for $k-$means Clustering

Inputs:
    $I = \{i_1, \ldots, i_k\}$ (Instances to be clustered)
    n : Number of clusters
Outputs:
    $C = \{c_1, \ldots, c_n\}$ (Cluster Centroids)
    $m : I \rightarrow C$ (Cluster Membership)
Procedure KMeans
    Set C to initial value (e.g. random selection of I)
    For each $i_j \in I$
        $m(i_j) = \underset{k \in \{1,\ldots,n\}}{argmin}\ distance(i_j, c_k)$
    End
    While m has changed
        For each $j \in \{1, \ldots, n\}$
            Recompute i_j as centroid of $\{i \mid m(i) = j\}$
        End
        For each $i_j \in P$
            $m(i_j) = \underset{k \in \{1,\ldots,n\}}{argmin}\ distance(i_j, c_k)$
        End
    End
    return C
End

Table 2. Confusion Matrix

		Actual	
		Genuine	Non-Genuine
Tested	Genuine	38	1
	Non-Genuine	2	1559

is reduced and hence the system possesses better tradeoff. To check the authenticity, we are following random sampling technique, which is the most promising technique to make statistical inference. The test-statistics used here helps in efficient authentication.

The simulation results on ECG waves of 40 individuals resulted into a confusion matrix as shown in Table 2. With the help of confusion matrix, the false acceptance ratio is found to be 2.56% and false rejection ratio is 0.13%. And the overall performance of the system is found to be 99.81%.

6 Conclusions

In this paper, we propose an efficient way for authentication using electrocardiogram signals of a person. The proposed technique uses statistical features like mean and variance of individual subwaves P, QRS and T in a 10-second sample. The features of a single wave are calculated for the person who undergoes authentication process. To infer whether the newly extracted single wave belongs to the same person or not, we use hypothesis testing based on Z, χ^2 and t tests. The proposed technique is more efficient as it uses statistical features instead of geometrical features. Also, the computational complexity is reduced by using a small set of parameters compared to earlier algorithms. This new technique produced impressive results with FAR of about 2.56% and FRR of 0.13%. The overall performance of the system is 99.81%.

References

1. Jain, A.K., Ross, A., Prabhakar, S.: An Introduction to Biometric Recognition. IEEE Trans. on Circuits Sys. 14(1), 4–20 (2004)
2. Boles, W., Chu, S.: Personal Identification using Images of the Human Palm. In: Proc. IEEE TENCON Conf. (1997)
3. Hegde, C., Manu, S., Shenoy, P.D., Venugopal, K.R., Patnaik, L.M.: Secure Authentication using Image Processing and Visual Cryptography for Banking Applications. In: Proc. Int. Conf. on Advanced Computing (ADCOM 2008), pp. 65–72 (2008)
4. Boles, W.: A Security System based on Human Iris Identification using Wavelet Transforms. In: Proc. First Int. Conf. Knowledge-Based Intelligent Electron. Syst. (1997)
5. Samal, A., Iyengar, P.: Automatic Recognition and Analysis of Human Faces and Facial Expressions: A Survey. Pattern Recognition 25(1), 65–77 (1992)
6. Hegde, C., Srinath, U.S., Aravind Kumar, R., Rashmi, D.R., Sathish, S., Shenoy, P.D., Venugopal, K.R., Patnaik, L.M.: Ear Pattern Recognition using Centroids and Cross-Points for Robust Authentication. In: Proc. Second Int. Conf. on Intelligent Human and Computer Interaction (IHCI 2010), pp. 378–384 (2010)
7. Dumn, D.: Using a Multi-layer Perceptron Neural for Human Voice Identification. In: Proc. Fourth Int. Conf. Signal Process. Applicat. Technol. (1993)
8. Hegde, C., Rahul Prabhu, H., Sagar, D.S., Vishnu Prasad, K., Shenoy, P.D., Venugopal, K.R., Patnaik, L.M.: Authentication of Damaged Hand Vein Patterns by Modularization. In: IEEE TENCON (2009)
9. Simon, B.P., Eswaran, C.: An ECG Classifier Designed using Modified Decision Based Neural Network. Computers and Biomedical Research 30, 257–272 (1997)
10. Biel, L., Pettersson, O., Philipson, L., Wide, P.: ECG Analysis: A New Approach in Human Identification. IEEE Trans. on Instrumentation and Measurement 50(3), 808–812 (2001)

11. Rijnbeek, P.R., Witsenburg, M., Schrama, E., Hess, J., Kors, J.A.: New Normal Limits for the Pediatric Electrocardiogram. European Heart Journal 22, 702–711 (1985)
12. Esbensen, K., Schonkopf, S., Midtgaard, T.: Multivariate Analysis in Practice, 1st edn., vol. 1 (1994)
13. Shen, T.W., Tompkins, W.J., Hu, Y.H.: One-Lead ECG for Identity Verification. In: Proc. of Second Joint Conf. of IEEE EMBS/BMES, pp. 62–63 (2002)
14. Kung, S.Y., Taur, J.S.: Decision-based Neural Networks with Signal/Image Classification Applications. IEEE Trans. on Neural Networks 6(1), 170–181 (1995)
15. Pan, J., Tompkins, W.J.: A Real Time QRS Detection Algorithm. IEEE Trans. on Biomedical Engineering 33(3), 230–236 (1985)
16. Singh, Y.N., Gupta, P.: Biometrics method for human identification using electrocardiogram. In: Tistarelli, M., Nixon, M.S. (eds.) ICB 2009. LNCS, vol. 5558, pp. 1270–1279. Springer, Heidelberg (2009)
17. Laguna, P., Mark, R.G., Goldberger, A.L., Moody, G.B.: A Database for Evaluation of Algorithms for Measurement of QT and Other Waveform Intervals in the ECG Computers in Cardiology, pp. 673–676 (1997)

Qualitative Survey-Based Content Analysis and Validation of Measures of Software Development Team Performance

Amar Nath Chatterjee and Duraipandian Israel

XLRI Jamshedpur, India
`amar_chatterjee1@yahoo.com, disrael@xlri.ac.in`

Abstract. Performance measurement of software development teams is an elusive and complex issue. Many IT organizations have tried evolving their own measures without focus on reliability and validity. There is yet no widely accepted scale for measuring software development team performance (SDTP). An examination of available measurement models of team performance/team effectiveness throws up gaps that call for identification and synchronization of dimensions. Based on expert surveys, this paper attempts to identify and short-list a set of content-validated dimensions of SDTP. First, SPSS Text Analysis package was used to content-analyze 94 industry experts' textual responses to an open-ended qualitative survey questionnaire, which led to extraction, categorization and short-listing of 34 measures of SDTP. Then followed another round of expert survey (N=30) that led to a distilled set of 20 content-validated measures of SDTP, based on Content Validity Ratios. This list of measures should help future research for SDTP scale development.

Keywords: Software Development Team Performance (SDTP), Measure, Dimension, Latent Construct, Multi-dimensional Construct (MDC), Unidimensional Construct (UDC), Instrument, Scale Development, SPSS Text Analyzer, Qualitative Data Analysis, Reliability, Validity, Content Validity Ratio (CVR).

[In this paper, Software (SW), Information Systems (IS) and Information Technology (IT) have been used broadly and interchangeably. Similarly, Scale and Instrument have also been used interchangeably; as also, measure and dimension (of a construct)].

1 Introduction

Nearly half of all software development projects worldwide are either over budget, or behind schedule, or deliver fewer features than specified. More than thirty-percent of the projects end up being cancelled or terminated prematurely. Among the completed projects, only about sixty percent satisfy originally specified features and functions [17]. According to the Standish Group's CHAOS Reports, the overall success rate of IT projects worldwide doubled from 16.2% in 1994 to 34% in 2003 but declined to

S. Dua, S. Sahni, and D.P. Goyal (Eds.): ICISTM 2011, CCIS 141, pp. 299–310, 2011.
© Springer-Verlag Berlin Heidelberg 2011

less than 28% in 2004 [30, 15] before reviving again to 35% in 2006 [29, 1]. SW development teams work under pressure from both customers and own management to deliver better, faster and cheaper outputs. While sales of software have registered a healthy growth [1], the still-poor and often fluctuating success rates of SW projects make Software Development Team Performance (SDTP) a concern of abiding priority.

2 Literature Review on Team Performance

A work team in an organization is defined as a cohesive production unit of two or more people with different tasks, a social identity and a structure of its own and characterized by interactions among its members who work together adaptively to achieve a set of common, shared and specified goals aligned to overall organizational goal of achieving excellence [11, 4]. The purposes behind measuring team performance could range from certification, diagnosis, feedback for training, selection and training evaluation to organizational research [4]. There could be other purposes like benchmarking, prediction, synchronization of other prevailing comparable scales etc. A challenge in teamwork measurement is the issue of giving due stress on both team process and team outcome [4].

Many researchers have found it difficult to standardize objective measures of team performance and instead, resorted to its perceptual evaluation [8, 9, 10]. Chia-Chen Kuo [7] compiled an integrated list of team effectiveness dimensions proposed by the scholars of the two decades of 1980's and 1990's (Table-1).

There are other researchers who have proposed yet other sets of measures for team performance though they have addressed it sometime as 'effectiveness', sometime as 'productivity', or, 'efficiency', or, 'satisfaction' or various other terms [23, 18]. Most such efforts have relied mainly on perceptual measures [34, 32]. While some have defined team effectiveness through a variety of team outcomes such as costs, time spent, product quality etc. [36], others have used an array of team process variables such as communication, leadership, motivation, and so on [24]. Although there are overlaps among the variables proposed by different scholars, there is still no consensus as to a definition of team performance variables [19].

More recently, Rico et al. [28] have adopted a *broad brush* approach to team performance by choosing effectiveness measures, such as timeliness, productivity, security, product/service quality, and accuracy of decisions. In comparison, Lee & Xia [22] have adopted a more focused view in SDTP context, wherein three basic dimensions are on-time completion, on-budget performance, and fulfillment of SW functionality.

Summing up, performance measurement is often an elusive and paradoxical issue. Questions like what should constitute and why study performance, have elicited widely divergent and often inconsistent responses from scholars and practitioners alike. For one, there is significant multiplicity of performance measures [19, 37, 7]; and for another, many of them prove rather weak as their validity degenerates with time [6].

Table 1. Integration of dimensions of team effectiveness. Source: [7]

Scholars:	Team effectiveness dimension
Ancona & Caldwell (1991-92-97); Tjosvold (1988), Song, Montoya-Weiss & Sohmidt (1997):	Team unification.
Cunmings (1981): *Employee value, Management value, Organizational value.*	
Glodstein (1984):	Performance, Satisfaction.
Hackman (1987):	Productivity, Society index.
Jehn & Shan (1997):	Team commitment.
KirKman & Rosen (1999): *Performance dimension* → Productivity, Active behavior, Customer service. *Attitude dimension* → Work satisfaction, Organizational commitment, Team commitment.	
Kolodny & Kiggundu (1980):	Productivity, Maintenance of labor power, Satisfaction.
Pearce & Ravlin (1987):	Productivity, Absence from work, Softies, Fluidity, Satisfaction, Innovation.
Shea & Guzzo (1987):	Productivity.
Sundstrom, DeMeuse & Futrell (1990):	Performance, Productivity.
Weldon & Weingart (1993):	Performance.

3 Software Development Projects Perspective

A SW development project is a time-bound initiative, including activities that require coordination and control, to develop, deliver, implement and often maintain a mandated software application (product or service) as uniquely specified by the customer. Project management is deployment of people, knowledge, skills, tools, and techniques to carry out multiple inter-dependent activities designed to meet the objectives of a given project [5, pp 735]. Of late, some of the organizational scientists and OB practitioners have tried to quantify IT or SW industry-specific team performance [25, 32, 34, 23, 26, 10]. There are other notable scholars too who have put in significant efforts in topics involving measurement of SDTP [27, 12, 39, 22].

SW development projects represent knowledge work contexts, where individual team members bring in specific slices of seminal skill-sets in their endeavour to collaboratively integrate them towards a specified common objective [10]. Team members need to perform inter-reliant and synchronized tasks. Intellective tasks and the need for enormous amount of coordination make them different from other teams in action in organizations [10]. Software being essentially an intangible entity, it cannot be measured along conventional length-mass-time dimensions. This poses an intriguing challenge in the measurement of SDTP, deserving to be treated separately.

According to Thite [39], theoretical success definitions for software development projects have traditionally followed the project dimensions of time, cost and quality. If a project is delivered on time, within budget and with agreed functionality/response, then a project would be classed as a success. Pinto and Mantel [27], on the other hand, had identified three facets of project outcome, namely, the implementation process itself, the perceived value of the project, and client satisfaction. In SW projects, *Effectiveness* of the team can be judged by observing characteristics such as motivation level, task commitment, and levels of performance, ability to withstand stress, more innovative solutions, and decreased development time [3]. Among the factors affecting *productivity*, stress has the most severe implications. Each factor that determines the productivity of a team acts as stress booster for many of its members resulting in delays and cost escalations [31]. Project *performance* is not determined merely by the type of management tools and processes used, but depends largely on the way in which these tools are integrated with the work process and the project team to support the activities toward scope, quality, time and cost objectives [38, 22].

4 Gaps

A previous study [6] referred to 32 different research papers on team performance measurement models authored by more than 70 scholars in a span of 30 years (1977-2006) - 21 of them addressed work teams in general and the balance 11 were more focused on SW development project teams. The study found expected overlaps in dimensions/ or measures proposed by various scholars, such as *productivity* and *satisfaction*, for instance, featuring across multiple models by multiple scholars (as also observed in Table 1). Many of the measurement models have dimensions overlapping with one another to varying extents, and simultaneously, many of them also differ from one another in both focus and comprehensiveness. [*A more comprehensive and chronological compilation of all the 32 papers/models can be had from the first author upon e-mail request.*]

Considering the points discussed above on performances of work-teams in general and of SW development teams in particular, as also highlighted in a number of previous studies [16, 7, 8, 9, 6], we perceived gaps in three broad areas, namely: multiplicity of instruments and non-uniformity of constituent measures or dimensions among different instruments (as evident in Table-1); non-specificity of vocabulary, wherein terms, such as *productivity* or *effectiveness* or *efficiency* etc., are used by different scholars to mean different things, as deliberated in details in a past study [6]; and not the least, obsolescence of the instruments in the face of rapidly changing technology, methodology, and business model with passage of time. These gaps led us to strongly perceive a need for a fresh development of a new scale for SDTP, which can synchronize its many facets and reasonably counter obsolescence. We planned to address this need by first ensuring that the proposed new scale encompasses an optimal mix of valid and relevant facets of SDTP – which defines the scope of this paper.

5 Toward a More Rigorous Scale Development

Recent studies of organizational teams have relied on the expert judgment of experienced stakeholders as the best way to assess team performance [10]. This reemphasizes the fact that for the domain of SDTP, objective measures are problematic and instead, judgment or opinions of Subject-Matter-Experts (SMEs) or experienced stakeholders are more useful, valid and accurate. Following the principles of validation prescribed by scholars [21, 14], we conducted field-surveys of SW development industry experts in order to come up with a set of content-validated measures that can act as facets or reflective unidimensional constructs (UDCs) for subsequent development of a scale for measuring SDTP, treating the latter as a *latent* multidimensional construct (MDC). By definition, a latent MDC "exists" at a deeper level in relation to its UDCs and lies in the common area of their mutual overlap [20]. This paper is limits its scope to the identification and validation of a set of UDCs representing the domain of SDTP.

6 Research Design

We followed a sequential mixed-method approach wherein qualitative data analysis was followed by quantitative surveys. The steps consisted of iterative conducting of and analyses of data from two rounds of expert surveys in order to arrive at a content-validated list of reflective dimensions of SDTP. Both rounds are described below.

6.1 Expert Survey Round-1: Exploratory Phase

This was a qualitative survey to ask an open-ended question to a pool of conveniently sampled software experts. In question asked was: "Could you please think through your past experience as an expert in the Software industry and prescribe as many specific measures as come to your mind, for evaluating the performance of a typical software development project team?" Procedure followed was either face-to-face or telephonic interview or through email and capturing the experts' responses. A total of 160 experts were contacted across India and abroad, of whom, 96 responded (60%). 94 responses were accepted for analysis after rejecting two responses due to incomplete respondent information. The process of administering the survey and collecting the textual responses spanned over four months during April-September 2009.

Of the 94 accepted responses, 84 experts were from large IT organizations, and the rest 10 were from small and medium-sized IT organizations engaged in SW development in India. Seventy-eight of them were male and sixteen female. Thirty-three were from Delhi/NCR, 25 from Mumbai/Pune, 9 each from Bangalore and Kolkata, 7 from Chennai, 4 from Hyderabad, and 8 were from overseas, namely, USA (2), UK (1), Oman (1), Emirates (1), Japan (2) and South Africa (1). However, many of those responding from Indian locations were with past overseas tenures ranging from 3 months to 5 years. Together, the 94 experts represented an experience pool of 957 person-years. The range of their relevant work experiences was from 2 to 30 years; the mean, mode and median being 10.18, 5 and 8 years respectively. Further, taking 6 years' work experience as a commonly observed cut-off point for an average software professional to move from the role of a typical team members (TM) to the supervisory role of a project leader (PL) in leading IT organizations like Infosys [2], it was observed that 33 of the experts (35%) were in the range of 2-6 years (i.e. ~TMs) and 61 of them (65%) above 6 years (i.e. ~PLs or above).

Results of Survey Round-1. The textual responses given by the 94 experts to the open-ended question were inputted into an Excel file as required by SPSS text analysis for surveys™ ver. 2.0 and then content-analyzed by using the software package [33]. The text analysis consisted of extraction of key terms, semantic categorization, and iterative refinement of results. Initially, 235 terms of varying frequencies of occurrence were extracted from the responses obtained from the experts, which could be grouped into 50 categories. Each such term and category were closely and critically looked at, to examine if it merited recognition as an important candidate for individual retention or for being merged with another similar term or category for the domain of SDTP; also, if it was very apparent from the inputs received from the experts that a new category should get added, the same was added manually. Rules of extraction and categorization, dictionary of synonyms and cut-off points in frequency of occurrence of key terms were altered and the process repeated several times, so as to keep the number of resultant categories of measures neither too large to be meaningful nor too few to be representative enough of the domain of SDTP. This process successively generated category-pool sizes of 50, 43, 40, 37, 35, 33, and a final pool of 34 measures. The final output of the SPSS text analysis package is summarized in Table 2 below showing the 34 categories created, along with the number of respondents endorsing and number of items absorbed by each category, in descending order of response volume.

6.2 Expert Survey Round-2: Content Validation

Round-2 began where round-1 ended - with the list of 34 measures of SDTP as its starting point. The central purpose of this round of survey was to adopt a quantitative approach, as prescribed by Lawshe [21], to determining content validity of the 34 measures of SDTP derived in survey round-1. Questionnaire for this round was designed in the form of an Excel sheet with five columns (in addition to the 'serial number' column), the first of which listed the same 34 measures as in Table 2 below, but with their sequence randomly scrambled and a short description inserted as comment for each measure, based on textual inputs received from the experts in survey round-1, and literature references in certain cases. The scrambling of sequence was done to avoid administering them in a likely discernable order.

Table 2. Results of content analysis of expert responses, round-1 using SPSS [33]

Category Name / SDTP Measure	# Items	# Respondents	Srl. No.
Time Overrun / Schedule Slippage	16	50	1
Customer Satisfaction (CSI)	8	48	2
Cost-Budget Overrun / Effort Slippage	19	45	3
CRM & Value-Add to Strategic Clients	26	43	4
Software Quality	20	37	5
Defect Density / Defect Containment	23	36	6
Revenue Realization and Profitability	15	31	7
Team Learning, Skill Development, Competency Growth	30	28	8
Repeat Business / Reference Orders	22	25	9
Requirements capture, Scope Fulfillment, Change Mgmt	18	23	10
Efficiency of SW Engg & Quality of Documentation	22	22	11
Customer Appreciations / Complaints / Feedback	20	22	12
Team Comm-Morale-Motivation-Synergy	28	21	13
Employee Satisfaction (ESI)	12	21	14
Reusable Artifacts/Tools Used/Created-Knowledge Mgmt	16	20	15
Process & Standard Compliance	5	19	16
Review Effectiveness	11	16	17
Quality of Design, Coding, Testing & Integration	12	14	18
Project Mgmt - Planning, Monitoring and Control	23	14	19
Productivity	8	11	20
Best Practices, Lessons Learnt, Innovations introduced	5	9	21
Difficulty Level due to SW Size, Complexity, Security, Reliability, Uptime, Response-time etc.	17	8	22
Alignment with Orgn's Vision-Mission-Goals & Orgnl Support	7	7	23
Cost of Quality (COQ)	2	7	24
Billing & Payment Collection	7	7	25
User-friendliness of the SW produced	10	6	26
Support in Deployment-Data Migration, Implementation, User Training	6	6	27
Senior-Junior Ratio	5	6	28
Attrition in Team	4	4	29
Ageing of Unresolved Issues fm Audits & Mgmt Reviews	5	4	30
Project Type	3	4	31
Onsite-Offshore Manpower Ratio (overseas projects)	4	4	32
Piloting New Tech, Domain or Business Models	4	2	33
Professionalism shown by the Team	1	1	34

The heading of the first column in the questionnaire was "Suggested name of the Measure (or attribute) for SDTP" and the respondent was asked to first read the description of the respective measure inserted as comment in the first column before responding to the question in the second column. The second column of the survey questionnaire was: "Do you think this measure is relevant for SDTP? Please respond 'Yes' or 'No' or 'Can't say' from the drop-down list for each measure." [*The actual format of the round-2 questionnaire can be had from the first author upon e-mail request.*]

In survey round-2, a total of 53 IT experts were initially contacted through email seeking their consent for participation. Out of fifty-three, 33 consented (62%) within a week, and were sent the questionnaire as email attachment. Actual responses came from 30 (91%). Of the 30 respondents, 23 were from large-sized firms, and 6 from medium-sized firms, and one on sabbatical from a small firm; 24 were male, and 6 female; 13 were from Delhi/NCR, 6 from Kolkata, 4 from Chennai, 2 each from Mumbai, Hyderabad and Bangalore, and 1 from overseas. Their work-experience ranged from 3 to 30 years, with a total experience pool of 379 person-years, mean/mode/median and standard deviation (SD) being 12.63/6/13 and 6.79 years respectively, and variability (SD/mean) 54%. Further, taking 6 years' work experience as an average threshold for a fresh software team member (TM) to move to the role of a project leader (PL), as argued earlier [2], it was observed that 10 respondents (33%) were in the range of 3-6 years (~TMs, N=10) and 20 of them (67%) were above 6 years (~PLs or above, N=20).

Results of Survey Round-2. Responses of the 30 experts were compiled and tabulated in one single Excel sheet, along with their comments, if any. For every measure, the expert responses varied from a clear "Yes" to a clear "No", or a "Can't say". Taking number of experts saying "Yes" to a particular measure as "Y" and total number of respondents as "N", the content validity ratio (CVR) was computed for each measure by using the formula prescribed by Lawshe [21]:

$$\boxed{CVR = (Y\text{-}N/2)/(N/2)}$$

CVRs for all 34 measures were separately computed for TMs (N=10), for PLs (N=20) and for all combined (N=30). CVRs below 62% (for N=10), below 42% (for N=20), and below 33% (for N=30), and also all CVRs below 50% were differently flagged to facilitate decisions on discretionary dropping or retention of measures based on Lawshe's [21] guidelines on a measure-to-measure basis. In addition, "Best Practices & Lessons Learnt" was renamed as "Process Knowledge Management (KM)", "Reusable Artifacts Used/Created" was renamed as "Product KM", and a new measure called "Greenfield KM" was introduced by us to address the entire KM-space. A few of the measures like "Time Overrun/Schedule Slippage" and "Cost-Budget Overrun / Effort Slippage" were marginally reworded to capture clearer meaning and a few others were semantically merged for better parsimony. Multiple measures pertaining to 'quality' were grouped into two broad heads, namely "Process Quality" and "Product Quality". All done, the outcome of survey round-2 is a shortlist of 20 assorted measures/UDCs of SDTP, listed below:

(1) Employee Satisfaction; (2) Process Knowledge Management; (3) Product Knowledge Management; (4) Greenfield Knowledge Management; (5) Adherence to Time Schedule; (6) Difficulty Level; (7) Process Quality; (8) Team Morale-Motivation-Synergy, and Alignment & Communication; (9) Team Learning, Skill Development & Competency Growth; (10) Customer Satisfaction; (11) Adherence to Cost/Effort Budget; (12) Scope/ Requirements capture & fulfillment, and Change Management; (13) Professionalism shown by the Team; (14) Cost of Quality (COQ); (15) Client Relationship Management (CRM) and Value Add to Client; (16) Quality of Deliverables, Design, Coding, Testing and Integration; (17) Project Management – Planning, Monitoring & Control; (18) Defect Density & Defect Containment; (19) Productivity; (20) Speed of resolving Review Issues.

The final output of this paper, however, is a table giving operational definition of each of the above 20 measures compiled from the textual inputs and comments received from the SW experts in the two survey rounds. [*This final output table and the aiding worksheet showing measure-wise CVRs can be had from the first author upon email request*].

7 Conclusions, Implications and Limitations

Need for measuring team performance is not one of recent origin. However, there have been almost as many viewpoints as there have been researchers. Often, the definitions and interpretations of basic terms like performance, productivity, efficiency and effectiveness etc. are varied. These concerns become more acute when it comes to measurement of SDTP due to intangible and dynamic nature of SW products and services. This paper is a step towards bridging these gaps by synthesizing and synchronizing various dimensions in order to evolve a set of UDCs of SDTP. It is expected to pave the way for further refinement and future development of a reliable and valid scale for SDTP.

One inescapable limitation of the study is that both SW development methodologies and SW industry's business models have been ever so rapidly changing and so, whichever scale, and dimensions thereof, are found good at a given point of time, are unlikely to continue to fit the bill for a very long period of time, thereby needing periodic reviews.

Another concern on the validity of the results could stem from the apparently limited pool-size and heterogeneity of the SW experts surveyed in both rounds, in spite of the fact that going by the literature, both rounds had statistically adequate sample sizes and demographic diversity.

Between the two rounds, six experts were common. We argued in favour of going with the overlap because, first, at least 3 months had elapsed between the two rounds; second, overlap of six was small; third, the nature of the two rounds were very different- one was a qualitative survey using an open-ended questionnaire, and the other was a multiple-choice pointed questionnaire for quantitative analysis of CVRs; and fourth, the order of the output of round-1 was scrambled before administering them in round-2.

Most of the domain experts surveyed were from top 2/3IT companies in India, except a few who could be reached overseas. This may raise doubts on its cross-cultural applicability. However, most of the experts surveyed have been operating in highly globalised environments and for global customers, many often visiting client-sites abroad. Thus, the impact of cross-cultural limitations on the study is expected to be minimal.

References

1. Balijepalli, V., Mahapatra, R., Nerur, S., Price, K.H.: Are two heads better than one for software development? The productivity paradox of pair programming. MIS Quarterly 33(1), 91–118 (2009)
2. Bose, P., Mishra, B.R.: Master technology first; be a manager later, says Infosys. Business Standard XXXV, 229 (2009)
3. Bradley, J.H., Hebert, F.J.: The effect of personality type on team performance. Journal of Management Development 16(5), 337–353 (1997)
4. Brannick, M.T., Prince, C.: An overview of team performance measurement. In: Brannick, M.T., Salas, E., Prince, C. (eds.) Team Performance Assessment and Measurement: Theory, Methods, and Applications, ch. 1, pp. 3–16. Lawrence Erlbaum Associates, NJ (1997)
5. Brown, C.V., DeHayes, D.W., Hoffer, J.A., Martin, E.W., Perkins, W.C.: Managing Information Technology, 6th edn. Pearson / Prentice Hall, NJ (2009)
6. Chatterjee, A.N.: A critical evaluation of the instruments for measurement of software development team performance. Global Digital Business Review 3(1), 134–143 (2008) ISSN 1931-8146
7. Kuo, C.-C.: Research on impacts of team leadership on team effectiveness. The Journal of American Academy of Business, 266–277 (September 2004)
8. Doolen, T.L., Hacker, M.E., Aken, E.M.V.: The impact of organizational context on work team effectiveness: A study of production team. IEEE Transactions on Engineering Management 50(3), 285–296 (2003)
9. Doolen, T.L., Hacker, M.E., Aken, E.M.V.: Managing organizational context for engineering team effectiveness. Team Performance Management 12(5/6), 138–154 (2006)
10. Faraj, S., Sambamurthy, V.: Leadership of information systems development projects. IEEE Transactions on Engineering Management 53(2), 238–249 (2006)
11. Forsyth, D.R.: Group Dynamics, 3rd edn. Wadsworth Publishing Company, CA (1998)
12. Gorla, N., Lam, Y.W.: Who should work with whom? Building effective software project teams. Communications of the ACM 47(6), 79–82 (2004)
13. Hackman, J.R.: The design of work teams. In: Lorsch, J.W. (ed.) Handbook of Organizational Behavior, pp. 315–342. Prentice-Hall, Englewood Cliffs (1987)
14. Hardesty, D.M., Bearden, W.O.: The use of expert judges in scale development – implications for improving face validity of measures of unobservable constructs. Journal of Business Research 57(2), 98–107 (2004)
15. Hayes, F.: Chaos is back. Computer World 38(45), 70 (2004)
16. Hoegl, M., Gemuenden, H.: Teamwork Quality and the Success of Innovative Projects: A Theoretical Concept and Empirical Evidence. Organizational Science 12(4), 435–449 (2001)
17. Hui, A.K.T., Liu, D.B.: A Bayesian Belief Network Model and tool to evaluate risk and impact in software development projects, pp. 297–301. IEEE, Los Alamitos (2004)

18. Imbrie, P.K., Immekus, J.C., Maller, S.J.: Work in progress – a model to evaluate team effectiveness. In: 35th ASEE/IEEE Frontiers in Education Conference. Session T4F, pp. 12–13 (2005)

19. Kraiger, K., Wenzel, L.H.: Conceptual development and empirical evaluation of measures of shared mental models as indicators of team effectiveness. In: Brannick, M.T., Salas, E., Prince, C. (eds.) Team Performance Assessment and Measurement: Theory, Methods, and Applications, ch. 4, pp. 63–84. Lawrence Erlbaum Associates, NJ (1997)

20. Law, K.S., Wong, C.: Multidimensional constructs in structural equation analysis: an illustration using the job perception and job satisfaction constructs. Journal of Management 25(2), 143–160 (1999)

21. Lawshe, C.H.: A quantitative approach to content validity. Personnel Psychology 28, 563–575 (1975)

22. Lee, G., Xia, W.: Toward agile: an integrated analysis of quantitative and qualitative field data on software development agility. MIS Quarterly 34(1), 87–114 (2010)

23. May, T.A.: Demographics and IT team performance. Computer World 38(20), 21–22 (2004)

24. McIntyre, R.M., Salas, E.: Team performance in complex environments: what we have learned so far. In: Guzzo, R., Salas, E. (eds.) Team Effectiveness and Decision Making in Organizations, pp. 333–380. Jossey-Bass, San Francisco (1995)

25. Nidumolu, S.R.: The effect of coordination and uncertainty on software project performance: residual performance risk as an intervening variable. Information Systems Research 6(3), 191–219 (1995)

26. Pattit, J.M., Wilemon, D.: Creating high-performing software development teams. R&D Management 35(4), 375–393 (2005)

27. Pinto, J.K., Mantel Jr., S.J.: The causes of project failure. IEEE Transactions on Engineering Management 37(4), 269–276 (1990)

28. Rico, R., Manzanares, M.S., Gil, F., Gibson, C.: Team Implicit Coordination Process: A Team Knowledge-Based Approach. Academy of Management Review 33(1), 163–184 (2008)

29. Rubinstein, D.: Standish Group Report: There's Less Development Chaos Today. Software Development Times (2007),
http://www.sdtimes.com/content/article.aspx?ArticleID=30247
(accessed on October 24, 2009)

30. Sandblad, B., Gulliksen, J., Aborg, C., Boivie, I., Persson, J., Goransson, B., Kavathatzopoulos, I., Blomkvist, S., Cajander, A.: Work environment and computer systems development. Behaviour & Information Technology 22(6), 375–387 (2003)

31. Sethi, V., King, R.C., Quick, J.C.: What causes stress in information system professionals? Communications of the ACM 47(3), 99–102 (2004)

32. Sonnentag, S., Frese, M., Brodbeck, F.C., Heinbokel, T.: Use of design methods, team leaders' goal orientation, and team effectiveness: a follow-up study in software development projects. International Journal of Human-computer Interaction 9(4), 443–454 (1997)

33. SPSS Inc.: SPSS Text Analysis for Surveys™ 2.0 User's Guide. SPSS Inc., Chicago (2006) ISBN 1-56827-379-7

34. Stevens, K.T.: The effects of roles and personality characteristics on software development team effectiveness, Dissertation submitted to the Faculty of Virginia Polytechnic Institute and State University for Ph.D (approved), p. 127 (1998)

35. Sundstrom, E., De Meuse, K.P., Futrell, D.: Work teams: applications and effectiveness. American Psychology 45(2), 120–133 (1990)

36. Tannenbaum, S.L., Beard, R.l., Salas, E.: Team building and its influence on team effectiveness: an examination of conceptual and empirical developments. In: Kelly, K. (ed.) Issues, Theory, and Research in Industrial / Organizational Psychology, pp. 117–153. Elsevier, Amsterdam (1992)
37. Tesluk, P., Mathieu, J.E., Zaccaro, S.J., Marks, M.: Task and aggregation issues in the analysis and assessment of team performance. In: Brannick, M.T., Salas, E., Prince, C. (eds.) Team Performance Assessment and Measurement: Theory, Methods, and Applications, ch. 10, pp. 197–224. Lawrence Erlbaum Associates, NJ (1997)
38. Thamhain, H.J.: Effective project leadership in complex self-directed team environments. In: Proceedings of the 32nd Hawaii International Conference on System Sciences (HICSS 1999), p. 12. IEEE, Los Alamitos (1999)
39. Thite, M.: Leadership: A critical success factor in IT project management. Management of Engineering and Technology. In: Portland International Conference on Technology and Innovation Management (PICMET 1999), vol. 1, pp. 298–303 (1999)

Computational Modeling of Collaborative Resources Sharing in Grid System

Ayodeji Oluwatope, Duada Iyanda,
Ganiyu Aderounmu, and Rotimi Adagunodo

Grid Lab, Center of Excellence in Software Engineering
Department of Computer Science and Engineering
Obafemi Awolowo University, Ile-Ife.Nigeria
{aoluwato,gaderoun,eadagun}@oauife.edu.ng

Abstract. In grid computing, Grid users who submit jobs or tasks and resources providers who provide resources have different motivations when they join the Grid system. However, due to autonomy both the Grid users' and resource providers' objectives often conflict. This paper proposes autonomous hybrid resource management algorithm aim at optimizing the resource utilization of resources providers using "what-you-give-is-what-you-get" Service Level Agreements resource allocation policy. Utility functions are used to achieve the objectives of Grid resource and application. The algorithm was formulated as joint optimization of utilities of Grid applications and Grid resources, which combine the resource contributed, incentive score, trustworthiness and reputation score to compute resource utilization. Simulations were conducted to study the performance of the algorithm using GridSim v5.0. The simulation results revealed that the algorithm yields significantly good result because no user can consume more than what it contribute under different scenarios; hence the problem of free riding has been addressed through this algorithm.

Keywords: Resource scheduling, Grid System, Computational modeling.

1 Introduction

Grid is a heterogeneous environment which allows collective resources operates different operating system and hardware, while cluster is a homogenous environment. Grid is therefore a system that coordinates resources that are not subject to centralized control, using an open standard, general-purpose protocol and interfaces to deliver non-trivial qualities of service, [10].

The Grid allow users to handle the exponentially growing database and to speed up their calculation in data processing by using existing sources. Besides, Grid enable users to share inexpensive access to computing power, storage systems, data sources, applications, visualization devices, scientific instruments, sensors and human resources across a distance department and organization in a secure and highly efficient manner. With these characteristics, Grid affords researchers capablility to collaborate more easily and gain access to more computing power, enabling more studies to be run and difficult problems solved [11].

S. Dua, S. Sahni, and D.P. Goyal (Eds.): ICISTM 2011, CCIS 141, pp. 311–321, 2011.

How these resources are shared in terms of CPU cycles, storage capacity, software licenses, and so on is dependent upon the availability of resources located in some foreign administrative domain(s). Resource providing and usage is settled by policy rules. Some challenging questions are: *what are the terms over which a party would allow others to use its resources? What are the conditions for utilizing third parties computational power* etc? Grid computing proffers solutions to these questions based on the concept of virtual organization, [28].

A virtual organization,VO, is a widely spread concept used in several fields of computer science (e.g. Agents Collaboration and Negotiation protocols, Collaborative Tools and Social Networks and so on). In Grid, a VO is conceived as coordinated resource sharing and problem solving infrastructure spanning multi-institutions which give users transparent access to these resources[12, 21]. The key concept here is its ability to negotiate resource-sharing arrangements among a set of participating parties (providers and consumers). This sharing is highly controlled by the resource providers and consumers *who define clearly and carefully what is shared, who is allowed to share, and the conditions under which sharing occur* [10, 12, 27]. Users have resource-consuming activities or jobs/tasks that must be mapped to specific resource through a resource allocation mechanism. The resource allocation mechanism may choose among alternative mappings in order to optimize some utility metric within the bounds permitted by the VO policy environment. It is envisioned that deployment of Grid technology will grow from its current modest scale to eventually overlay the global Web, hence, requiring allocation mechanism to be highly scalable and robust [4]. A specific problem that underlies the Grid concept is coordinating resource sharing and problem solving in dynamic, multi-institutional virtual organizations. Grid users are basically concerned with the performance of their applications, for example the total cost to run a particular application, while resource providers usually pay more attention to the performance of their resources, such as resource utilization in a particular period, [22]. As observed by [7] scheduling algorithms in Grid system can be classified into two categories i.e. application-centric scheduling and resource-centric scheduling. Also according to [3, 5]and [32], challenges of collaborative resource sharing include resource free-riding (RFR), denial of distributed service attack (DoDS attack), contention management, Quality of Service (QoS), failure detection and recovery, heterogeneity, un-trustedness, selfishness, resource allocation inefficiency, users un-satisfaction, deadline missing etc. As a result of these identified tribulations among others, there is the need to develop a resource sharing scheme that averts these problems in order to enhance the performance of the Grid system and consequently encourages formation of virtual organization for collaborative resource sharing. Therefore, this paper addresses the problem of resource free-riding among competing organizations and it focuses on modeling of a fair resource sharing technique in Grid.

2 Previous Works

Different economic-based models such as currency based -microeconomics and macroeconomics principles for resource management; have been extensively proposed and studied in the pas]. In this model, a consumer is allowed to negotiate an arbitrarily complex provision of resources and pay for it accordingly in some currency. This scheme suppresses negotiation problem between resource consumers and providers and requires well established infrastructure for secure e-commerce and service auditing [1, 13, 31, 2]. In [31] ticket approach that builds on the concept of tickets was introduced. This model combats challenge of resource free riding but not all frauds are detectable and traceable, hence free riding was noticeable.

Li Chunlin and Li Layuan in [22] proposed a system-centric scheduling policy based on maximizing users sum of utility and incentives for Grid user and resource provider interplay. The model however is plague with system imbalance. Foster, I et al in [11] introduced the matching of the QoS request and service between the tasks and hosts based on the conventional Min–Min algorithm. It suffered poor QoS scope definition due its binary nature of decision between high and low QoS tasks. Other parameters such as number of satisfied users and total utility of the meta-task are not addressed. In [19] challenges of inconsistency in distributed virtual environment were solved with Time-Stamped Anti-Entropy Algorithm. The algorithm fails to update users with information about changes in the distributed virtual environment as users. In [6], free riding problem was addressed again using peer auditing scheme. In the scheme, users audit each other's contributions and allowed resource consumption as a direct proportion of each other's contribution according to the agreed consumption ratio. However, there are instances of users bypassing auditing through hacking or conspiracy and issue of unreliable audit reports. [25] proposed random audits scheme to tackle over consumption of resources by malicious participants. The drawback of the scheme is its inability to include scenarios such as contention for temporal resources like bandwidth. Utility score was introduced in [20] to eliminate resource allocation problem and free riding in file sharing system. In this framework, users provide their utilities to the system and incentives to improve their utility score to the system; hence the equilibrium of the system is maintained. Its drawback is evident in a theoretical framework which fails to consider a specific scheme and a solution which does not provide an incentive scheme capable of luring users to play along with the system. Wolski et al [30] proposed a scheme in which applications must buy the resources from resource suppliers using an agreed-upon currency utility function. This work fails to consider the global optimization of utility of all Grid users and providers. Fu, et al [13] applied trustworthiness as resource anti-boasting and anti-inflation mechanism. [26] used reputation score in their work to tackle the problem of false IDs attacks- Debit and Credit Reputation Score (DCRC) and Credit Only Reputation Score (CORC). In [24] the problem of unfair resources sharing in distributed environments was addressed with a computational model for trust and reputation. In this model, reputation evaluation was used to calculate the trustworthiness of every organization in a self-policing way. Reputation is well modeled but the risk of over-consumption by malicious users was not put into consideration. [23] applied prisoner's-dilemma (PD) incentive to combat free riding in distributed systems. The solution is plagued with failure to capture other characteristics of the free riders including sybilproofness. [14] introduced mutual authorization and authentication scheme to tackle quality of service (QoS) in grid environment. Its lacks support for advance reservation, checkpoint and restart protocols which invariably results in considerable delay. In this paper, we propose

Grid resource sharing technique using the combination of trustworthiness, reputation and incentive scoring. The rest of the paper is organized as follows: section two presents the proposed scheme modeling, section three discussed the simulation setup and result, and section four concludes the paper and provides future research directions.

3   The Proposed Scheme Modeling

In an attempt to address the conflicting objectives in Grid resource sharing, we propose a hybrid resource sharing scheme that performs dual functions of enforcing cooperation among participants during resources sharing in VO and exterminating free-riding concurrently. Fair resource sharing scheme was formulated using the concept of resource contribution, reputation, incentive and trustworthiness scoring. Modeling of an Infra-Grid was carried out using GridSim v5.0, a Java based simulator. Behavioral characteristics of resources users were modeled as a deterministic utility function. To formulate a fair resource sharing scheme, utility functions were derived and transformed to scoring functions using Euler's transformation. Scoring functions were modeled as a resource allocation and management algorithm. To evaluate the scheme, an infra-Grid simulation model was built using Gridsim v5.0 toolkit with "*what-you-give-is-what-you-get*" Service Level Agreements resource allocation policy. The performance of algorithm was investigated using resource utilization ratio under different scenarios.

3.1   Formulation of Collaborative Resource Sharing Scheme

The necessary and sufficient requirements for collaborative resource sharing and estimation of optimal resource consumption of each organization participating in the proposed VO model are: *resource contributed incentive score, trustworthiness score, and reputation score of the respective participating organizations.* These four entities are interrelated. There exit a relationship among these entities such that the incentive score of an organization is derived from the resources contributed by such organization to the VO, the trustworthiness score is derived from the derivative of both the resource contributed by an organization to the total resource in the VO and its incentive score, while reputation score is a function of the incentive score and trustworthiness score. The resource utilization of every participating party is then derived from incentive, trustworthiness and reputation scores of respective participants. The algorithm of the proposed scheme is as follow.

```
1.   Initialize resource contribution
2.      While (contribution is made)
3.          Calculate incentive
4.          For(each Organization)
5.   Compute incentive according to
6.   Service Level Agreement (SLA)
7.      I_i = f (R_i)
8.          End For
9.                  For (each Organization)
```

```
10.                    Compute Trustworthiness
11.              T_i = f (I_i, R_i).
12.                    End For
13.         For (each Organization)
14.                    Compute Reputation
15.                 Re_i, = f (I_i, T_i)
16.             End For
17.                 For (each Organization)
18.         Estimated optimal Resource utilization
19.                    R_c = f (I_i, T_i, R_i )
20.                 End For
21.           End While
22.      End
```

The algorithm to generate the requested number of resources is described as follows:

```
input : The vector pRr of the resource request distribution
output: Returns the number of resources requested
1.   r  ← Random(0.0 to 1.0);
2.   for i ←  0 to ElementsIn(pRr) do
3.     if i is 0 then
4.         if r < pRr 0 then
5.                 Rr ←   1;
6.         else
7.           if r > p Rr_{i-1} and r ≤ p Rr_i  then
8.                 Rr ←   2^i;
9.             end if
10.       end if
11.     end if
12.  end for
13.   return Rr;
14.   end
```

The algorithm for generating execution time request is as follows.

```
input :  The  vector  pT  with  the  execution  time  request
distribution
output: Returns the execution time to request
1.     r  ← Random(0.0 to 1.0);
2.   for i  ← 0 to ElementsIn(pT ) do
3.     if i is 0 then
4.           if r < pT_0 then
5.                 T ←   1;
6.         else
7.               if r > pT_{i-1} and r ≤ pT_i then
8.                 T←    2^i;
9.             end if
10.       end if
11.       end if
12.   end for
13.   T  ← T × C_t ;
13.   return T;
14.   end
```

The algorithm to generate the time until the next job submission is given as:

```
input   :   baseMean,   meanModifier,          offHourModifier,
offHourStart, offHourEnd and simTime which is the current
time of the simulation
output: Returns the time to pass from now until the next job
is to be submitted to the VO
1.      mean← baseMean;
2.      mean   ← mean × meanModifier;
3.  if simTime ≥ offHourStart and simTime <
            offHourEnd then
4.      mean ←   mean × peakModifier;
5.      r ←   Random(0.0 to 1.0);
6.          t ←   r × mean;
7.      if t > mean × 10 then
8.              t ←   mean × 10;
9.          else
10.                 return t;
11.         end if
12.  end if
13. end
```

3.2   Modeling of the Proposed Scheme

The description of some basic assumptions and notations required for the model and simulation are explained below.

  i.   The VO is assumed to be heterogeneous and dynamic.
 ii.   Every participating organization contributes resource to and consumes resources from the VO Grid.
iii.   The total resource, R_{tot} in the VO Grid is an aggregate of the resources contributed by the participating organizations.
 iv.   The incentive score, I_i, of the participating organizations can be computed as ratio of such organization's resource contribution, R_i, in the entire VO Grid, $I_i = f(R_i)$
  v.   The trustworthiness score, T_i, of individual organization is a function of both the contributed resources and the incentive score of such organization, $T_i = f(I_i, R_i)$.
 vi.   The reputation score, Re_i, of individual organization is a function of both the incentive and trustworthiness scores, $Re_i, = f(I_i, T_i)$
vii.   The estimated optimal resource to be consumed, R_c, by each participating organization is a function of incentive, trustworthiness and reputation scores, $R_c = f(I_i, T_i, R_i)$

The proposed Infra-Grid was modeled using three independent organizations, named P, Q, and R to form a VO Grid WXYZ as depicts in Figure 1 and figure 2depicts its topology consisting of leaf router, edge router and core router. The rectangle WXYZ is the VO Grid formed by these organizations rectangles ABCD, EFGH and IJKL. Each organization contributes some of its resources to the VO Grid for collaborative sharing among the participating parties. The three organizations also have some of their resources that are not part of the VO, these are being managed privately outside the VO by the respective organization.

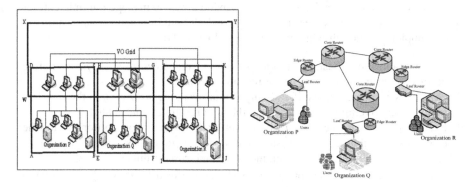

Fig. 1. Architecture of the Proposed Infra-Grid   **Fig. 2.** Topology of the Proposed Infra-Grid

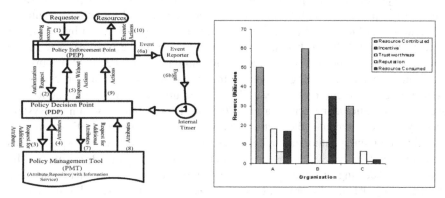

Fig. 3. Policy engine for the proposed Scheme

Fig. 4. Simulation result of resource utilization ratio of the participating organizations

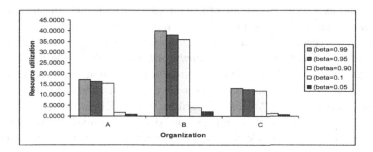

Fig. 5. Simulation result of resource utilization of the participating organizations under different values of beta (the resource utilization determinant)

3.3 Policy Engine for the Proposed Scheme

Policies are statements that specify how the VO deals with requests, what type of content can be moved out to an organization, what resources can be shared and the degree of sharing, and so on. Thus, policies provide a way to manage multiple entities deploying complex technologies. In the proposed VO environment, organizations P, Q and R implement the following policies tagged P1 to P4:

P1: Each organization must contribute part of its resources to the system before permission is granted for the use of resources from the system.

P2: Grant new resource request if and only if total resource utilization in the past few minutes (say five minutes) is less than the total resources contributed.

P3: Each organization receives resource utilization credit proportionate to the resources contributed to the system based on (wygiwyg) using the concepts of resources contribution, reputation, incentive and trustworthiness scores.

P4: The 1/N policy that is each organization should perform 1/Nth of job submitted to the system at a particular time "t".

The proposed policy engine in figure 3 consists of; the policy management tools (PMT), the policy decision point (PDP) and the policy enforcement points (PEPs). PDP retrieves policies from the system, interprets them (based on policy condition), and decides on which set of policies to be enforced (i.e. policy rules) by the policy enforcement points (PEPs). It obtains information such as the type and size of the resource contributed needed to enforce P1, to enforce P2 and P3, PDP obtains and utilizes historic data including incentive, trustworthiness and reputation scores on resource usage. PDP enforces P4 by collecting and utilising information about work done by each participating organization. PEPs are logical entities within the system boundary, which execute the actions defined in the policies. PEPs enforces P1,P2,&P4 by issuing authorization decisions to accept or reject certain organizations /operations on resources when the usage limit is reached. Policy P2, is enforced by periodically executing management tasks to restore the system to a state consistent with the policy. P1 and P2 are configuration policies, while policies P3 and P4 are conditional policies. In figure 3, steps 1 to 5 describe interactions among the requestor, PEP, PDP and PMT to negotiate resource access control. The "event interaction" which is steps 6 to 10 describe event recording process, action list generation and execution. An event is the aggregation of relevant attributes sent to a PDP when a given situation occurs. An event reporter supplies convention attributes when an event is generated. Each attribute is described using the "attribute" element. The "attribute" element must have an *AttributeId, the attribute* (the name), a *Type attribute* (the type) and a *Value element* (the value). An action is a management task to be performed on a resource. it is defined with a function name and a set of function arguments. Actions consists *"action-id"* attribute, which defines the operation or task to be executed fired. A set of arguments consist of *"ArgumentId"* and *"DataType"* attributes and a *"Value"* sub-element. The PEP maintains a mapping from *action-id* names to executable code representing particular management tasks.

4 Simulation Setup and Results

We described the proposed resources sharing algorithm using Java-based discrete-event Grid simulation toolkit, GridSim v5.0. To start the SimJava Kernel, GridSim.init( ) method was instantiated. This is required in order to start the SimJava simulation kernel. The Gridsim entities created are the Grid resources, the Grid users, the Grid network and the tasks or jobs. Three Grid resources, Grid users and routers were created. Each router was 1Gbps in capacity running First-In-First-Out (FIFO) scheduling scheme. Three tasks were created and mapped to the list of users and resources. The resources considered for the simulation are hard disk space, memory and bandwidth. Figure 3 depicts the policy engine as implemented in the simulation.

4.1 Results Discussion

The simulation was conducted using GridSim v5.0. The resources used for simulation are; harddisk space, memory and bandwidth. The performance of the algorithm was observed under free-riding and its associated problems such as; deal of distributed service attack (DoDS attack), boaster, per collision and multiple identity attacks. Figure 4 shows the initial parameter values per organisation. It is refereed to beta in the Simulations (resource utilization determinant). Figure 5 demonstrates a situation such that resources consumed are commensurate with the resources contributed, thereby eliminating free-riding. This therefore confirmed our asumption, every user was allowed to utilize resource accodrding to the *"what –you-give-is-what-you-get"* SLAs policy.

We discover that the model robustness depends on the truthful computations of reputation and trustworthiness scores, and resource contribution statistics. In addition, when a new organization joins the system, its reputation score is essentially is set to 0 but grows with time and such an organinization cannot consume any resources. In fact, new incomers in this mechanism have to first contribute and "prove their worthiness which addresses the issue of conspiracy among participating organisations. This mechanism is quick to response such that as soon as an organization builds up some reputation, it begin to consume resources from the VO.

5 Conclusion and Further Work

Problems of free-riding, denial of distributed service attack, deadline missing, allocation efficiency, and user satisfaction pose challenges to Grid computing research community and virtual organization environments. These challenges necessitate the need to develop pragmatic and fair resource sharing scheme. As an attempt to solve the conflicting objectives of Grid resource sharing, this work proposed an autonomous hybrid resource management algorithm using the concept of resources contribution, reputation, incentive and trustworthiness scoring. Simulation result of the modeled Infra-Grid revealed that our technique based on *"what-you-give-is-what-you-get"* Service Level Agreements resource allocation policy addresses challenges of free riding and its associated problems. Further investigation of the performance of the scheme on resource allocation efficiency, user satisfaction, deadline miss ratio, throughput ratio, algorithm sensitivity test, model analysis, and benchmark will form the direction of further work.

Acknowledgement

This work is partly sponsored by the STEP-B (The STEP-B Project is a Nigerian project on Science and Technology Education at the Post-Basic Level (STEP-B)) funding through a World Bank grant for the establishment of a Center of Excellence in Software Engineering in Obafemi Awolowo University, Ile-Ife.NIGERIA

References

1. Amir, Y., Awerbuch, B., Borgstrom, R.: A Cost-Benefit Framework for Online Management of Meta-Computing Systems. In: Proceedings of the First International Conference on Information and Computational Economy (1998)
2. Andrade, N., et al.: Our-Grid: An Approach to Easily Assemble Grids with Equitable Resource Sharing. In: Proceedings of the 9th Workshop on Job Scheduling Strategies for Parallel Processing (June 2003)
3. Sulistio, A., Poduval, G., Buyya, R., Tham, C.-K.: On Incorporating Differentiated Levels of Network Service into GridSim. Future Generation Computer Systems (FGCS) 23(4), 606–615 (2007)
4. Galstyan, A., Czajkowski, K., Lerman, K.: Resource Allocation in the Grid with Learning Agents. Journal of Grid Computing 3, 91–100 (2005)
5. Caminero, A., Sulistio, A., Caminero, B., Carrion, C., Buyya, R.: Extending GridSim with an Architecture for Failure Detection. In: Proceeding of the 13th International Conference on Parallel and Distributed Systems (ICPADS 2007), December 5-7, pp. 1–8 (2007)
6. Wallach, D.S., et al.: Enforcing fair sharing of distributed resources. In: Kaashoek, M.F., Stoica, I. (eds.) IPTPS 2003. LNCS, vol. 2735. Springer, Heidelberg (2003)
7. Dong, F., Akl, S.G.: Scheduling algorithms for grid computing: State of the art and open problems, Technical report, Queen's University School of Computing (January 2006)
8. Elmroth, E., Gardfjall, P.: Design and Evaluation of a Decentralized System for Grid-Wide Fair Share Scheduling. In: Proceeding of the 1st International Conference on e-Science and Grid Computing, Melbourne, Australia, pp. 1–9 (2005)
9. Foster, I., Kesselman, C.: The Grid: Blueprint for a Future Computing Infrastructure. Computational Grids 2(1), 15–52 (1998)
10. Foster, I.: What is the Grid? A Three Point Checklist. Daily News and Information for the Global Grid Community 1(6), 55–65 (2002)
11. Foster, I., Kesselman, C., Nick, J., Tuecke, S.: Grid Services for Distributed System Integration. Computer 35(6), 37–46 (2002)
12. Foster, I., Kesselman, C.: The Grid2 Blueprint for a New Computing Infrastructure. Morgan Kaufmann Publishers, San Francisco (2003)
13. Fu, Y., Chase, J., Chun, B., Schwab, S., Vahdat, A.: SHARP: An Architecture for secure resource peering. In: Proceedings of the 19th ACM Symposium on Operating Systems Principles, SOSP-19 (2003)
14. Gagliardiand, F., et al.: Building an Infrastructure for Scientific Grid Computing: Status and Goals of the EGEE project. Philosophical Transactions-A of the Royal Society: Mathematical, Physical and Engineering Sciences (2005)
15. Ghosh, S., Rajkumar, R., Hansen, J., Lehoczky, J.: Integrated Resource Management and Scheduling with Multi-Resource Constraints. In: Proceedings of 25th IEEE Real-Time Systems Symposium (2004)
16. Grimshaw, A.: What is a Grid? Grid Today 1(26), 200–220 (2002)

17. Aydt, H.: Simulation-aided decision support for cooperative resource sharing. Master's thesis, Royal Institute of Technology, Sweden (2006)
18. He, X.S., Sun, X.H., Laszewski, G.V.: QoS Guided Min–Min Heuristic for Grid Task Scheduling. Journal of Computer Science & Technology 18(4), 442–451 (2003)
19. Krauter, K., Buyya, R., Maheswaran, M.A.: Taxonomy and Survey of Grid Resource Management Systems for Distributed Computing. International Journal of Software Practice and Experience 32(2), 135–164 (2002)
20. Kwok, M., et al.: Scalability Analysis of the Hierarchical Architecture for Distributed Virtual Environment. IEEE Transactions on Parallel and Distributed Systems 19(3), 408–417 (2008)
21. Lakshmish, R., Liu, L.: Free riding: A New Challenge to File Sharing Systems. In: Proceedings of the 36th Hawaii International Conference on System Sciences (HICSS 2003), pp. 1–10 (2003)
22. Smarr, L., Catlett, C.E.: Metacomputing. Communication ACM 35(6), 44–52 (1992)
23. Chunlin, L., Layuan, L.: A System-Centric Scheduling Policy for Optimizing Objectives of Application and Resource in Grid Computing. Elsevier: Computers & Industrial Engineering 57(9), 1052–1061 (2009)
24. Feldman, M., Chuang, J.: Overcoming Free-Riding Behavior in Peer-to-Peer Systems. ACM SIGecom Exchanges 5(4), 41–50 (2005)
25. Mui, L., Mohtashemi, M.: A Computational Model of Trust and Reputation. In: Proceedings of the 35th Hawaii International Conference on System Science (HICSS 2002), pp. 1–9 (2002)
26. Ngan, T.-W., Wallach, D.S., Druschel, P.: Enforcing Fair Sharing of Peer-to-Peer Resources. In: Kaashoek, M.F., Stoica, I. (eds.) IPTPS 2003. LNCS, vol. 2735, pp. 1–6. Springer, Heidelberg (2003)
27. Gupta, M., Judge, P., Ammar, M.: A Reputation System for Peer-to-Peer Networks. In: Proceedings of the 13th International Workshop on Network and Operating System Support for Digital Audio and Video (NOSSDAV 2003), Monterey, California, June 1-3, pp. 1–9 (2003)
28. Singh, G., Kesselman, C., Deelman, E.: A Provisioning Model and its Comparison with Best-Effort for Performance-cost Optimization in Grids. In: Proceedings of International Symposium on High Performance Distributed Computing (HPDC 2007), Monterey Bay, California, pp. 117–126 (2007)
29. Chen, S., Luo, T., Liu, W., Song, J., Gao, F.: A Framework for Managing Access of Large-Scale Distributed Resources in a Collaborative Platform. Data Science Journal 7(1), 137–147 (2008)
30. Westerinen, A., Schnizlein, J., Strassner, J., Scherling, M., Quinn, B., Herzog, S., Huynh, A., Carlson, M., Perry, J., Waldbusser, S.: Terminology for Policy Based Management. RFC Editor, United States (2001)
31. Wolski, R., Plank, J., Brevik, J.: Analyzing market-based resource allocation strategies for the computational grid. International Journal of High-performance Computing Applications 15(3), 258–281 (2001)
32. Zhao, T., Karamcheti, V.: Enforcing Resource Sharing Agreements Among Distributed Server Clusters. In: Proceedings of the 16th International Parallel and Distributed Processing Symposium (IPDPS 2002), April 2002, pp. 1–10 (2002)
33. Zhengqiang, L., Shi, W.: Enforcing cooperative resource sharing in untrusted p2p computing environment. Mobile Network and Application 10(2), 971–983 (2005)

A Novel Approach for Combining Experts Rating Scores

Tom Au, Rong Duan, Guangqin Ma,
and Rensheng Wang

AT&T Labs, Inc. – Research, USA
{sau,rongduan,gma,rw218j}@att.com

Abstract. Based on the same information, subjects are classified into two categories by many experts, independently. The overall accuracy of prediction differs from expert to expert. Most of the time, the overall accuracy can be improved by taking the vote of the experts committee, say by simply averaging the ratings of the experts. In this study, we introduced the ROC invariant representation of experts rating scores and proposed the use of beta distribution for characterizing experts rating scores for each subject. The momentum estimators of the two shape parameters of beta distribution can be used as additional features to the experts rating scores or equivalents. To increase the diversity of selections of combined score, we applied a boosting procedure to a set of nested regression models. With the proposed approach, we were able to win the large AUC task during the 2009 Australia Data Mining Analytical Challenge. The advantages of this approach are less computing intensive, easy to implement and apparent to user, and most of all, it produces much better result than the simple averaging, say. For an application with a base consists of hundreds of millions of subjects, 1% improvement in predictive accuracy will mean a lot. Our method which requires less efforts and resources will be one more plus to practitioners.

Keywords: Model Ensemble, ROC Curves, Data Mining.

1 Introduction

Hundreds of experts are asked to independently build models to predict given subjects into one of two classes. While determine which expert's prediction to use, researchers has found that by simple averaging all or top expert's results may yield even more accurate prediction than any particular expert could produce alone. For example, the recently concluded Netflix Prize competition was won by teams that somehow combined their individual model predictions [1]. If the top two teams (Bellkor's Pragmatic Chaos and The Ensemble) make a 50/50 blend of their result, they would achieve even more accuracy than any one of them alone could achieve [2]. As prediction accuracy is paramount, to find a way better than simple averaging to combining the experts' predictions is desired.

S. Dua, S. Sahni, and D.P. Goyal (Eds.): ICISTM 2011, CCIS 141, pp. 322–329, 2011.
© Springer-Verlag Berlin Heidelberg 2011

In data mining literature, ensemble is often used for a collection of models by combining their predictions to achieve the optimum performance subjecting to a given metric such as accuracy. Methods like bagging [3] and boosting [5] are proposed to generate weights for averaging or voting the collection of models. Many researchers just view the collection of models as a set of given features, and apply learning techniques like regressions, supporting vector machine, artificial neural nets, memory-based learning, decision tree, boosting stumps, etc., to train or learn a sparse representation of the collection of models [6]. From machine learning point of view, Dieterich [4] summarized these approach into five categories: Bayesian Voting, manipulating the training examples, manipulating the input features, manipulating the output targets and injecting randomness.

In this study, we are focusing on the binary classification problem. Given hundreds of classifiers created by many experts, we will view these classifiers not only as input features for all subjects, but also a depiction of rating score distribution for (or spectrum of experts views on) each subject. We use both the collection of these classifiers, and a set of created feature depicting individual subject's rating score distribution as our modeling input to boost the combined prediction power. In practice, this approach for model ensemble had taken us to the top for the Large AUC Task during the 2009 Australia Data Mining Analytic Challenge.

In section 2, we layout the data structure and discuss their ROC invariant transformations; in section 3, we propose a set of features created based on both percent rank score and normal rank score of the classifiers; in section 4, we suggest a regression, boosting and model assembling strategies. Section 5 concludes our analysis. We use the Large AUC task training data from the 2009 Australia Data Mining Analytic Challenge (AUSDM09) as examples throughout this paper.

2   Classifiers and Their Invariant Property

We consider the following data structure

$$\{(Y_i, f_{i,1}, f_{i,2}, \dots f_{i,m}), i = 1,2,3, \dots, n\}, \tag{1}$$

where $Y_i = 1 \; or \; 0$ indicating the category that subject i belongs to or not, n is the number of subjects, $f_{i,j}$ is the rating score assigned to subject i by expert j, $j = 1,2,\dots,m$. Without loss of generality, we assume that $Y = 1$ associates with higher rating scores.

For binary classification problem, the performance of a classifier can be measured by the receiver characteristic curve (ROC) and the area under the curve (AUC). Two classifiers have different ROC curves not because they have different values, but because they rank subjects risk of being 1 (or 0) differently. Therefore, the ROC curve for a classifier is an invariant under a monotone transformation of the decision variables (i.e., $f_{i,j}$), for a given j.

For classifier j, we sort $\{(Y_i, f_{i,j}), i = 1,2, ..., n\}$ in terms of $f_{i,j}$ in ascending order, and let $\{r_{i,j}, i = 1,2, ..., n\}$ be the corresponding percentage rank scores of $f_{i,j}$. Note that $0 \leq r_{i,j} \leq 1$. Aligning the percentage ranks scores by the target Y_i, we yield the following transformed data

$$\{(Y_i, r_{i,1}, r_{i,2}, ... r_{i,m}), i = 1,2,3, ..., n\} \tag{2.2}$$

Based on the rank scores $\{r_{i,j}, i = 1,2, ..., n\}$, another transformation similar to standardizing the input features can be made as the following:

$$z_{i,j} = \Phi^{-1}\left(\left(n * r_{i,j} - \frac{3}{8}\right) \Big/ \left(n + \frac{1}{4}\right)\right), i = 1,2,...,n,$$

where $\{z_{i,j}\}$ are called the Blom [7] normal rank scores, and $\Phi()$ is the cumulative standard normal distribution function. The original data set (1) in terms of Blom normal rank score is

$$\{(Y_i, z_{i,1}, z_{i,2}, ... z_{i,m}), i = 1,2,3, ..., n\} \tag{2.3}$$

An interesting property of the ROC curve is that it is invariant under order preserving transformations of the decision variables and the classifiers. Based on the invariant argument, the ROC curves based on $f_{i,j}$, $r_{i,j}$ and $z_{i,j}$ are the same for a given expert j. We loosely call the three data sets (1), (2) and (3) the ROC invariant data sets. The classifiers in data set (1) may measured on different scales, but in data sets (2) and (3) the classifiers are on the $[0,1]$ and $\Phi^{-1}()$ scales, respectively.

3   Characterizing Individual Rating Scores

Each classifier assigns a rating score to an individual subject, with hundreds of classifiers, the confidence of each individual subject's true status of being 1 or 0 is revealed in certain extent. If subject i's true status is $Y_i = 1$, then one would expect that relatively higher rating score is assigned to it, and vice versa.

An immediate candidate model for describing the characteristics of the score distribution is the beta distribution with two shape parameters. Our percentage rank data set (2) meets the domain requirement of beta distribution, so for individual i, we assume that $\{r_{i,j}, j = 1,2,...,m\}$ is from the beta distribution $beta(\alpha_i, \beta_i)$. If the percentage rank score is well fitted by the beta distribution, then the shape of the score distribution will be well depicted by the two shape parameters α_i and β_i. When $\alpha_i < \beta_i$, the distribution of the scores is skewed to the left (in the sense that smaller values become more likely); when $\alpha_i > \beta_i$, the distribution of the scores is skewed to the right; when $\alpha_i = \beta_i$ the score distribution is symmetric about 0.5. Following the above intuition, we use data set (2) to create two features based on the momentum estimators of the two shape parameters:

$$a_i = \bar{r}_i \left(\frac{\bar{r}_i (1 - \bar{r}_i)}{s_i^2} - 1 \right),$$

$$b_i = (1 - \bar{r}_i) \left(\frac{\bar{r}_i(1 - \bar{r}_i)}{s_i^2} - 1 \right),$$

where $\bar{r}_i$ and s_i^2 are the mean and variance of $(r_{i,1}, r_{i,2}, \dots r_{i,m})$, respectively.

To see how sometimes these two created features can summarize the score distribution characteristics, we use the first 200 expert's ratings of the AUSDM09 data as an example. In Figure 1, we graphed four individuals percentage rank scores histograms and the fitted beta distribution $beta(a_i, b_i)$. The top left panel figure of Figure 1 shows that the scores are skewed to the right (more high scores), and the estimated $(a_i, b_i) = (17.50, 10.47)$, and the actual $Y_i = 1$; the top right panel of Figure 1 shows that the scores are skewed to the left (more low scores), and the estimated $(a_i, b_i) = (7.69, 12.51)$, and the actual $Y_i = 0$; the bottom left panel figure of Figure 1 shows that the scores are very skewed to the right, and the estimated $(a_i, b_i) = (23.05, 0.30)$, and the actual $Y_i = 1$; the bottom right panel figure of Figure 1 is peculiar at first sight, but with close inspection, one can see that the scores are more skewed to the left, and the estimated $(a_i, b_i) = (0.52, 0.76)$, and the actual $Y_i = 0$.

In Figure 2, we plotted the (a_i, b_i) for the 15000 observations, and it showed how the two created features may separate the target 1 from target 0, most of the times.

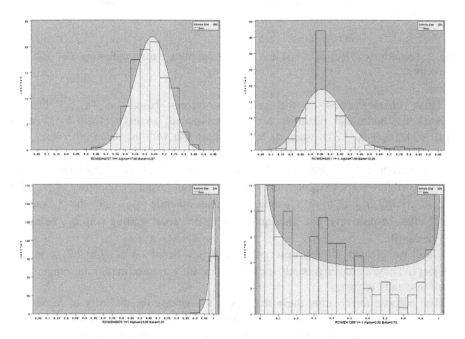

Fig. 1. Percentage rank score distributions for selected four individual subjects. The scores are shown in histograms and the continuous curves are fitted beta densities.

For those individuals with experts score distributed symmetrically around the middle, that is, percentage scores is symmetrically around 0.5 (points close to the diagonal line in Figure 2), it may be hard to classify them into 1 or 0. So we may need to create more features besides the two shape parameters.

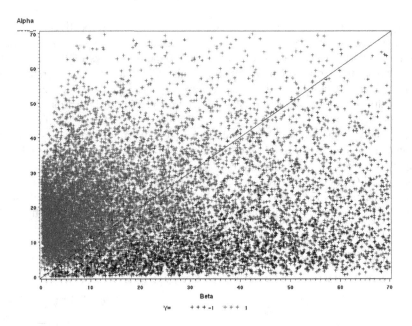

Fig. 2. Scatter plot of $\{(a_i, b_i), i = 1, 2, \ldots, n\}$, red and black crosses indicating Y=1 and Y=0, respectively

Other characteristics of the percentage scores based on (a_i, b_i) can be created as additional features are the estimated mean $(a_i/(a_i + b_i))$, mode $((a_i - 1)/(a_i + b_i - 2))$, when $a_i, b_i > 1$ (replace it with the mean otherwise), and variance $(a_i b_i/(a_i + b_i)^2(a_i + b_i + 1))$, of the beta distribution.

The percentile like features can also be used to summarize the characteristics of the score distribution, for example, for subject i, we can create 21 quantiles of $\{z_{i,j}, j = 1, 2, \ldots, m\}$: $q_{i,1}, q_{i,2}, \ldots, q_{i,21}$. Quantile $q_{i,1}$ and quantile $q_{i,21}$ are the 1 and 99 percentiles, respectively; the other quantiles are 5% apart starting with $q_{i,2}$ being the 5th percentile.

Last, we created all the descriptive statistics based on $z_{i,j}, j = 1, 2, \ldots, m$, for subject i. These descriptive statistics include mean, variance, max, min, range, IQR, skewness and kurtosis.

The purpose for creating these features is to increase the diversity of features in model ensemble. We also want to emphasize on that the beta distribution related features and the quantile like features are non-linear function of the experts' scores.

4   Regression and Boosting

So far, for each subject, we have created Blom normal scores, characteristics of scores based on beta distribution, quantiles, and descriptive statistics of scores based on the m experts.

The next step is to use these created features as explanatory variables (X) to perform a LogitBoost [8] for target $\{Y_i\}$. At the regression stage a backward selection is used. More precisely, let $\frac{1}{2} ligit\big(P(Y = 1|X)\big) = F(X)$, where $F(X)$ is a linear combination of explanatory variables. Then we start with

1. Assign each individual X with the same weight $\omega(X) = \frac{1}{N}$, N is the number of observation in the data, $F(X) = 0$ and $P(Y = 1|X) = \frac{1}{2}$;
2. Repeat from $m = 1,2,...,M$:
   a. Compute the working response and weights
   $$f_m(X) = \frac{Y - P(Y=1|X)}{P(Y=1|X)\big(1 - P(Y=1|X)\big)}$$
   $$\omega(X) = P(Y = 1|X)\big(1 - P(Y = 1|X)\big);$$
   b. Fit the function $f_m(X)$ by weighted least-squares regression to X using weights $\omega(X)$ and backward selection;
   c. Update $F(X) \leftarrow F(X) + \frac{1}{2}f_m(X)$ and
   $$P(Y = 1|X) \leftarrow e^{F(X)}/\big(e^{F(X)} + e^{-F(X)}\big);$$
3. Output $F(X) = \sum_{m=1}^{M} f_m(X)$ and
   $$P(Y = 1|X) \leftarrow e^{F(X)}/\big(e^{F(X)} + e^{-F(X)}\big).$$

We can set out to build four nested models and hope each model will capture some aspects of the subjects' true status:

- Model 1: The Blom's Rank Scores $\{z_{i,j}\}$;
- Model 2: Model 1 + the a_i and b_i;
- Model 3: Model 2 + the 21 quantiles $q_{i,1}, q_{i,2}, ..., q_{i,21}$;
- Model 4: Model 3 + the rest of the created features.

For each of the above model, we perform the regression and boosting procedure on the training data and score the testing data. The iteration stops where the optimum AUC is yield for the testing data. One can use the combinations of these models as the final scoring procedure; or instead of using simple average; the best weights for the weighted average of these scores can be searched.

To examine the above strategy, we use the target variable and the first 200 experts rating from AUSDM09. We split the data randomly into 50/50 training and testing data sets. The training data contains 25035 observations and the testing data contains 24965 observations. For each model, we stop the boosting procedure at the fifth iteration. In terms of the area under ROC curve (AUC), Figure 3 and Figure 4 show the individual model performance and the combined performance, respectively.

In this example, we see that the average of top 10 experts score produces higher AUC than that of simply average all the 200 experts scores; the Model 1 gives higher AUC than that of averaging the top 10 experts. We also see that Model 2, by adding

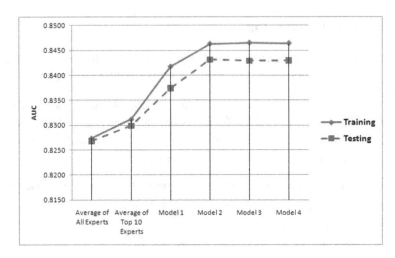

Fig. 3. Individual Model Performance

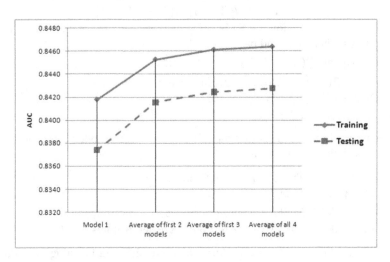

Fig. 4. Combined Performance

the two estimated shape parameters to the linear combination of the experts score model, has improved the overall prediction accuracy by 1% point. The improvement with rest of the features created is less than 1%.

5   Conclusions

We proposed to manipulate expert's ratings by creating extra features to describe their distribution characteristics. For example, we fitted a beta distribution to the distribution of experts' scores for each subject, and used the two shape parameters as features to describe experts scores distribution, then applied the simple boosting procedure to different regression models. The four regression models may or may not improve the prediction power individually, but by averaging them, we may improve the overall prediction. We view our approach as a way to create more divers models based on experts rating scores, and in return, it may yield chances of having a more (than averaging) powerful ensemble result. Another important aspect of our approach is interesting in that it is straight forward and no sparse representation attempt.

The proposed approach does not forbid one to use a sparse representation of the experts' data. One can create a sparse representation of the experts' data and add the created features to it.

References

1. Bell, R., Koren, Y., Volinsky, C.: The BellKor 2008 Solution to the Netflix Prize (2008), http://www2.Research.att.com/~volinsky/Netflix/Bellkor2008.pdf
2. The Ensemble. Netflix Prize Conclusion, http://www.the-ensemble.com
3. Breiman, L.: Bagging predictors. Machine Learning 24, 123–140 (1996)
4. Dietterich, T.G.: Ensemble methods in machine learning. In: Kittler, J., Roli, F. (eds.) MCS 2000. LNCS, vol. 1857, pp. 1–5. Springer, Heidelberg (2000)
5. Sullivan, J., Langford, J., Blum, A.: Featureboost: A meta-learning algorithm that improves model robustness. In: Proceedings of the Seventeenth International Conference on Machine Learning (2000)
6. Caruana, R., Niculescu-Mizil, A., Crew, G., Ksikes, A.: Ensemble Selection for Libraries of Models. In: Proceedings of International Conference on Machine Learning, vol. 64 (2004)
7. Blom, G.: Bagging predictors. In: Statistical Estimates and Transformed Beta Variables. John Wiley and Sons, New York (1958)
8. Friedman, J., Haste, T., Tibshirani, R.: Additive Logistic Regression: A Statistical View of Boosting. The Annals of Statistics 28(2), 337–407 (2000)

A Collaborative Filtering Based Approach for Recommending Elective Courses

Sanjog Ray and Anuj Sharma

Information Systems Area, Indian Institute of Management Indore,
Indore, Madhya Pradesh, India
{sanjogr,f09anujs}@iimidr.ac.in

Abstract. In management education programmes today, students face a difficult time in choosing electives as the number of electives available are many. As the range and diversity of different elective courses available for selection have increased, course recommendation systems that help students in making choices about courses have become more relevant. In this paper we extend the concept of collaborative filtering approach to develop a course recommendation system. The proposed approach provides student an accurate prediction of the grade they may get if they choose a particular course, which will be helpful when they decide on selecting elective courses, as grade is an important parameter for a student while deciding on an elective course. We experimentally evaluate the collaborative filtering approach on a real life data set and show that the proposed system is effective in terms of accuracy.

Keywords: Course Recommender System, Collaborative Filtering, User based Collaborative Filtering, Item based Collaborative Filtering.

1 Introduction

The structure, content and delivery of post graduate programmes within management schools in India has been reorganized from time to time and is now delivered on the basis of a fully modular, semester and credit-based curriculum in many of the top business school. While this new semester or trimester (terms as in the Indian Institutes of Management) approach provides increased choices and flexibility and allows students the ability to personalize their studies, challenges have arisen with regard to enabling students to appreciate the range and diversity of modules (courses) in each term or semester that are available to them. In particular, the current enrolment system makes it difficult for students to locate course options that might best fit their individual niche interest. Due to the diversity of different electives available, students find it difficult and time consuming to select the courses they will like and at the same time can get relatively better grades.

Students pursuing higher education degrees are faced with two challenges: a myriad of courses from which to choose, and a lack of awareness about which courses to follow and in what order. It is according to their friends and colleagues' recommendations that the many of them select their courses and register accordingly. It would be useful to help students in finding courses of interest by the intermediary of a recommender system.

S. Dua, S. Sahni, and D.P. Goyal (Eds.): ICISTM 2011, CCIS 141, pp. 330–339, 2011.
© Springer-Verlag Berlin Heidelberg 2011

Recommender systems implement advance data analysis techniques to help users find the items of their interest by producing a predicted likeliness score or a list of top-N recommended items for a given active user. Item recommendations can be made using different methods where each method is having different results. Collaborative filtering (CF) based algorithms provides item recommendations or predictions based on the opinions of other like-minded users.

In other domains, the benefits of deploying recommendation systems to assist users in finding relevant items is well understood and researcher are finding different usage of recommender systems in generating recommendations for different category of items [1, 2]. More recently, research has been conducted into developing such technology for course recommender systems.

In this paper, we present our work on developing collaborative filtering based course recommendation system for integration into business school's existing enrolment system either online or offline. We believe collaborative filtering approach provides student an accurate prediction of the grade they may get if they choose a particular subject based on their performance in earlier courses. The prediction of the grade will be very helpful when students decide on selecting elective courses, as grade is an important parameter for a student while deciding on elective courses.

This paper is organized as follows. Section 2 describes related work done in the area of course recommendation and motivation for the proposed work. Section 3 describes different collaborative recommendation algorithms which can be used to facilitate the course recommendation and enrolment process. These algorithms are empirically evaluated in Section 4. The results are discussed in Section 5 and conclusions are presented in Section 6.

2 Related Work

From the last decade, Recommendation System (RSs) have been widely developed, implemented and accepted for various categories of application like recommendation of products (e.g., books, music, movies) and of services (e.g., restaurants, hotels, websites), likewise research has been conducted into developing such technology for course recommender systems.

SCR [3], which is an acronym for Student Course Recommender, suggests courses by using a strategy based on Bayesian Network Modeling. The SCR network learns from the information stored about the students who have used the system. It requires the presence of enough cases in the student database. Therefore, if a user has not started or completed any courses, and is not pursuing any degree at the university, SCR cannot give him any course recommendation.

The Course Recommender System [4] is based on the several different collaborative filtering algorithms like user-based [5], item-based [6], OC1 [7], and a modified variant of C4.5 [8]. The system can predict the usefulness of courses to a particular student based on other users' course ratings. To get accurate recommendations, one must evaluate as many courses as possible. Based on the evaluation results, the author suggests C4.5 as the best algorithm for course recommendation. The system cannot predict recommendations for students who have not taken any courses at the University.

PEL-IRT stands for Personalized E-Learning system using item response theory [9]. It recommends appropriate course material to students, taking into account both course material difficulty and student ability. When using PEL-IRT, students can select course categories and units and can use appropriate keywords to search interesting course material. Once the course material has been recommended to students and they have browsed through, the system asks them to answer two questionnaires. This explicit feedback is used by PEL-IRT to re-evaluate the students' abilities and adjust the course material difficulty used in the recommendation.

The AACORN system, which stands for Academic Advisor Course Recommendation Engine, applies a case-based reasoning approach to course recommendation, has been proposed in [10]. The AACORN system recommends courses to graduate students at De-Paul CTI. The system uses the experience of past students and their course histories as the basis for course advising. In order to determine the similarities between course histories, the system uses a metric commonly used in bio-informatics called the edit distance. The system requires a partial history of the courses followed by a student before it can provide useful recommendations.

CourseAgent is a community-based recommendation system that employs a social navigation approach to deliver recommendations for courses based on students' assessment of their particular career goals [11]. The main theme of this approach is to obtain students' explicit feedback implicitly, as part of their natural interaction with the system. The basic and obvious benefit of the system to the students is as a course management system that keeps information about courses they have taken and facilitates communication with their advisors.

The RARE, a course recommender system based on association rules combines association rules together with user preference data to recommend relevant courses [12]. RARE was used on real data coming from the department of Computer Science at the Universit´e de Montr´eal. It analyses the past behavior of students concerning their course choices. More explicitly, it formalizes association rules that were implicit before. These rules enable the system to predict recommendations for new students. A solution to the cold start problem, which is a central question in recommender systems, is also proposed in RARE.

The Course Recommender System [13] is based on variation on the widely-used item-based collaborative filtering algorithm. The objective of module recommender system is to facilitate and enhance the on-line module selection process by recommending elective modules to students based on the core modules that they have selected. Evaluation using historical enrolment data shows very encouraging performance in terms of both recall and coverage.

Some recent research is focused on using course recommender systems in niche area like for civil engineering professional courses [14] and for university physical education [15].

From the review of the literature, it is evident that recommendation technology applied in education field can facilitate the teaching and learning processes. Considering the significance and seriousness of education, the help of recommendation system can improve efficiency and increase veracity of learners in the actual situation.

Comparing to other approaches like SCR [3] based on bayesian network modeling, RARE based on association rules, and AACORN [10] based on case-based reasoning, the proposed approach uses collaborative filtering as in Course Recommender System [4] but using students' grades that is indicator of performance in earlier courses. The other systems like PEL-IRT [9] and CourseAgent [11] are explicit feedback based system but the proposed approach in this paper does not need any feedback from students. Given the challenges and constraints of integrating this technology into an existing live environment, the proposed work is in its initial stages but the vast literature suggests that this domain offers great potential and scope for future research and development.

3 Collaborative Filtering Methods

Collaborative filtering methods are those methods that generate personalized recommendations based on user preferences data or judgment data on different items present in the system. Judgment data is primarily in form of ratings assigned by users to different items. Ratings data can be explicitly or implicitly obtained. Explicit ratings are those given directly by the user, while implicit data can be collected indirectly by studying data about the user from different sources like purchase data, browsing behavior etc.

Collaborative filtering (CF) was first introduced by to recommend jokes to users [16]. Since then many systems have used collaborative filtering to automate predictions and its applications in commercial recommender systems has resulted in much success [17]. Because of its minimal information requirements and high accuracy in recommendations, variations of CF based recommender systems have been successfully implemented in Amazon.com [1], TiVo [18], Cdnow.com [19], Netflix.com [20] etc. The most widely used approach in collaborative filtering are the nearest neighbors approach. We describe below two of the most popular collaborative filtering approaches, user-based collaborative filtering and item-based collaborative filtering.

3.1 User-Based Collaborative Filtering

User-based collaborative filtering was first introduced by GroupLens research systems [21, 22] to provide personalized predictions for Usenet news articles. The basis implementation details of user–based CF remains the same as proposed in [22]. CF systems are primarily used to solve the prediction problem or the top-N prediction problem. For an active user U_a in the set of users U, the prediction problem is to predict the rating active user will give to an item I_t from the set of all items that U_a has not yet rated. The steps followed in user-based CF to make a prediction for user U_a are as follows:

Step 1: Similarity between the active user U_a and every other user is calculated.
Step 2: Based on their similarity value with user U_a, set of k users, most similar to active user U_a is then selected.
Step 3: Finally, prediction for item I_t is generated by taking the weighted average of the ratings given by the k similar neighbors to item I_t.

In step 1 to calculate the similarity between users Pearson-r correlation coefficient is used. Let the set of items rated by both users u and v be denoted by I, then similarity coefficient ($Sim_{u,v}$) between them is calculated as

$$Sim_{u,v} = \frac{\sum_{i \in I}(r_{u,i} - \bar{r}_u)(r_{v,i} - \bar{r}_v)}{\sqrt{\sum_{i \in I}(r_{u,i} - \bar{r}_u)^2} \sqrt{\sum_{i \in I}(r_{v,i} - \bar{r}_v)^2}} \quad (1)$$

Here $r_{u,i}$ denotes the rating of user u for item i, and $\bar{r}_u$ is the average rating given by user u calculated over all items rated by u. Similarly, $\bar{r}_v$ denotes the rating of user v for item i, and $\bar{r}_v$ is the average rating given by user v calculated over all items rated by v. In some cases to calculate similarity cosine vector similarity is used. In [5] through experimental evaluation they have shown the using person correlations results in better accuracy. To improve accuracy of predictions, various methods have been proposed to improve the way similarity is measured between users. Significance weighting [23] and case amplification [5] are two methods that have been experimentally shown to impact accuracy positively. Case amplification transforms the correlation value when used as weight-age in step 3. Correlations closer to one are emphasized and those less than zero are devalued. Significance weighing is applied to devalue correlation value calculated between two users based on a small number of co-rated items. When the numbers of co-rated item are above a particular threshold it has no impact on actual correlations value. In [23] the threshold has been shown to be 50, but threshold value may differ among datasets, so it should be experimentally determined.

Once similarities are calculated, a set of users most similar to the active user U_a are selected in step 2. There are two ways in which a set of similar users can be selected. One is to select all users whose correlation with user U_a lie above a certain specified correlation value or select a set of top-k users, similarity wise. Experimentally it has been shown that top-k approach performs better than the threshold approach [23]. Value of k is obtained by conducting experiments on the data as it depends on the data set used.

In step 3 to compute the prediction for an item i for target user u, an adjusted weight-age sum formula is used to take into account the fact that different users have different rating distributions.

$$P_{u,i} = \bar{r}_u + \frac{\sum_{v \in V} Sim_{u,v}(r_{v,i} - \bar{r}_v)}{\sum_{v \in V}|Sim_{u,v}|} \quad (2)$$

Where, v represents the set of k similar users. While calculating prediction, only those users in set v, who have rated item I, are considered.

3.2 Item-Based Collaborative Filtering

The main difference between item-based CF [6, 24] and user-based CF is that item-based CF generates predictions based on a model of item-item similarity rather than user-user similarity. In item-based collaborative filtering, first, similarities between the various items are computed. Then from the set of items rated by the target user, k items most similar to the target item are selected. For computing the prediction for the target item, weighted average is taken of the target user's ratings on the k similar items earlier selected. Weight-age used is the similarity coefficient value between the target item and the k similar items rated by the target user. To compute item-item similarity adjusted cosine similarity is used.

Let the set of users who rated both items i and j be denoted by U, then similarity coefficient ($Sim_{i,j}$) between them is calculated as

$$Sim_{i,j} = \frac{\sum_{u \in U}(r_{u,i} - \bar{r_u})(r_{u,j} - \bar{r_u})}{\sqrt{\sum_{u \in U}(r_{u,i} - \bar{r_u})^2} \sqrt{\sum_{u \in U}(r_{u,j} - \bar{r_u})^2}} \tag{3}$$

Here $r_{u,i}$ denote the rating of user u for item i, and $\bar{r_u}$ is the average rating given by user u calculated over all items rated by u. Similarly, $r_{u,j}$ denotes the rating of user u for item j.

To compute the predicted rating for a target item i for target user u, we use the following formula.

$$P_{u,i} = \frac{\sum_{j \in I} Sim_{i,j} * r_{u,j}}{\sum_{j \in I}|Sim_{i,j}|} \tag{4}$$

In Equation 4, I represent the set of k most similar items to target item i that have already been rated by target user u. As earlier mentioned, $r_{u,j}$ denotes the rating of user u for item j.

4 Experimental Evaluation

We performed the experimental evaluation on the anonymized data set of 255 students and the grades they scored in 25 subjects. Among the 25 subjects, semester 1 and semester 2 comprised of 9 subjects each, while the third semester comprised of 7 subjects. Out of the total 6375 data rows, training set comprised of 6200 data rows and test set comprised of 175 data rows. To create the test data set we randomly selected 25 students and separated their grade data for the third semester from the dataset. We evaluated the algorithms for different ratios of test /train data i.e., x values. Both item-based and user-algorithms were implemented as described in the earlier section. Fig. 1 shows the proposed approach for recommending courses.

To measure the recommendations quality we use most prominent and widely used metric for predictive accuracy mean absolute error (MAE) [17], [23], [25]. Mean absolute error (often referred to as MAE) measures the average absolute deviation between a predicted rating and the user's true rating. The MAE is computed by first summing the absolute errors of the N corresponding ratings-prediction pairs and then computing the average. Lower MAE values indicate higher accuracy. MAE is the most widely used metric because the mechanics of its computation are simple and easy to understand. In our experiments we measured the MAE for user-based and item-based algorithms for neighborhood size 5, 10, 15, 20 and all users respectively. In our experiment only those users with positive correlation values with the test user were considered for selecting K-nearest neighbors.

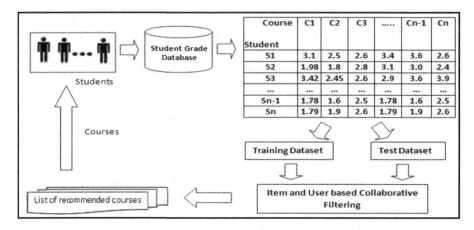

Fig. 1. The Proposed Course Recommendation Approach

5   Experimental Results

In this section, we present the results of our experimental evaluation. Figs. 2 and 3 shows the MAE values for user-based CF and item-based CF for different values of k i.e. neighborhood size and different values of x i.e. test/train data ratio.   Overall prediction accuracy of both the algorithms is very high as MAE values for all possible neighborhood size falls in the range of 0.33 to 0.38. As we can observe from the results, there is not much difference in accuracy between item-based CF and user based CF algorithms. In item based CF we also observe that for larger values of k, MAE values hardly change. The reason for this may be the small number items present in the dataset. Both the algorithms perform worse at $k=5$ and MAE values don't vary much after $k=10$. In Fig. 4, we compare user-based CF and item based CF different values of x for neighborhood size $(k) = 10$. While item-based CF perform better for $x=10\%$, user-based CF performs slightly better from higher values of x.

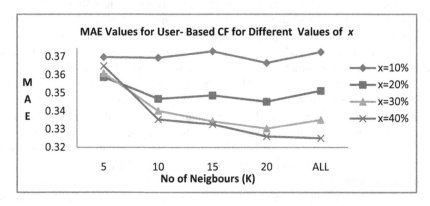

Fig. 2. MAE Values for User-Based CF for Different Values of x

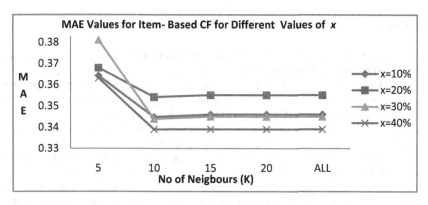

Fig. 3. MAE Values for Item-Based CF for Different Values of x

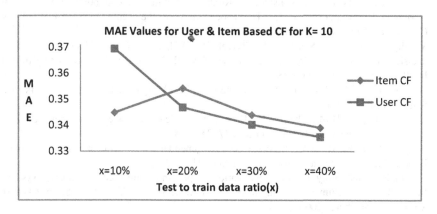

Fig. 4. MAE Values for User and Item CF for Neighborhood Size k =10

6 Conclusions

In this paper, we have compared two collaborative filtering approaches for predicting the grade a student will obtain in different courses based on their performance in earlier courses. Overall, the results of our experimentation on a real life dataset are very encouraging. We believe collaborative filtering approach provides student an accurate prediction of the grade they may get if they choose a particular subject, which will be very helpful when they decide on selecting elective courses, as grade is an important parameter for a student while deciding on elective courses.

For future work, research can done in developing integration strategies for approaches that can accurately predict student performance in courses and approaches that help a select a subject or courses based on student interests and learning objectives. These approaches can be used to provide valuable advice to students during career guidance advice and courses selection process.

References

1. Linden, G., Smith, B., York, J.: Amazon.com recommendations: item-to-item collaborative filtering. IEEE Internet Computing 7(1), 76–80 (2003)
2. Miller, B., Albert, I., Lam, S., Konstan, J., Riedl, J.: Movielens unplugged: experiences with a recommender system on four mobile devices. In: The Proceedings of the Seventeenth Annual Human-Computer Interaction Conference (2003)
3. Ekdahl, M., Lindström, S., Svensson, C.: A Student Course Recommender. Master of Science Programme Thesis, Lulea University of Technology, Department of Computer Science and Electrical Engineering/Division of Computer Science and Networking (2002)
4. Simbi, P.: Course Recommender. Senior Thesis, Princeton University (2003)
5. Breese, J., Heckerman, D., Kadie, C.: Empirical analysis of predictive algorithms for collaborative filtering. In: The Proceedings of the Fourteenth Conference on Uncertainty in Artificial Intelligence, pp. 43–52 (1998)
6. Sarwar, B., Karypis, G., Konstan, J., Riedl, J.: Item-based collaborative filtering of recommendation algorithms. In: The Proceedings of the Tenth International WWW Conference, Hong Kong (2001)
7. Murthy, S., Kasif, S., Salzberg, S., Beigel, R.: OC1: randomized induction of oblique decision trees. In: The Proceedings of the Eleventh National Conference on Artificial Intelligence, pp. 322–327 (1993)
8. Quinlan, J.R.: C4.5: Programs for Machine Learning. Morgan Kaufmann Publishers Inc., San Francisco (1993)
9. Chen, C.M., Lee, H.M., Chen, Y.H.: Personalized e-learning system using item response theory. Computers and Education 44(3), 237–255 (2005)
10. Sandvig, J., Burke, R.: AACORN: A CBR recommender for academic advising. Technical Report TR05-015, DePaul University, Chicago, USA (2006)
11. Farzan, R., Brusilovsky, P.: Social navigation support in a course recommendation system. In: Wade, V.P., Ashman, H., Smyth, B. (eds.) AH 2006. LNCS, vol. 4018, pp. 91–100. Springer, Heidelberg (2006)
12. Bendakir, N., Aimeur, E.: Using association rules for course recommendation. In: Proceedings of the AAAI Workshop on Educational Data Mining, pp. 31–40 (2006)
13. O'Mahony, M.P., Smyth, B.: A recommender system for on-line course enrollment: an initial study. In: The Proceedings of the ACM Conference on Recommender Systems, pp. 133–136. ACM, New York (2007)
14. Zhang, X.: Civil Engineering professional courses collaborative recommendation system based on network. In: The Proceedings of the First International Conference on Information Science and Engineering (2009)
15. Liu, J., Wang, X., Liu, X., Yang, F.: Analysis and design of personalized recommendation system for university physical education. In: The Proceedings of the International Conference on Networking and Digital Society (2010)
16. Goldberg, D., Nichols, D., Oki, B.M., Terry, D.: Using collaborative filtering to weave an information tapestry. Communications of ACM 35(12), 61–70 (1992)
17. Good, N., Schafer, J.B., Konstan, J.A., Borchers, A., Sarwar, B., Herlocker, J.L., Riedl, J.: Combining collaborative filtering with personal agents for better recommendations. In: The Proceedings of the Sixteenth National Conference on Artificial Intelligence and The Eleventh Innovative Applications of Artificial Intelligence Conference Innovative Applications of Artificial Intelligence, Orlando, Florida, United States, pp. 439–446 (1999)

18. Ali, K., Van Stam, W.: TiVo: making show recommendations using a distributed collaborative filtering architecture. In: The Proceedings of ACM SIGKDD International Conference on Knowledge Discovery and Data Mining (2004)
19. CDNow retail website, http://www.cdnow.com
20. Netflix, Inc., http://www.netflix.com
21. Resnick, P., Iakovou, N., Suchak, M., Bergstrom, P., Riedl, J.: GroupLens: an open architecture for collaborative filtering of netnews. In: The Proceedings of the ACM Conference on Computer Supported Cooperative Work, pp. 175–186. ACM, Chapel Hill (1994)
22. Konstan, J.A., Miller, B.N., Maltz, D., Herlocker, J.L., Gordon, L.R., Riedl, J.: GroupLens: Applying collaborative filtering to usenet news. Communications of the ACM 40(3), 77–87 (1997)
23. Herlocker, J.L., Konstan, J.A., Borchers, A., Riedl, J.: An algorithm framework for performing collaborative filtering. In: The Proceedings of SIGIR, pp. 77–87. ACM, New York (1999)
24. Deshpande, M., Karypis, G.: Item-based top-N recommendation algorithms. ACM Transactions on Information Systems 22(1), 142–177 (2004)
25. Shardanand, U., Maes, P.: Social information filtering: algorithms for automating word of mouth. Human Factors in Computing Systems (1995)

Architecture of a Library Management System Using Gaia Extended for Multi Agent Systems

Pooja Jain and Deepak Dahiya

Jaypee University of Information Technology, Waknaghat, Solan
{pooja.jain,deepak.dahiya}@juit.ac.in

Abstract. An intelligent and efficient library management system can be achieved with the help of a network of intelligent software agents. Agent-oriented techniques can be used to analyze design and build complex software systems. The designing of the interacting agents is done with the help of Gaia, extended for the multiagent systems.

Keywords: Gaia, multi agent systems, Library management system.

1 Introduction

Agent oriented techniques can effectively improve the current practices in software engineering. A multi agent system (MAS) is a set of interacting sub-organizations with a subset of agents possibly belonging to one or more organizations. These agents collaborate, coordinate and cooperate with each other to accomplish a complex task. The design of such a MAS can be effectively done by using extended Gaia [7]. By following the Gaia methodology, one can make full use of an agent-oriented approach in terms of system development and system use [8].

This paper deals with a multi agent library management system. The designing of the various agents is done through Gaia. The remainder of the paper is structured as follows. Section two talks about the various works done in the field of multiagent systems and Gaia. Section three talks about the current work done. Section four deals with the analysis phase of agent design of the library management system using Gaia. Section five talks briefly about the architectural design. Section six deals with the detailed design. Section seven deals with the partial implementation of the Library management system using JADE. Section eight talks about the results and the future scope of this paper.

2 Related Work

Gaia [1] was the first complete methodology proposed for the analysis and design of MAS. The Gaia methodology guides the system developers to define the agent structure (micro-level) and agent organization structure (macro-level) in two phases – analysis and design phase. In the analysis phase, the roles model addresses the micro-level aspects, and the interaction model and the environment model address the

S. Dua, S. Sahni, and D.P. Goyal (Eds.): ICISTM 2011, CCIS 141, pp. 340–349, 2011.

macro-level aspects. The major tasks are to define a collection of roles of agents, the relationships between agents and between agents and their environments. In the design phase, the agent model addresses the macro-level aspects and the services model addresses the micro-level aspects. The major tasks are to establish a mapping from roles to agent types and specify the functions and services provided by individual agent types. The output of the Gaia process is a collection of practical designable and reusable models [8].

However, the original version of Gaia suffered from the limitations of being suitable for the analysis and design of closed MAS. Several extensions to the basic Gaia methodology have been proposed to overcome these limitations. It was required to produce an ordered sequence of steps, an identifiable set of models, and an indication of the interrelationships between the models, showing how and when to exploit which models and abstractions in the development of a MAS.

The new, extended version of Gaia exploits the new organizational abstractions and significantly extends the range of applications to which Gaia can be applied.

3 Current Work: Agent Design of a Library Management System

The library system that's being considered here is a college library. The users are the students and the faculty of the college. The library management system (LMS) is a MAS comprising of various independent, reactive and autonomous agents.

A library has a large number of books related to different fields and topics. The user can search for the books as per his choice. Depending upon his search, a list of books is displayed. The list will have the attributes like, title of the book, author/authors, publishing house and the number of available copies. When the user wants to have a particular book, the system will check the number of books already issued to the user, as there is a limit on the number of books that can be issued to a user. If the user has the sufficient balance, then the book will be issued to him, and the number of copies of that particular book will be reduced by one.

If the user asks for a book which is present in the library, but is currently unavailable, then the user can see the names of the people to whom the copies have been issued. The system will give an option, whether the user wants to send an email to them, to return the book. Depending upon the choice of the user, an email can be send to some/all of the people having the copies of the desired book.

This LMS can be considered as an intelligent system, due to its one unique feature. If suppose, the user searched for a topic and that topic are not available in any of the books present in the library, then the user will be given a choice by the system for the web search. The web search is done by a topical web crawler. A web crawler is a program or an automated script which browses the World Wide Web in a methodical automated manner [4, 5, and 6]. A focused crawler or topical crawler is a web crawler that attempts to download only web pages that are relevant to a topic or set of topics. Topical crawling generally assumes that only the topic is given, while focused crawling also assumes that some labeled examples of relevant and not relevant pages are available.

When the user comes to return the book, then the system accepts the book and increases the number of copies of that particular book by one. The books can be recommended by the faculty as well. When a recommendation request reaches the LMS, it prepares a quote and sends it to the accounts department, so that the book can be purchased.

4   Analysis Phase

The main goal of the analysis phase is to organize the collected specifications and requirements for the system-to-be into an environmental model, preliminary role and interaction models, and a set of organizational rules, for each of the (sub organizations) composing the overall system [2].

4.1 The Organizations

The first phase in Gaia analysis is concerned with determining whether multiple organizations have to coexist in the system and become autonomous interacting MASs. In our system, the organizations can be easily identified as:-

- The one that takes care of issue and return of books
- The one that keeps all the information about the available books

4.2 The Environmental Model

The environment is treated in terms of abstract computation resources, such as variables or tuples, made available to the agents for sensing [11]. The environment model for the library management system can be depicted as:

Table 1. Environment model

reads	book_catalogue	the collection of all the books can be read
	copies	the number of copies of the book available can be read
changes	book_catalogue	changes whenever a new book is added or deleted
	copies	the number of copies of the book will be changed by the agent whenever the book is issued or returned
reads	recommended_books	the list of books recommended by the faculty
reads	book_balance	the total of the number of books issues to a user can be read
changes	reg_details	register the students and the faculty

The *book_catalogue* is represented by a data structure including information such as author of the book; title of the book, the year in which the book was published, the publishing house etc. *Copies* is the total number of copies of a book available in the library.

4.3 The Preliminary Role Model

Given the identification of the basic skills and of their basic interaction needs, respectively, the analysis phase can provide a preliminary definition of the organization's roles and protocols.

To represent (preliminary) roles, Gaia adopts an abstract, semiformal, description to express their capabilities and expected behaviors. These are represented by two main attribute classes, respectively: (i) permissions and (ii) responsibilities [2].

In the library management system, the roles can be identified as:

1. Book_manager
2. Register
3. Authorization
4. Solve_query
5. Display
6. Issuer
7. Web_crawler
8. Recommend
9. Return
10. Send_email

Book_manager role keeps track of all the books available in the library. It also keeps all the information about the *book_catalogue* and the copies of a particular book present. It will also keep the entire information about a particular user, like the books issued in his name and his transaction history of 3 months.

The *register* role is responsible to register the new students and new faculty. The user will give his enrollment number, name and mother's name and accordingly he will be allotted a default user id and a password. When the user logs to the system, it will give an option of exiting user or a new user. If the user is new then the registration process is handled by this agent otherwise the request is send to the authorization agent [11].

The *authorization* role is used to get the username and the password from the user. The details are matched with the database and correspondingly authorization is done. Once the authorization is done, a message is passed to the *solve_query* agent.

The *solve_query* agent takes the query from the user and solves it. It will send a message to the *display* role, which will display all the books/journals/magazines matching the search results. If the *display* agent is unable to display anything, or in other words, the book/topic searched by the user is not available in the library, then the *solve_query* role will give the option to the user for searching the topic from the web. If the user agrees, a message is sent to the *web_crawler* role, which will take care of the topic to be searched and store the results in a folder. This folder is private for the user and he can see the downloaded material at later point of time. This data is available only for a month. The data of this folder is also displayed by the *display* agent.

If the *display* agent is able to display the books and its details in coordination with the *book_manager* agent, then the user can select any book which he wants. Once the user has selected a particular book, the *Issuer* role checks the book balance of the user and accordingly issues the book.

The *display* role is used to display the results to the user. These results can be-

1. Books/ Journals/ magazines search result
2. Search results stored by the web crawler

Once a book is issued by the *issuer* agent, a mail is send to the user through the *send_email* role confirming the issue of the book. It will also show the list of all the books currently issued to the user, collaborating with the *book_manager* role.

The *recommend* role keeps a track of recommended books. The books can be recommended by the faculty or a request can be made by the student, as well, for the purchase of a book. Then this role will form a quotation of the book and send an email to the accounts department through the *send_email* agent.

Return role concerns the returning of the books by the users. As the user returns the book, the "copies" attribute is incremented by one. Once the book is returned by the user, a mail is send to the user through the *send_email* agent confirming the return of the book. It will also show the list of all the books currently issued to the user, collaborating with the *book_manager* role. The permissions of the different roles can be referred from another paper by the author [11].

4.4 Responsibilities

These attributes determine the expected behavior of a role and, as such, are perhaps the key attribute associated with a role. Responsibilities are divided into two types: liveness properties and safety properties.

Liveness expression of an agent is [11]:

Issuer = (Read_book_catalogue, Read_user_book_balance, update_copies, update_user_balace)

The safety requirements of the various agents are:

- Student_bal   NOT > limit
- Faculty_bal   NOT > limit

The role schema of some of the roles is:

4.5 The Preliminary Interaction Model

This model captures the dependencies and relationships between the various roles in the MAS organization, in terms of one protocol definition for each type of inter role interaction [2]. The protocol definition of some is as follows:-

The most basic liveness rule can be that a book can be returned only when it has been issued. This can be depicted as

Issuer⟶Return

Another can be that a book can be issued only when the user has been authorized by the authorization role.

Authorization⟶Issuer

Table 2. Role Schema

Role Schema: book_manager	Role Schema: Issuer
Description: This preliminary role involves keeping a track of all the books available in the library. It uses a data structure called book_catalogue that contains all the necessary fields of a book. It also stores all the information about each user. His details like name, email id etc and the number and name of the books issued in his name.	Description: This preliminary role involves issuing the book to the user on the condition that the copy of the book is available and the user has sufficient balance to issue a book
Protocols and activities Change_book_catalogue, read_book_catalogue, change_user_details, read_user_details	Protocols and activities Read_book_catalogue, Read_user_book_balance, update_copies, update_user_balace
Responsibilities Liveness: Book_manager= (Change_book_catalogue, read_book_catalogue, change_user_details, read_user_details) Safety: • Student_bal NOT > limit • Faculty_bal NOT > limit	Responsibilities Liveness: Issuer = (Read_book_catalogue, Read_user_book_balance, update_copies,update_user_balace) Safety: • Student_bal NOT > limit • Faculty_bal NOT > limit

5 Architectural Design

In the architectural design phase, the organizational structure is defined in terms of its topology and regime control, and the interaction model and role model are completed [9]. The output of the Gaia analysis phase systematically documents all the functional (and to some extent non functional) characteristics that the LMS has to express, together with the characteristics of the operational environment in which the MAS will be situated.

While the analysis phase is mainly aimed at understanding what the MAS will have to be, the design phase tells the actual characteristics of the MAS. Many factors influence the actual design of the agents and the interaction between them. Like in the case of LMS, it may happen that the limit of books to be issued is different for different faculty. For example, the limit for a lecturer can be less than the limit for an associate professor, which in turn may have a limit less than the professor. Sometimes it may also happen in some of the colleges that the user doesn't need a *user_id* and a *password*, since the book is issued by a librarian. The user himself doesn't need to log onto the system. Only on the basis of the roll number of the student and employee code of the faculty, the books are issued. In such a case, the authorization role will not come into the picture. So, the architectural design will entirely depend upon the actual implementation of the library management system, i.e. on the basis of the requirement specifications given by a particular college.

6 Detailed Design

The last phase, detailed design, is concerned with detailing the agent model, which consists of mapping the identified roles to agent classes and instances, and the services model, which are the blocks of activities in which the agents will engage [9]. In the design phase, the abstract constructs from the analysis stage, such as roles, are mapped to concrete constructs, such as agent types, that will be realized at runtime. The agent model outlines the agent types in the system. The services model outlines the services required by the roles assigned to the agent types. The acquaintance model depicts communication links between agent types [10].

6.1 Definition of the Agent Model

There is one to one correspondence between the roles and the agents. In the LMS, it can be said that the different agents can be

Book_manager, Register, Authorization, Solve_query, Display, Issuer, Web_crawler, Recommend, Return, Send_email

It is desirable to have a *book_manager* agent that takes care of the *book_manager* role and keeps the entire information about all the books in the library as well as the complete information about the users.

The *register* agent takes care of the register role and registers a new user in the system. The *authorization* agent is responsible of validating the user_id and the password of the users, so that the books can be issued and returned. Similarly, other agents can be designed on the basis of the roles, already described above.

The agent model can be depicted as:-

$$
\begin{array}{lll}
\text{Book\_manager}^1 & \xrightarrow{\text{Play}} & \text{Book\_manager} \\
\text{Register}^1 & \xrightarrow{\text{Play}} & \text{Register} \\
\text{Display}^2 & \xrightarrow{\text{Play}} & \text{Display}
\end{array}
$$

6.2 Service Model

The aim of the Gaia services model is to identify the services associated with each agent class or, equivalently, with each of the roles to be played by the agent classes. Thus, the services model applies both in the case of static assignment of roles to agent classes as well as in the case where agents can dynamically assume roles. For each service performed by an agent, it is necessary to document its properties. The inputs, outputs, preconditions and post conditions should be known for all the services performed by an agent. Inputs and outputs can be obtained from the protocols model as described above. Pre- and post conditions represent constraints on the execution and completion, respectively, of services [2]. In the case of *book_manager* agent, the input is the output generated by the *solve_query* agent and the output of the *book_manager* agent is the list of books searched for.

7 Implementation

After the successful completion of the Gaia design process, developers are provided with a well-defined set of agent classes to implement and instantiate, according to the defined agent and services model.

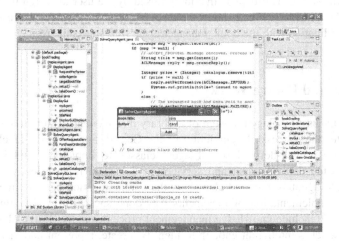

Fig. 1. Solve_Query Agent

Fig. 2. Display Agent

The MAS for the library management system is implemented using JADE (Java Agent Development framework). JADE is probably the most widespread agent-oriented middleware in use today. It is a completely distributed middleware system with a flexible infrastructure allowing easy extension with add-on modules. The framework facilitates the development of complete agent-based applications by means of a run-time environment implementing the life-cycle support features required by agents, the core logic of agents themselves, and a rich suite of graphical tools [12].

The user searches for a book through the *solve_query* agent and the *display* agent displays the results. The book is issued to the user and the *"copies"* of the book and the number of books issued to the user is manipulated accordingly. The solve_query agent is depicted in the Fig. 1. It depicts the query raised by the user to search for the books on Java by Daryl Morey. Fig. 2 shows the display agent that displays the results of the query in coordination with the book_manager agent. It displays the results as "java one copy present."

8 Results and Future Work

This paper discusses the agent design of a library management system to be used in a college. After the successful completion of the Gaia design process, developers are provided with a well-defined set of agent classes to implement and instantiate, according to the defined agent and services model. Gaia is an effective means to design a multi agent system and also show the communication between the different agents. The implementation of the agents was done successfully. The future work encompasses to include knowledge management processes in the design phase so that the final product is intelligent enough to identify, create, represent and distribute the knowledge.

References

1. Wooldrige, M., Jennings, N.R., Kinny, D.: The Gaia methodology for agent-oriented analysis and design. In: Autonomous Agents and Multi-Agent Systems, vol. 3, pp. 285–312. Kluwer Academic Publishers, The Netherlands (2000)
2. Wooldridge, M., Zambonelli, F., Jennings, N.R.: Developing multiagent systems: the Gaia methodology. ACM Transactions on Software Engineering and Methodology 12(3), 317–330 (2003)
3. Arenas, A.E., García-Ojeda, J.C., de Pérez Alcázar, J.: On combining organisational modelling and graphical languages for the development of multiagent systems. Integrated Computer-Aided Engineering 11(2), 151–163 (2004)
4. Shettar, R., Shobha, G.: Web crawler on client machine. In: The International Multi-Conference of Engineers and Computer Scientists II, IMECS 2008, Hong Kong, March 19-21 (2008)
5. Liu, X., Tan, D.: A web crawler (2000)
6. Hughes, B.: Web crawling. Department of Computer Science and Software Engineering, University of Melbourne
7. Huang, W., El-Darzi, E., Jin, L.: Extending the Gaia methodology for the design and development of agent-based software systems. In: The Proceedings of the Thirty-First IEEE Annual International Computer Software and Applications Conference (COMPSAC 2007), Beijing, China, pp. 159–168 (2007)
8. Liu, S., Joy, M., Griffiths, N.: GAOOLE: a Gaia design of agent-based online collaborative learning environment

9. Tiba, F.T.K., Capretz, M.A.M.: An overview of the analysis and design of SIGMA: supervisory intelligent multi-agent system architecture. In: Information and Communication Technologies, pp. 3052–3057 (2006)

10. Juan, T., Pearce, A., Sterling, L.: ROADMAP: extending the Gaia methodology for complex open systems. In: The Proceedings of the First International Joint Conference on Autonomous Agents and Multi-Agent Systems, Bologna, Italy, pp. 3–10 (July 2002)

11. Jain, P., Dahiya, D.: Knowledge management systems design using extended Gaia. International Journal of Computer Networks & Communications (IJCNC) (paper accepted to the published in January 2011)

12. Bellifemine, F., Caire, G., Greenwood, D.: Developing multi agent systems with JADE, Wiley Series in Agent Technology

Interorganizational Information Systems Diffusion: A Social Network Perspective

Laurence Saglietto[1] and Federico Pigni[2]

[1] GREDEG - UMR 6227
CNRS et Université de Nice Sophia-Antipolis, France
[2] Grenoble Ecole de Management
Ecole de Management des Systèmes d'Information, France
sagliett@gredeg.cnrs.fr, federico.pigni@grenoble-em.com

Abstract. The adoption process of interorganizational information systems has been widely studied. However, little is known on the forces actually acting against their diffusion and deployment. In this paper we suggest the use of elements derived from social network studies to explore the context and environment in which the diffusion of an IOIS takes place. We imply that the adoption of an IOIS depends both on the nature of the innovation and on its diffusion processes intended as the flow of communication within a system of interdependent actors leading to a dynamic process of social influence. Using the concept of avalanches and bifurcations as predictors of change in a nonlinear diffusion of innovation process, we explore the case of the adoption of a new HR management system after the merger of two large pharmaceutical groups providing evidence that this methodology could shed lights on elements only implied in previous studies.

Keywords: Diffusion of innovation, social network, interorganizational systems (IOS).

1 Presentation of the Subject

Several conceptual frameworks have been proposed in literature to explain the diffusion of an IOIS eventually based only on a "processual" view of the adoption process, studying the main determinants and success factors, or focusing on the study of the resulting organizational changes. We suggest that the key dimension of the relational network at the base of the diffusion process is often implied and only partially explored in previous works. The relational network is the result of the collective action emerging from the characteristics of the relationships among the individuals involved in the adoption process. Indeed, it has been shown that the adoption of an innovation depends not only on the nature of the innovation itself but also on the diffusion processes [1]. The diffusion process consists of the flow of communication within a system of interdependent actors leading to a dynamic process of social influence [2]. According to Rogers [3], "the diffusion of an innovation has traditionally been defined as a process by which that innovation is communicated

S. Dua, S. Sahni, and D.P. Goyal (Eds.): ICISTM 2011, CCIS 141, pp. 350–354, 2011.

through certain channels over time among the members of a social system." In IOIS studies Roger's theory of diffusion of innovations and the related theories of technology acceptance and adoption are still the dominant paradigm. The studies belonging to this stream of research aims at identifying the antecedents and the factors affecting IOIS adoption [4]. Steyer and Zimmermann [5] propose an approach where the influence of the adoption spreads in "the form of avalanches" assigning relevance to the network structure. This propagation is then a model of diffusion based on the flow of the influence through a social network. Specifically, this model does not result from transmission from one actor to another (insufficient to cause the adoption of an innovation), but from the transmission to an individual from its overall environmental relationships (cumulative effect). The authors show that the success of the adoption and diffusion process relies not only on the number or mass of adopters but also on their position within the network structure. We adapt this model to analyze the various events that determine changes in the direction of the diffusion. Such events are often viewed as sudden breaks or "bifurcation," situations hardly predictable that potentially could have important implications [6]. Thépot [7] defines bifurcations as "those presently so small actions that can turn into reality in the future" that is to say, they are indeed "enormous for their virtual consequences". Bifurcations are then identified from the analysis of the emerging different paths and channels of information, in turn enabling the study of their impacts on the behavior of the network players. Bifurcations, themselves generate new paths. They correspond to the transition from one situation to another: a social situation involving an element of unpredictability. In other words, these bifurcations, these critical moments are to be intended for the analysis as the cross product of a subjective decision (a transaction, a negotiation, conflict, failure) and the objectivity of a path constraint. This study aims at exploring, structure of the network that hosts the diffusion and adoption process, investigating the social structure providing the context and environment of the diffusion process of an IOS. Theoretically, we develop our arguments by relating avalanches diffusion structures and bifurcations to critically analyze the diffusion process. We illustrate the phenomenon with a case study.

2 Case Setting

2.1 The Context

This study frames the diffusion of an IOIS for the integrated management of human resources. More specifically, this study was performed at the time of the merger of two different multinational companies in the pharmaceutical sector. We seek to appreciate "the avalanche path" and the link between the results of the adoption process and the structure of the enacted social network, when bifurcations occur.

2.2 Methodology

We followed the entire project as it was defined for the adoption of the software as part of the project group HR/IA (Global Human Resources Information Applications) specifically formed.

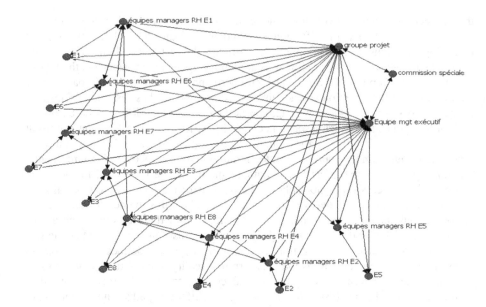

Fig. 1. The Relational Network of Project Founder

Legend : HR = Human Resources, E*n*= Entreprises

The release was planned on a period of 16 months to unify and bring the 70 existing human resource systems in the new group (about 75,000 people spread over 11 major countries - note that many branches were lost after the merger, so only 40 branches were subsequently involved in the project). Our analysis is based on the research methodology of case studies described by Yin [8], using a triangulation of sources of information. Our analysis is based on internal documentation, and on the materials collected and the experience gained in participating in the project as observer.

2.3 Analysis

The description of the diffusion of innovation process, analysis of phase 1, highlights the key elements for the understanding of the problem, such as the importance of the group, that is to say, "the ensemble", more than that of individuals.

The relational network of the project is composed of different groups (HR/IA project team, the executive management group, the special committee, the HR & IS teams, the companies). The collective logic adopted by the surveyed companies appears to be based primarily on social relations that reveal the effective underlying exchanges, the influence of status and control. This allows overcoming the issues generally emerging from an individualistic conception of the resulting adoption and influencing the change management process. In moving towards a relational logic, the individual reactions are indeed bypassed. Companies are the latest stage in the diffusion process, and therefore is where the accumulated tension and resistance to social change can be found. The second phase of the diffusion process will be devoted to their understanding as they will enact local social networks. At the heart of the various local networks the success depended on the position and power of the agents of change. They granted the success of the diffusion process and affected in "avalanches" all the other groups simultaneously. The objective of this wide distribution was to reach the 40 newly formed subsidiaries. Diffusion has followed an avalanche model since dissemination of information has undergone a multichannel transmission. Indeed, subsidiaries have received the same information from their teams of HR managers, the Project Group HR/IA, and the executive management team. From the project Group HR/ IA to the subsidiaries, intermediates (teams of HR managers) have also received an information in avalanche, from the project HR/IA, the special committee and the executive management team. These intermediates, top management and subsidiaries, have therefore an indispensable position for information exchange. However, significant symptoms (loss of control, no Perceived need, uncertainty and fear of the unknown, real or perceived threats, concerns regarding competence and resources) have led to bifurcation when their treatments have been established. For example, they brought an additional player in the network: coaches to guide and manage the changes, and they have also led to the review of the content of published information and their distribution support. Consequently, the communication of information through these meetings allowed on the one hand a collective appropriation of the project and on the other hand a loss of control on behalf of the management. Some bifurcations at the local relational networks led to changes in the orientation of the trajectory of diffusion. When the treatment was then applied it generated uncertainty both in the timing and in the outcomes of the diffusion. The relational pressure played a key role in these bifurcations influencing the diffusion paths by recombining and contributing to endanger the time schedule of the project. The geographical and relational proximity of the HR teams with the different companies was effective in the diffusion of the innovation. Connectivity and relational cohesion contributed to explain the performance through the specific morphology of the networks. The analysis of bifurcations implicated the assessment of the organizational change at both the global and local levels and respecting the schedule assigned to the team to accomplish its mission.

3 Conclusions

The present study suggests some theoretical and empirical implications. From a theoretical perspective it provides a first improvement of the avalanche model of Steyers and Zimmerman [5] introducing a predictor of change existing in a nonlinear diffusion of innovation. Second, this analysis has highlighted the importance of an approach in terms of social networks to explain the adoption of an IOIS, which provides additional insights to this issue to date. Ultimately, this helps to overcome the bias of the pro-innovation that characterizes the traditional studies in IOIS, while repositioning the IOIS artefact in a characterized social context. Methodologically, this paper is a first attempt to use social network arguments to support the study or IOIS diffusion going beyond the classical diffusion theories.

From a more empirical perspective, this study provides managers with new insight into change management. Differently from current practices, it is observed a more critical role played by time responsiveness to the success of change management process. Moreover, project initiators should concentrate on the delicate management of the interfaces between networks (in this case global and local): it is at this level that the forces acting against the diffusion process become apparent. Similarly, the choice of the project team, and the implied social network, has to be carefully considered.

References

1. McGrath, C., Krackhardt, D.: Network conditions for organizational change. Journal of Applied Behavioral Science 39, 324–336 (2003)
2. Burkhardt, M., Brass, D.: Changing patterns or patterns of change. Administrative Science Quaterly 35, 104–127 (1990)
3. Rogers, E.M.: Diffusion of innovations. The Free Press, New York (1995)
4. Robey, D., Im, G., Wareham, J.D.: Theoretical foundations of empirical research on interorganizational systems: Assessing Past Contributions and Guiding Future Directions. Journal of the Association for Information Systems 9, 497–518 (2008)
5. Steyer, A., Zimmermann, J.-B.: Social influence and diffusion of innovation. Mathematics and Social Sciences 168, 43–57 (2004)
6. Lucas, Y.: Représentation et rôle des réseaux en sociologie. In: Conference Représentation et Rôle des Réseaux en Sociologie (2007)
7. Thépot, J.: Jacques Lesourne. Revue Française de Gestion 3, 9–15 (2005)
8. Yin, R.K.: The case study anthology. SAGE Publications, Thousand Oaks (2004)

Cognitive Models and Its Current Challenges

Mukta Goyal, Alka Choubey, and Divakar Yadav

Jaypee Institute of Information Technology University,
Noida, India
{mukta.goyal,alka.choubey,divakar.yadav}@jiit.ac.in

Abstract. Cognitive learning and learner's mind processes play important roles in education. It is a process of acquiring new habits, knowledge and skills which together enable the students to do something they could not have done before. Some of the theories are bloom's taxonomy, information processing theory and critical thinking. Students understand and remember better if they can fit their learning into a framework and are motivated to learn if that framework fits into what they understand as their ultimate goal. Here we reviewed some of the cognitive models which may support students to learn the learning content through the learning management system. In this context, the student performs the user role while using the cognitive model as a tool to learn and take part in the subject in a more effective way.

Keywords: Cognitive model, rote learning, e-learning analytical learning, innovative learning, example-based learning.

1 Introduction

Behaviorism, Cognitivism and Constructivism are the three fundamental learning theories having their own potential impact on any adaptive e-Learning system. Behavior [15] theorists define learning as nothing more than the acquisition of new behavior. Learning emphasizes the constructivist approach, which argues that learners must actively "construct" knowledge by drawing out experiences that have meaning and importance to them [9].

Cognitivism [15] makes mental process as the primary object of study and tries to discover and model the mental processes on the part of the learner during the learning process.

Cognitive strategies have played a major role to design the learning system. The goal of this paper is to organize the content that facilitates the different type of students in the learning process.

2 Cognitive Models

Learning can be enhanced with the use of following six cognitive models.

S. Dua, S. Sahni, and D.P. Goyal (Eds.): ICISTM 2011, CCIS 141, pp. 355–358, 2011.
© Springer-Verlag Berlin Heidelberg 2011

2.1 Rote Learning

Rote learning is a learning technique which avoids understanding of a subject and instead focuses on memorization. Rote learning is widely used in the phonics in reading, the periodic table in chemistry, multiplication tables in mathematics, anatomy in medicine, cases or statutes in law, basic formulas in any science, etc. Research [1] has shown that when properly applied, rote learning is a consistently effective teaching method. The danger of rote teaching is that it allows, even encourages, us to avoid imagine. Indoctrination potential is another possible danger which is created by rote teaching [2].

2.2 Analytical Learning

Analytical learning uses prior knowledge and deductive reasoning to augment information given by training examples. It helps middle school students who are learning mathematics by constructing unique solutions to complex problems [3]. Another way of solving problem is to apply Jung's [4] theory of personality types and identifies specific techniques to support individual differences .Moore [5] presented new information on visualization framework that supports the analytical reasoning process. SMILE (State machine Interactive Learning Environment) learning model can be used in merging the different solutions of a problem based on the personality of different persons [16].

2.3 Example Based Learning

Example-based learning is a form of learning that exploits a very strong, or even perfect, domain theory to make generalizations or form concepts from training examples. It is known to be as robust learning system. It is a combination of supervised and unsupervised learning methodology. It provides various learning technique like instinctual learning. Example based learning proposed a methodology which is provided by previous research and problems simulations. This learning methodology merges previous models of problem solving and further critical and most desirable changes. Earlier research in the field of Artificial Intelligence has provided various examples focused on systems based on basic ideas of human long-term learning. This model is composed of six basic structures: a) instinctual learning, b) short-term learning, c) long-term learning, d) motor memory, e) short-term memory, f) long-term memory/knowledge base.

2.4 Dynamic Learning

Dynamic learning is an attempt to demonstrate how learner's mind processes dynamically, interact with surrounding environments by establishing two novel theoretical models of perception and imagination systems as shown Fig. 1 and Fig. 2, respectively. Comparison of these two models shows that removal of the imagination element from the model of imagination system results in loss of the awareness as demonstrated in the model of perception system, indicating the importance of the imagination in learning processes.

The term dynamic is added to distinguish the construct from traditional, centralized groups of learners found in many classrooms. In a dynamic community, all members share control, and everyone learns, including the teacher or group leader. Teaching according to the learner's aptitude" increases his or her performance therefore dynamic learning can help a lot in learning mathematics [7]. Modular Object-Oriented Dynamic Learning Environment (MOODLE) is another learning environment which is highly reliable and functional alternative to popular commercial products like WebCT and Blackboard [6].

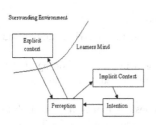

Fig. 1. Trigonal cyclic model of perception system

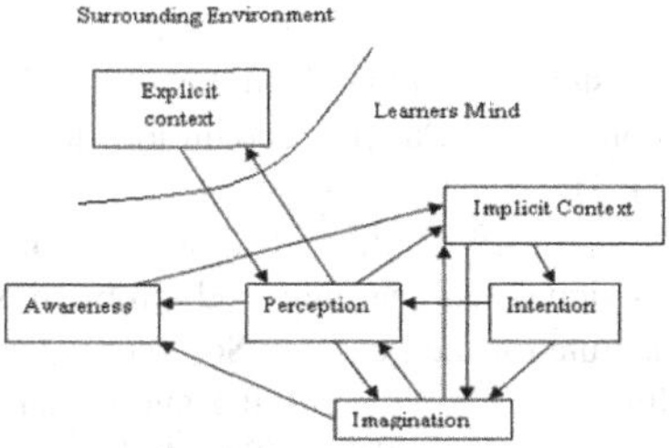

Fig. 2. Trigonal bipyramidal cyclic model of imagination system

2.5 Common Based Learning

Common Sense learning judges the things given by other sense. Commonsense comes out from ideologies, cultures, mentalities, collective memories, past experiences and everyday knowledge. Commonsense Knowledge [8], [10], [11] is used to help instructors for the following: a) identify topics of general interest to be taught, b) identify facts that are inadequate in order to fill in gaps in knowledge, c) fit the instructional material content to the learner's previous knowledge, d) provide a suitable vocabulary to be used in the instructional material, and e) minimize the time used to prepare lessons.

2.6 Innovation Learning

Innovation is defined as the bringing of new methods, ideas, etc. To innovate simply means to make changes [14]. Innovative learning environments can be described through the four levels of models [13] given as: a) The overall external structural framework such as national school policies, including curricula, funding, etc., b) the organizational/institutional setting in which learning takes place, c) The learning environment in which learners and teachers interact and in which pedagogical theories and practices and ICTs are used in different ways to improve learning, and d) actors/individuals, viewed as the individuals involved in the learning activities.

3 Conclusions

Today rapid growth of web based courses for education and training impose challenges to e-learning systems to generate content according to the level of the learner. Identification of documents through these cognitive models may enable e-learning systems to match learner needs.

Acknowledgements

The authors wish to thank M.Hima Bindu from Jaypee Institute of Information Technology, Noida, India for her support.

References

1. Webster, M.: Online: Rote Learning Memorization (1990)
2. Leonard, U.: Pattern recognition: learning and thought. Prentice-Hall, Englewood Cliffs (1973)
3. Susan, M.W., Ray, B., Reiser, J.: An environment for supporting independent individualized learning and problem solving. School of Education and Social Policy Institute for the Learning Sciences
4. Huitt, W.: Problem solving and decision making: consideration of individual differences using the Myers-Briggs Type Indicator. Journal of Psychological Type 24, 33–44
5. Yedendra, B., Shrinivasanand, J., Wijk, J.: Supporting the analytical reasoning process in information visualization. In: CHI, Florence, Italy (2008)
6. Dale, N., Pavlo., A., Serkan, T.: Modular object-oriented dynamic learning environment: what open source has to offer. Iowa State University, Iowa
7. Pofen, W., Wenlung, C., Weichung, W., Pi-hsia, H.: Dynamic learning system and its effects. In: International Conference on Computers in Education (ICCE 2002) (2002)
8. Luis, V.A., Mihir, K., Manuel, B.: Verbosity: a game for collecting common-sense facts. In: CHI, Montréal, Québec, Canada (2006)
9. Maria, R.: Learning by doing and learning through play: an exploration of interactivity in virtual environments for children. ACM, New York (2004)
10. Junia, C.A., Aparecido, F.P., Vania, P.A., Muriel, G., Silvia, Z.M., Américo, T.N., Henry, L.: Applying common sense to distance learning: the case of home care education. ACM, New York (2006)
11. Aparecido, C., Junia, C.A., Silvia, H.Z.: Learning activities on health care supported by common sense knowledge. In: SAC 2008, Fortaleza, Ceará, Brazil (2008)
12. Nalin, S.: Creating innovative new media programs: need, challenges, and development framework. Computer Science and Mathematics
13. Ramboll, M.: Study on innovative learning environments in school education (2003)
14. Bartlett, S.: Developing a Bartlett strategy for innovative learning. In: Bartlett, School of Planning
15. Conlan, O., Hockemeyer, C., Wade, V., Albert, D.: Metadata driven approaches to facilitate adaptivity in personalized eLearning systems. Journal of the Japanese Society for Information and Systems in Education (2003)
16. Ilja, L., Lieberman, E.: Developing analytical and synthetic thinking in technology education

Resolving Productivity Paradox of Fully and Partially IT Oriented Banks: Evidence from India

Dhiraj Sharma

School of Management Studies, Punjabi University,
Patiala 147002, Punjab, India
dhiraj_sharma22@rediffmail.com

Abstract. Technology has become the buzzword in the Indian banking sector these days. It has been, now more than 15 years since the Indian banking sector is liberalized, privatized and globalized. All the bank groups in India are rapidly embracing Information Technology (IT). However, it is a matter of debate whether Technology provides better financial results. There is no conclusive evidence that IT improves financial performance. It is generally believed that the technology provides efficiency, hence improves working and performance of an organization (Productivity Paradox). This implies that with the technology induction, the financial performance of an organization should also improve. Findings of the paper show that the fully IT oriented Indian banks are financially better off than the partially IT oriented Indian banks. This implies that technology does provide an edge which leads to better financial performance.

Keywords: Information technology, financial performance, Indian commercial banks, profitability analysis.

1 Introduction

There is no conclusive evidence that investment in Information Technology (IT) improves financial performance of a business. It has been a matter of debate whether Technology provides better financial results. The scholars call it *Productivity Paradox*. However, additional IT investments may negatively contribute to financial productivity [7]. On the similar lines, studies by [8] and [4] have also concluded that there is insignificant correlation between IT spending and profitability measures, which means IT investment is unproductive. There are, on the other hand, studies which show that there is no correlation between IT investment and financial productivity [1]. Thus, it is difficult to determine whether banking technology has a significant impact on bank performance. And there are studies which have found significant contributions from IT toward financial growth [5], [3]. There are studies which have drawn on statistical correlation between IT investment and profitability or stock value for their analyses and they have concluded that impact of IT on productivity is positive [2, 9].

S. Dua, S. Sahni, and D.P. Goyal (Eds.): ICISTM 2011, CCIS 141, pp. 359–362, 2011.

2 Research Methodology

The major objective of the paper is to study the inter-group comparison of financial performance of fully and partially IT oriented Indian banks from the year 1997 to the year 2008. The total time period for the study has been divided into two sub-periods: Pre-Technology Induction Era (from 1997-98 to 2000-01) and Post Technology Induction Era (from 2001 afterwards). The universe of the study is Indian Scheduled Commercial Banking. As per Reserve Bank of India (RBI), Partially IT-oriented Banks include - Group I – Public Sector Banks (20 Banks) and Group II – State Bank of India and its 7 Associates (08 Banks)]. Fully IT-oriented Banks are – Group III – Private Sector Banks (25 Banks) and Group IV – Foreign Banks (29 Banks). From each group of banks, top five banks (in terms of highest business per employee in the year 2008) have been taken.

3 Results and Discussion

Spread which is the difference between interests earned and interest paid by the banks plays a major role in determining the profitability of banks. Table 1 show that Interest rates in Indian banking sector have declined from the low to high technology period. Thus, the spread of banks have declined because of lower interest rates in the recent years. There is significant difference in the fully and partially IT oriented banks in case of Interest Earned and Interest paid in the two periods, however, the difference in spread is not statistically significant as the interest rates have overall declined in the high technology period. Overall, the t-test exhibits insignificant difference in the means of two periods for the Indian banking industry.

Burden is the difference between non-interest expenditure and non-interest income of the banks. Burden is usually taken in the negative sense since non-interest expenses tend to exceed non-interest income in the banking industry. Like the Spread, the average burden has also decreased from 2.14% to 1.81 % for Indian banking industry (Table 1). There is a decline in the Burden as a Percentage of Average Assets of Indian banks which is a positive sign. Overall, the t-test exhibits significant difference in the means of two periods at 1 % LOS for the Indian banking industry.

The profits of Indian banks have increased significantly from low technology era to high technology era (Table 1). The t-test exhibits significant difference in the means of two periods at 1 % LOS for the Indian banking industry. There is significant difference in the means in case of partially IT banks and insignificant difference in the means of fully IT banks. This is because that the fully IT banks (Private and Foreign Banks) from the very inception are Hi-tech banks thus in terms of technology induction there cannot be any significant difference in the low and high periods.

Table 1. Spread, Burden, Profitability Analysis of Indian Commercial Banks

Group	Average	Spread Ratios			Burden Ratios			Profitability Ratios		
		IE%AAs	IP%AAs	S%AAs	NIE%AAs	NII%AAs	B%AAs	NP%TI	NP%TD	NP%AAs
G-I	X_1	9.24	6.36	2.87	3.55	1.07	2.48	3.75	0.44	0.39
	X_2	7.30	4.44	2.85	3.24	1.25	1.99	10.21	1.01	0.86
	Mean Gap	1.94	1.92	0.02	0.31	0.18	0.49	6.46	0.57	0.48
	S.E.	0.80	0.75	0.19	0.41	0.39	0.15	1.49	0.16	0.14
	t-Value	3.89	4.09	0.13	1.22	0.73	5.28	6.93	5.67	5.29
	LOS	**	**	--------	-----	--------	**	**	**	**
G-II	X_1	8.90	6.01	2.91	3.68	1.37	2.31	5.86	0.77	0.60
	X_2	7.62	4.59	3.04	3.58	1.40	2.18	9.50	1.00	0.86
	Mean Gap	1.28	1.41	0.13	0.10	0.03	0.13	3.64	0.24	0.26
	S.E.	0.58	0.68	0.28	0.20	0.32	0.35	1.29	0.16	0.12
	t-Value	3.50	3.30	0.72	0.80	0.14	0.60	4.50	2.39	3.41
	LOS	**	**	--------	-----	--------		**	*	**
G-III	X_1	8.69	6.59	2.08	2.71	1.73	0.99	10.48	1.46	0.64
	X_2	6.25	4.12	2.11	2.92	1.86	1.06	13.00	1.52	0.78
	Mean Gap	2.44	2.47	0.03	0.21	0.13	0.07	2.52	0.06	0.14
	S.E.	1.09	1.04	0.49	0.42	0.43	0.26	3.41	0.44	0.31
	t-Value	3.58	3.77	0.09	0.78	0.50	0.45	1.18	0.22	0.71
	LOS	**	**	-------	-------	--------	------	------	-----	-------
G-IV	X_1	9.74	6.31	3.43	5.39	2.59	2.80	4.60	1.00	0.62
	X_2	6.69	3.34	3.34	4.79	2.77	2.02	14.46	2.37	1.32
	Mean Gap	3.05	2.97	0.09	0.60	0.18	0.79	9.86	1.36	0.70
	S.E.	1.00	1.11	0.54	0.35	0.21	0.22	6.56	1.17	0.68
	t-Value	4.88	4.27	0.27	2.72	1.34	5.69	2.40	1.87	1.63
	LOS	**	**	--------	*	--------	**	*	------	--------
Indian Banking Industry	X_1	9.14	6.31	2.83	3.83	1.69	2.14	6.17	0.91	0.56
	X_2	6.96	4.12	2.81	3.63	1.82	1.81	11.79	1.47	0.95
	Mean Gap	2.17	2.19	0.02	0.20	0.13	1.33	5.62	0.56	0.39
	S.E.	0.71	0.75	0.11	0.16	0.23	0.11	1.72	0.25	0.17
	t-Value	4.40	4.03	0.77	1.73	1.27	4.95	4.31	2.84	2.79
	LOS	**	**	--------	--------	--------	**	**	*	*

Source: Computed from the data published by Performance Highlights of Indian Banks, Indian Bank Association, 1997-2008.

Note: IE%AAs : Interest Earned as percent of Average Assets; IP%AAs : Interest Paid as percent of Average Assets; S%AAs: Spread as percent of Average Assets; NIE%AAs : Non-Interest Expenditure as percent of Average Assets; NII%AAs : Non-Interest Income as percent of Average Assets; B%AAs : Burden as percent of Average Assets; NP%TI: Net Profit as percent of Total Income; NP%TI: Net Profit as percent of Total Depostis; NP%TI: Net Profit as percent of Average Assets, X1 – Average in Low-technology induction period; X2 – Average in High-technology induction period, * Mean is significant at the 0.05 level, **Mean is significant at the 0.01 level. LOS: Level of Significance.

4 Conclusions

In the present study, it is found that the partially IT oriented banks are less profitable than the fully IT oriented banks. However, in terms of overall productivity and profitability their performance is gradually improving over the recent years. Foreign banks and private banks are on the top in terms of the overall profitability parameters. This can be attributed to higher level of technology induction in comparison to public banks. Analyzing further, it is found that SBI and associate banks are ranked second after the foreign Banks in terms of the spread ratios but their higher Burden places them next to Private Banks in terms of profitability. The Private Banks are more profitable as they have the lowest financial burden in the two periods. Moreover, they have a high proportion of non-interest income and comparatively low level of non-interest expenditure as compared to SBI group of banks. The foreign banks have also higher Burden ratios but they have been able to maintain high Non-Interest income which is offsetting high Burden. The difference in the Burden and Non-Interest income is the main reason for the differences in the profitability among different banks groups. The Interest earned ratios are declining over the years for all groups of banks because over the last few years RBI has lowered the interest rates. Still foreign Banks were able to have highest Interest earned ratios in low technology era as compared to the Indian Banks. From the view point of temporal comparison and on the basis of usage of technology by various bank groups, the fully IT oriented banks have financially performed better than partially IT oriented banks in both low and high technology periods. This implies that technology does provide an edge which leads to better financial performance. That means, technology does align with financial performance. Finally, the Indian banking sector has performed well on various fronts in the high technology induction period but especially those banks have performed well which had a better technology base.

References

1. Barua, A., Baily, M., Kriebel, C., Mukhopadhyay, T.: Information technology and business value: an analytical and empirical investigation. University of Texas, Austin Working Paper, TX (May 1991)
2. Brynjolfsson, E.: The productivity paradox of information technology. Communications of the ACM 35, 66–67 (1993)
3. Brynjolfsson, E., Hit, L.: Paradox lost? Firm-Level Evidence on the Returns to Information Systems Spending. Management Science 42, 541–558 (1996)
4. Dos Santos, B.L., Peffers, K.G., Mauer, D.C.: The impact of information technology investment announcements on the market value of the firm. Information Systems Research 4(1), 1–23 (1993)
5. Lichtenberg, F.: The output contributions of computer equipment and personnel: a firm level analysis. Economics of Innovation and New Technology 3(4) (1995)
6. Mariappan, V.: Changing the way of banking in India. Vinimaya XXVI (2), 26–34 (2006)
7. Morrison, C.J., Berndt, E.R.: Assessing the productivity of information technology equipment in the U.S. manufacturing industries. National Bureau of Economic Research Working Paper 3582 (January 1990)
8. Strassman, P.A.: The business value of computers. Information Economics Press, New Canaan (1990)
9. Wilson, D.: Assessing the impact of information technology on organizational performance. In: Banker, R., Kauffman, R., Mahmood, M.A. (eds.) Strategic Information Technology Management. Idea Group, Harrisburg (1993)

Inside Story of Business Process Reengineering in State Bank of Patiala (SBOP): Straight from the Horses' Mouth

Ratinder Kaur

School of Management Studies, Punjabi University,
Patiala-147002, Punjab, India
ratinder_talwar1@rediffmail.com

Abstract. In India, the changing dynamics of Indian economy have brought many reforms in financial sectors especially in banking and insurance sector. To meet new competitive challenges due to technology induction in banks and change in the customers' perspective forced organizations to rethink about their ways of doing business operations. Consolidations, amalgamations, pressures to reduce operating costs stressed banking community to adopt tools like Business Process Reengineering (BPR) in order to bring strategic benefits to organizations. State Bank of Patiala (SBOP) has initiated BPR initiatives on the lines and under the umbrella of State Bank of India (SBI). The study attempts to determine whether there is an improvement in the competitive measures of cost management, customer service, quality and productivity of bank under study. In India, no such study on BPR in banking specifically with reference to State Bank of Patiala has been found so far. Thus, present study contributes to the research in banking with regard to BPR.

Keywords: Business process reengineering (BPR), employees' efficiency, reduction in cost, higher productivity.

1 Introduction

In the face of competition and economic pressures, firms are changing their fundamental unit of analysis from the business function to business processes. Consolidations, amalgamations, pressure to reduce operating costs pressurized banking community to adopt tools like Business Process Reengineering (BPR) in order to bring strategic benefits to organizations. The purpose of business process re-engineering is basically to redesign and change the existing business practices in order to take advantages like, increase in employees' efficiency, reduction in cost, higher productivity, better efficiency etc. To have such advantages from Business Process Reengineering State Bank of Patiala (SBOP) has initiated BPR initiatives on the lines and under the umbrella of State Bank of India (SBI). The Bank launched various BPR initiatives starting from February 2005 to strengthen its ability to acquire new customers, build lasting relationships with existing customers and increase customer

S. Dua, S. Sahni, and D.P. Goyal (Eds.): ICISTM 2011, CCIS 141, pp. 363–366, 2011.

satisfaction through quality customer service. In this paper an effort has been made on the part of researcher to know the employees' perspective regarding the implementation of BPR initiatives in the bank under study.

2 Literature Review

Business process reengineering (BPR) is the "fundamental rethinking and radical redesign of business processes to achieve remarkable improvements in critical, contemporary measures of organization performance, such as cost, quality, service and speed". The Business Process Reengineering Method (BPR) is described by Hammer and Champy as the fundamental reconsideration and radical redesign of organizational processes, in order to achieve strong improvement in current profits, services and speed [1]. The purpose of reengineering is to "make all your processes the best-in-class." BPR is radical and therefore will require transformational changes in the organizations' work flows, systems, job roles and values [2]. Frederick Taylor suggested in the 1880's that managers could find out the best processes for getting their work done and reengineer them to enhance productivity. However, Holland and Kumar noted that 60–80% of BPR initiatives have been unsuccessful [3]. Business process is a structured set of activities designed to produce specific outputs for internal or external customers or markets. Each activity is a result of the efforts on the part of people who takes help from the other resources of production like technology, raw materials, methods and internal and external environment. A number of publications continue to report BPR activity. MacIntosh, for example, declares that BPR is 'alive and well' in his study of process activity within the public service sector [4].

3 BPR Initiatives at SBOP

State Bank of Patiala has introduced revolutionary dimensions in customer service by introducing Business process Reengineering (BPR) initiatives after achieving the landmark of cent per cent automation. BPR benefits customers through significantly reduced transaction time, flexibility in servicing and improved value chain of service. The bank has initiated various initiatives since 2005 including ATM Migration, Grahak Mitra (customer's friend), Loan Processing Centre (LPC), Currency Administration Cells (CAC), Central Pension Processing Centre (CPPC), and Relationship Managers (PB/ME) etc. The purpose of ATM initiative is basically, reduction in transaction cost, increase in customer service, increase in number of hits, convenience of ATM withdrawals, increase in ATM networks, better location of ATMs in cities, expedited card issuance for existing and new customers. On the other hand, Currency Administration Cell (CAC) act as "Centre of Excellence" that monitors and optimizes cash balances across the branches within their area of operation. The cells organize transportation of currency. The other initiative launched by bank is Drop Box which provides hassle-free cheque tendering round the clock throughout the year. Yet another initiative on the part of bank is Grahak Mitra (Customer's friend) which is created for the support of customers. Grahak Mitra

(GM) facilitates migration to alternate channels and focus on cross-selling. One more feather in BPR initiatives is Retail Assets Central Processing Centre (RACPC). The role of RACPCs is to provide best service and experience at the time of processing and granting loans. RACPCs emphasized reduction in the time taken to sanction and disburse retail loans across branches in Bank. Relationship Manager (RM), another effort for strengthening the relationship with existing customer and bringing new customers. The RM provides personalized services especially the VIP customers of the bank and offers value added services too. He generates new leads through existing high value customers after developing rapport with them.

4   Results and Discussion

BPR is not just getting work from processes or technology rather from the people who actually runs the show in organizations. Employees of the organization can throw better light on the insider's story about the status of BPR in Bank. Therefore, an attempt has been made on the part of researcher to know the success of BPR initiatives introduced in Bank. It was deemed appropriate to select 300 employees as sample size from the total universe of State Bank of Patiala's (SBOP) employees. The bank employees who are the actual users of the BPR were asked about their views about 'to what extent BPR increases employee's efficiency and productivity?' Age wise majority of the respondents of all the ages have agreed to the statement. On the basis of designation 69% of the Front line managers agree to the statement. On the other hand 76% of the middle level managers claim efficiency and productivity. Employees believe that BPR reduces manual labour and thus brings comfort and ease to the users. The bank employees, the actual users of the banking BPR, were asked 'to what extent BPR provides comfort and ease to the employees? Overall, the majority of respondents think that BPR does provide comfort and ease.

Majority of the respondents agree that with BPR the bank employee-customer relations have improved. Age wise majority of the highest and the lowest aged respondents are also of the same opinion. All the respondents of more than 55 years of age think BPR has improved the relations to a large extent. Majority of employees, on the basis of experience and designation also feel the same way. 62.56% respondents having more than 5 years experience also feel that BPR has improved the relations to a large extent. Overall, the majority of respondents (89.11 per cent) do believe that BPR has improved the employee customer relations. Most of the respondents agree that customer service speed has increased due to BPR. Overall, the majority of respondents (93.33%) believe that customer service is faster in post BPR period. Almost all the respondents of Banks agree that overall performance of banks has improved due to BPR induction. Age wise also, all the respondents of below 25 years of age and above 55 years of age feel that overall performance of banks has improved due to BPR induction. On the basis of experience majority of the lower and highly experienced respondents feel that overall performance of banks has improved due to BPR initiatives. On the basis of designation top level almost 100 percent managers claim that overall performance has increased.

The question was raised to judge employees' view about the extent of training and technical skills being provided by the bank. On the basis of experience 74.34%

respondents of below 01 year experience agreed that banks are providing technical skills or training to use BPR and 89.87% respondents of above 5 years of experience said that banks are providing technical skills or training to use BPR. However, 25.05% respondents of 3 to 5 years of experience disagree that banks are providing technical skills or training to use BPR. According to designation majority of banks feel that Bank is providing technical skills or training to facilitate BPR initiatives. The majority of all bank groups' respondents agreed that overall BPR has benefited the bank employees. Overall, the majority of the respondents (92.33%) feel BPR in bank has overall benefited the bank employees.

5 Conclusions

SBOP is the pioneer bank in implementing BPR in its improvement programme. Since the implementation of BPR in full swing from 2005 onwards, the performance of the bank has improved considerably. The majority of the bank employees feel BPR initiatives have benefited not only the banking performance but also the bank employees. The bank is also providing technical skills and required training to use BPR channels. Employees of the bank also feel that customer service speed has also increased due to BPR. Finally, BPR initiative has lead to improvement in internal efficiency and competitive advantage.

References

1. Hammer, M., Champy, J.: Reengineering the corporation: a manifesto for business revolution. Harper Collins, London (1993)
2. Dale, M.W.: The reengineering route to business transformation. The Journal of Strategic Change 3, 3–19 (1994)
3. Holland, D., Kumar, S.: Getting past the obstacles to successful reengineering. Business Horizons, 79–85 (1995)
4. Macintosh, R.: BPR: alive and well in the public sector. International Journal of Operations & Production Management 23(3), 327–344 (2003)

Author Index